Philosophy, Science, and History

Philosophy, Science, and History: A Guide and Reader is a compact overview of the history and philosophy of science that aims to introduce students to the groundwork of the field, and to stimulate innovative research. The general introduction focuses on scientific theory change, assessment, discovery, and pursuit. Part I of the Reader begins with classic texts in the history of logical empiricism, including Reichenbach's discovery-justification distinction. With careful reference to Kuhn's analysis of scientific revolutions, the section provides key texts analyzing the relationship of HOPOS to the history of science, including texts by Santayana, Rudwick, and Shapin and Schaffer. Part II provides texts illuminating central debates in the history of science and its philosophy. These include the history of natural philosophy (Descartes, Newton, Leibniz, Kant, Hume, and du Châtelet in a new translation); induction and the logic of discovery (including the Mill–Whewell debate, Duhem, and Hanson); and catastrophism versus uniformitarianism in natural history (Playfair on Hutton and Lyell; de Buffon, Cuvier, and Darwin).

The editor's introductions to each section provide a broader perspective informed by contemporary research in each area, including related topics. Each introduction furnishes proposals, including thematic bibliographies, for innovative research questions and projects in the classroom and in the field.

Lydia Patton is Associate Professor in the Department of Philosophy, Virginia Tech. Dr. Patton's research centers on the history and philosophy of science and epistemology. Recent work includes "Methodology of the Sciences," forthcoming, *Oxford Handbook of Nineteenth Century German Philosophy*; "Experiment and Theory Building" (*Synthese*); "Hermann Von Helmholtz" (*Stanford Encyclopedia of Philosophy*); and "Signs, Toy Models, and the A Priori" (*Studies in History and Philosophy of Science*).

Philosophy, Science, and History

A Guide and Reader

Edited by
Lydia Patton

Routledge
Taylor & Francis Group

NEW YORK AND LONDON

First published 2014
by Routledge
711 Third Avenue, New York, NY 10017

and by Routledge
2 Park Square, Milton Park, Abingdon, Oxon, OX14 4RN

Routledge is an imprint of the Taylor & Francis Group, an informa business

Library of Congress Cataloging in Publication Data
A catalog record for this book has been requested

ISBN: 978–0–415–89830–0 (hbk)
ISBN: 978–0–415–89831–7 (pbk)
ISBN: 978–0–203–80245–8 (ebk)

Typeset in Minion
by Swales & Willis Ltd, Exeter, Devon, UK

Printed and Bound in the United States of America by
Edwards Brothers Malloy

Contents

List of Figures

Chapter 19
"Critical Remarks," Gottfried Leibniz
These figures are redrawn from the Loemker edition.

Chapter 21
"On the Divisibility and Subtlety of Matter," Du Châtelet
Figures 7 and 10 are redrawn from the du Châtelet text.

Chapter 28
"On the Geological Succession of Organic Beings," Darwin
Diagrams 1 and 2 are from Chapter Four of the *Origin of Species*, 6th edition.

Acknowledgments

Above all, I would like to thank the Philosophy editor at Routledge, Andrew Beck, and the editorial assistant at Routledge, John Downes-Angus, for their patience and skill in helping to bring this text to completion.

Matthew Wisnioski suggested the Rudwick, Shapin and Schaffer texts for Part I of the book. Benjamin Jantzen suggested the theme of catastrophism and uniformitarianism for Part II, and contributed productive discussion and suggestions for texts and themes.

Alice Lay has contributed a great deal of work as a research assistant. She prepared lists of references, did research on the authors and texts, discussed plans for the work, and, in general, made significant contributions to the research for this book.

Hasok Chang's suggestions and recommendations have shaped the project and have improved it significantly; many of the most interesting and effective features of the work are a result of his recommendations.

Don Howard took time to discuss the project several times when it was first proposed, suggested directions the project could take, and had a significant influence on the initial trajectory of the work.

James W. McAllister provided an early and incisive review of the project, which had a great and salutary influence on its direction and shape, especially in the selection of texts and themes.

Several anonymous reviewers for Routledge made comments that stimulated revisions and improvements. Colin Morgan at Swales & Willis oversaw the final process of proofreading and preparation of the text and figures. Lorrie Spence assisted with scans for two of the figures.

Discussion with colleagues, including Richard Burian, Benjamin Jantzen, James Klagge, Deborah Mayo, Walter Ott, and Joseph Pitt, has been stimulating and invaluable. Elements of the project arise from talks and panels I have attended, and books I have read by numerous others in the HOPOS, STS, HPS, & HPS, and history of science communities, to whom I am grateful.

This book is dedicated to my family.

The author and publishers wish to thank the following for their permission to reprint copyrighted material:

Cambridge University Press for Descartes, Rene. 1985 [1647]. *Principles of Philosophy*, in *The Philosophical Writings of Descartes, vol. I*, translated by Cottingham, Stoothoff and Murdoch, Cambridge University Press. Part I, §1–8, §13–14, §18–21, §26–8; Part II, §4–26, §32–45, §54–64. Part III, §26–29, §119, §140; Lakatos, Imre, 1970. "Falsification and the Methodology of Scientific Research Programmes," from *Criticism and the Growth of Knowledge*, Cambridge: Cambridge University Press, pp. 132–138; Newton, Isaac. 1978 [original date unknown]. *Unpublished Scientific Papers of Isaac Newton*, ed. and trans. A. Rupert Hall and Marie Boas Hall. "On the Gravity and Equilibrium of Fluids," pp. 121–148.

Hackett Publishing for Kant, Immanuel. 2001 [1783]. "How is Pure Natural Science Possible?" *Prolegomena to any Future Metaphysics*, Translated by James W. Ellington. Indianapolis: Hackett, pp. 38–64.

MIT Press for Kuhn, Thomas. 1987. "What Are Scientific Revolutions?" in *The Probabilistic Revolution*, ed. Lorenz Krüger, Lorraine Daston, and Michael Heidelberger. Cambridge: The MIT Press, pp. 7–22.

Notre Dame University Press for Hans Reichenbach. 1938. *Experience and Prediction*, Chapter I, "Meaning," §1, "The Three Tasks of Epistemology," pp. 3–16. © University of Notre Dame Press; Comte de Buffon. 1981 [1749]. "Initial Discourse," trans. John Lyon in *From Natural History to the History of Nature*. University of Notre Dame Press, pp. 97–8, 100–104, 122–127. ©Notre Dame University Press.

Princeton University Press for Duhem, Pierre, 1954 [1914]. "Physical Theory and Experiment," from *The Aim and Structure of Physical Theory*, trans. Phillip Wiener. Princeton University Press, pp. 180–205 and pp. 208–211;

Shapin, Steven and Simon Schaffer. 1985. *Leviathan and the Air Pump.* Princeton: Princeton University Press, pp. 332–344.

University of Chicago Press for Kuhn, Thomas. 1977. "The Relations between the History and the Philosophy of Science," *The Essential Tension,* Chicago: University of Chicago Press, pp. 3–20; Kuhn, Thomas. 1977. "The History of Science," in *The Essential Tension,* Chicago: University of Chicago Press, pp. 105–126; Rudwick, Martin. 1985. *The Great Devonian Controversy.* Chicago: The University of Chicago Press. Chapter 1, pp. 3–16.

Every effort has been made to trace the copyright holders but if any have been inadvertently overlooked the publisher will be pleased to make the necessary arrangements at the first opportunity.

Editor's Introduction

The history of the philosophy of science (HOPOS) is an interdisciplinary endeavor. It may have begun as the project of tracing the developments of the philosophy of science from the discipline's origins in the Vienna Circle in the early twentieth century. Much recent work in the field, however, now predates, or is not strictly relevant to, the philosophy of science as a subdiscipline of philosophy. Instead, HOPOS is far-reaching and, from a philosophical perspective, staggeringly open-ended. At HOPOS conferences, which have taken place every two years since 1990, papers are presented on Carnap, Reichenbach, and Hempel, but also on Aristotle, Boyle, Malebranche, Leibniz, Feynman, Hilbert, Dedekind, Haeckel, Darwin, and many others from the history of science. Work in HOPOS encompasses even disciplines that might not have been considered scientific at the time, and disciplines, such as mathematics, that are even now not considered empirical sciences.

The flexibility of HOPOS as a disciplinary endeavor has made it strong. On the email list for HOPOS: The International Society for the History of Philosophy of Science, Alan Richardson, George Gale, and others have debated whether HOPOS can count as a *Fach* in the German sense, that is, whether it has its own special subject matter, sometimes referred to as a canon, and a distinctive approach to that subject matter. On whichever side one comes down in the debate on this issue, it is significant that the question came up, and that it was debated so carefully. The open-endedness of HOPOS has enriched and strengthened the community of scholars, but has also put in question the cohesion of that community and of the research tradition at its core.

In my view, this is a situation that HOPOS shares with science itself. The demarcation problem, of how to distinguish science from non-science, is not easy to solve. The difficulty of finding a solution surely is attributable at least in part to the fact that science evolves constantly, and thus that any attempt to list its characteristic features a priori might be falsified by the evolution of science in practice. To use an example that already is becoming well-worn, string theory may fail several demarcation criteria that have been used in the past. But it is not abandoned as non-science on those grounds, because it has virtues, including mathematical fruitfulness, that scientists value in practice. Some (Chang 2012, *Is Water H₂O?*) have suggested that, far from listing demarcation criteria, we ought to encourage pluralism about scientific methods, theories, and approaches.

Similarly, how ought we to define HOPOS, and to distinguish it from related intellectual pursuits? The account of HOPOS as history *of* the philosophy of science at least provided a time limit for the subject matter it investigated, and a clear notion of the cast of players. On this account, HOPOS is an inquiry into the history, influences, motivations, and arguments of the Vienna Circle and associated philosophers, who founded the philosophy of science. Questions relevant to this view of HOPOS include, then, the changes to the views and interrelationships of the Vienna Circle philosophers over time, the influence and impact of their views, the material and intellectual conditions that influenced their careers and thought, and the like.

Present-day HOPOS manifestly does not limit itself to anything like this list of topics, though they are central to the research of many HOPOS practitioners. If you glance at a recent HOPOS conference program, or the contents of a HOPOS journal such as *Studies in the History and Philosophy of Science*, you will see papers that investigate the origins of, influences on, and justifications of specific scientific results, papers that analyze the history and influence of instrumentation on scientific theories, papers that focus on scientific development and change, papers that focus on theory building or theory choice, and papers that appear to be mainly philosophical and that have very little historical content.

Work in HOPOS is also marked by methodological pluralism. Among the methods that can be observed in recent work in HOPOS, History and Philosophy of Science (HPS), Integrated History and Philosophy of Science (&HPS), Science and Technology Studies (or Science and Technology in Society (STS) and related fields can be found:

- Using the methods of history, sociology, and other social sciences to investigate the sciences. This work sometimes is done in other, related research programs as well, including Science and Technology Studies programs. For instance, in some HOPOS and STS programs, doctoral students who wish to focus on the historical practice of HOPOS are required to obtain a Master's degree in history.

- Using philosophical methods to analyze the scientific and philosophical contributions of scientists in history. This is sometimes referred to as HPS or &HPS. This strain of research advocates a thoroughgoing integration of the history and philosophy of science, rather than choosing primarily to take an historical approach and to allow some philosophical influences, or choosing to take a philosophical approach and to include a few case studies. &HPS advocates building a truly integrated and coherent discipline, rather than merely allowing distinct disciplines to inform each other.
- Contributing to the history of logical positivism and logical empiricism, which some consider the first tradition of philosophy of science properly speaking. This was among the impetuses for the institutional founding of HOPOS: The International Society for the History of the Philosophy of Science.
- Non-contemporary philosophical accounts of science, which might mean any text that takes a synthetic look at scientific traditions and their philosophical consequences. Galileo's *Two New Sciences*, Berkeley's *The Analyst*, Boyle's *Skeptical Chymist*, and Whewell's *Philosophy of the Inductive Sciences* would all count as early HOPOS texts in this sense.
- Examining accounts of scientific change, for instance, using scientific methods. Probing the empirical and deductive foundations of philosophy of science itself. This is arguably not HOPOS, but some work that appears indistinguishable from HOPOS work is done in this vein—for instance, using historical case studies to test conjectures or hypotheses about the origin or development of a theory and its philosophical consequences. Donovan, Laudan, and Laudan, *Scrutinizing Science*, is an example of this sort.
- Other methods and traditions may well be available to a student who looks further.

I do not seek to make any normative recommendations here. I find this methodological and thematic pluralism to be one of the stimulating and bracing elements of HOPOS research. And it for this reason that the publishers and I have chosen—at the recommendation of one reviewer—the open-ended title: *Philosophy, Science, and History: A Guide and Reader*. Moreover, HOPOS and the allied pursuits of &HPS, HPS, STS, and the like change as science changes, and as the political and social contexts change.

That said, it is eminently possible, and necessary, to tell the history of HOPOS itself as a discipline, and to put one's own research into this context. I have organized the Parts of this book with the aim in mind of pointing out some major features of the development of the discipline, especially those that may be useful to a student who is encountering the material or the standards for HOPOS work for the first time. The Introductions to the material, especially, aim to place classic texts and debates in a wider context and to give references to further reading.

One of my secondary aims was to introduce those new to the discipline, whether students or not, to the basic elements that one is expected to know if one is traveling in HOPOS circles. I know that I have left out some of the rarer and more interesting flora and fauna, but I have tried to make this a reliable field guide.

Beyond remarking on this broad shift toward pluralism in HOPOS study, and beyond making a few remarks about the layout of the work, I will let the texts and introductions speak for themselves. However, I will point out in closing that this work has been conceived and constructed as a guide to the field as I know it, and as my advisors on the project felt it would be most effective. As such, it focuses mainly on the Western tradition, and on work in English, with a limited number of references in French and German.

However, the field is changing, and is growing. There is an increasingly international base of HOPOS studies. Israel and Hungary have long been sources of significant HOPOS scholarship. Mexico, Brazil, India, Spain, Turkey, South Africa, Portugal, and surely others have communities of varying sizes working in HOPOS or HOPOS-related fields. In 2011, *Brazilian Studies in Philosophy and History of Science*, edited by Décio Krause and Antonio Videira, appeared as volume 290 in the Boston Studies in the Philosophy and History of Science series. The CFCUL in Lisbon is also pursuing a number of current projects; see http://cfcul.fc.ul.pt/projectos/. The history of Chinese and Arabic science and philosophy is too much neglected in Western scholarship, and is an area in which, one hopes, more work will be done in future.

A brief note on the texts, in closing. They were chosen with several variables in mind. One is whether the text is already available freely online: if an entire relevant text was available, often I limited myself to noting that fact in the introduction to the relevant section. Moreover, I include mostly what might be considered "primary" sources, with a few exceptions. My aim was not to select the best of contemporary scholarship in HOPOS, but rather, as I've said, to provide a guide to what one needs to know to work in the field, and to understand as much work in the field as possible. I have made an effort to give references to excellent more recent work, to key debates, and to ideas for further research or study, in the introductions to each part.

In so doing, I am sure that I have missed a great deal of excellent work, but I have given a representative sample. If this book is being used as a textbook, the professor for the course surely will find it useful to supplement this material depending on her or his aims for a given course; the introductions are provided as a springboard.

Note

1. With a few exceptions, the references in the footnotes to the primary texts can be found in the reference list in the Introduction to that Part.

Approaches to the History and Philosophy of Science

Themes

This part focuses on the methodology and argument of well-known approaches to studying the history and philosophy of science. To this end, I have excerpted those who defend the **rational reconstruction** of science as defining the limits of inquiry into what is properly called science (Hempel, Lakatos), and from those who argue that **historical and sociological** methods result in knowledge that is of independent value and validity (Rudwick, Shapin and Schaffer). The distinction between the **contexts of discovery and of justification** tracks this difference in approach (Reichenbach). Others argue that moments in science that are most productive, including **revolutionary changes to scientific paradigms**, are not susceptible to rational reconstruction at all (Kuhn), or give a naturalist argument that history is itself a blend of natural science and constructed categories (Santayana). Contemporary philosophers have developed research programs that incorporate analyses of **material culture, experiment, theory change**, and **historical epistemology and methodology**, which indicate potential for increased collaboration across the disciplines.

CHAPTER 2

Introduction to "Approaches to the History and Philosophy of Science"

Early History of Science

The history of science I will discuss here is the history of science as practiced in the contemporary discipline. Writing on the history of science more generally has a multifarious character. Practicing scientists who teach might incorporate materials on the history of science into their course syllabi, or might write books that deal with episodes in the history of science. Some write about the history of science for the popular press, as Dava Sobel did with *Galileo's Daughter* (1999). Historical episodes in science serve as the source material for fictional narratives, as in the case of John Banville's *Revolutions Trilogy* (2001). But here, we will deal with history of science as a modern discipline, as it is practiced in academic departments, including departments of HPS and STS.

Whewell's *History of the Inductive Sciences* (1837) often is counted as the first Western history of science. The second half of the nineteenth century saw a definite increase in studies of science's history, including Friedrich Albert Lange's *History of Materialism* (1866, a history of science and of philosophy), Ernst Mach's *Mechanics in Its Development* (1960/1883) and Paul Tannery's *For the History of Hellenic Science* (1887). One of the most well known of the pioneers of the discipline is the Belgian historian George Sarton (1884–1956), who founded the journal *Isis*, and who wrote several books on the study of the history of science (Sarton 1927–48).

The development of the history of science over the nineteenth century was influenced by trends and developments in history proper. As Laura

Snyder has emphasized recently, Whewell's coinage "scientist" only came to be used in its contemporary sense relatively late, in the nineteenth century (Snyder 2012). Moreover, the nineteenth century was alive with debate about what counted as true science. Early in the century, Wilhelm von Humboldt, who founded the university in Berlin, had encouraged a pluralistic view of science ("Wissenschaft"), as a rigorous methodology applied to any subject matter studied by the four university faculties: law, philosophy, theology, and medicine. This tolerance gave way, toward the end of the century, to a well-known debate between Wilhelm Dilthey and Wilhelm Windelband about differences between the methods of the human and the natural sciences (the *Geistes-* and the *Naturwissenschaften*).

Dilthey, in his major work *Introduction to the Human Sciences* (1883), argued that the methods of the human sciences were independent of the methods of the natural sciences.[1] Dilthey employed methods of interpretation, along the lines of those used by Johann Herder (1744–1803) and Johann Gustav Droysen (1808–1884), to support a distinction between explaining (erklären) a subject of interest, whether an event or a phenomenon, and understanding (verstehen) that subject.[2] The human sciences, including history, can contribute to understanding on several fronts: illuminating the context of an event, of course, but also showing how distinct areas of knowledge can be investigated together. Moreover, weighing the influence of one theory on another involves an element of interpretation.

In 1894, Wilhelm Windelband (1848–1915) gave a rectorial address at Strasbourg, "History and Natural Science," in which he responds to Dilthey's arguments. As a neo-Kantian, Windelband appeals to Kant's argument in the *Metaphysical Foundations of Natural Sciences* that only sciences with an a priori, mathematical basis of lawlike "principles" or rules have a truly scientific foundation. Windelband argues that the "human sciences" have a distinct aim and subject matter from the natural sciences. The natural sciences are "nomothetic," or law-governed, while the human sciences are "idiographic," describing particular events or specific phenomena. Not only can these analyses of the specific and particular not be generalized or brought under a universal law, Windelband argues, they cannot be objective, because they are relative to a specific historical or cultural context and thus essentially subjective.

In addition to the dialogue over the natural and human sciences, there are other trends in nineteenth-century history, philosophy, and sociology that had an impact on early approaches to the study of science. The topic of history more generally was central to key developments in philosophy. G. W. F. Hegel's (1770–1831) idealist approach to history was challenged by the young Hegelians, including Ludwig Feuerbach (1804–1872) and Karl Marx (1818–1883). By the end of the century in Germany, there was a more general division between naturalism, materialism, and empiricism on one hand, and idealism on the other.

The impact of materialist and positivist approaches to history is part of the background to the debate over the status of history as a science. If historical events can be reduced to material law, then history is a science, even on Kant's and Windelband's accounts. The English positivist historian Henry Thomas Buckle (1821–1862), among others, argued for this claim.

In France, Auguste Comte (1798–1857) wrote the *Course of Positive Philosophy*, a founding text of modern sociology. Comte's ideas became widely known, and his views on the stages of development of societies and the classification of the sciences were influential. Émile Bréhier, Émile Boutroux, and Victor Delbos would later criticize Comte's views on the necessary progress of philosophy in history (Chimisso 2001, 131). Gaston Bachelard (1884–1962) criticized Comte's view that science makes continuous progress. However, as Chimisso notes, even some of the historians and philosophers who criticized Comte's teleology adopted versions of his views about the "three-stage periodisation of the progress of the scientific mind," and about the stages of development of human society (*ibid.*).

In the United States, the Scottish intellectual tradition influenced university curricula; the Common Sense philosophers Thomas Reid (1710–1796) and Dugald Stewart (1753–1828) were particularly influential. However, there was also a strong naturalist bent to intellectual life, both in the philosophical and in the scientific senses of the word. Louis Agassiz (1807–1873), a Swiss biologist and student of Georges Cuvier and Alexander von Humboldt, gave a series of lectures in the mid-nineteenth century: at the Lowell Institute in Boston, and at Harvard, where he settled as a professor. Charles Sanders Peirce (1839–1914), a pioneering figure of American Pragmatism, was active as a writer during the second half of the nineteenth century, though he rarely taught on a university faculty. He did publish widely, and gave lectures. Ralph Waldo Emerson (1803–1882) was an influential figure on the American scene, though, like Peirce, he spent most of his career outside the university. Emerson's 1837 lecture "The American Scholar" is considered a watershed moment in American intellectual culture. The empiricist philosopher William James (1842–1910) spent his academic career at Harvard, teaching philosophy, physiology, and psychology. William James's philosophically oriented students at Harvard included Alain Locke (1885–1954), the "Father of the Harlem Renaissance" and author of theories of "value, pluralism, and cultural relativism"; C. I. Lewis (1883–1964), a leading figure in early analytic philosophy; and George Santayana (1863–1952), who developed a naturalist theory of history and of science (Carter 2012).

Santayana's mature naturalist philosophy is found in his *Scepticism and Animal Faith* (Santayana 1923; see also Saatkamp 2010). Santayana's earlier account of the relationship between history and science in *The Life of Reason*, excerpted here, begins with a straightforward statement of naturalism about historical methods, that "historical research" is "a part of physics." However, there are disanalogies between historical and physical methods, of necessity. We can't perform repeatable experiments on the events of history, as we can on the objects of physics.

Further, history involves first-hand observation and the long series of events of the past. In neither case can we amass all the data. It would be impossible, and pointless, to try to gather all the sense impressions of all observers of history, for instance. In analyzing events through their records and artefacts, one must choose a set of concepts and conventions to give structure to the historical record: "historical terms mark merely rhetorical unities, which have no dynamic cohesion." The *laws* of history are the laws of physics, but the *objects* of history, its conceptual categories, are "rhetorical" or conventional tools. Thus, Santayana found a way to argue that history is a law-governed science, but that its concepts are stipulated.

Logical Empiricism: Rational Reconstruction, the Unity of Science, and Demarcation

Over the first half of the twentieth century, the logical empiricists, many of whom emigrated to the United States following the rise of the Nazi Party in Germany, defended a number of theses in the philosophy of science.[3] Logical empiricism emerged in the 1920s from the Berlin Circle or "Society for empirical philosophy," logical empiricist members of which included Hans Reichenbach (1891–1953) and Carl Hempel (1905–1997), and the Vienna Circle or "Ernst Mach Society," logical empiricist members of which included Moritz Schlick (1882–1936), Rudolf Carnap (1891–1970), Philipp Frank (1884–1966), Otto Neurath (1882–1945), and Hans Hahn (1879–1934). Inspired by Frege, by Russell, and by Wittgenstein's 1922 *Tractatus Logico-Philosophicus*, the logical empiricists advocated giving an account of knowledge by means of logical analysis of language, with the aim of eliminating metaphysics, mysticism, and pseudoscience.[4]

Early in the logical empiricist movement, many of the philosophers in it endorsed a form of verificationism drawn from the *Tractatus*: that all complex or "molecular" propositions are truth functionally composed of "atomic" propositions, that "atomic" propositions correspond to single facts or states of affairs, and that it can be checked in experience whether these facts or states of affairs obtain or not (Hempel 1935, 50ff.). Verificationism was allied to a correspondence theory of truth: that "truth consists in a certain agreement or correspondence between a statement and the so-called 'facts' or 'reality'" (Hempel 1935, 49).

Later, Neurath and Carnap challenged strict verificationism and the correspondence theory, urging instead a form of coherence theory, the view that "truth is [. . .] a certain conformity of statements with each other" (Hempel 1935, 49), and for "confirmation" of a theory's claims by evidence, not verification, as the criterion of meaning. Confirmation bypasses the requirement that complex propositions be truth-functionally composed of atomic, verifiable elements. Instead, for a theory's claims to be confirmed, the hypotheses or predictions it makes must be either confirmed or rejected by the evidence. The hypothesis itself, and indeed the resulting claim, need not be a truth-functional composition of the atomic facts or states of affairs provided by the evidence.

Whether they adopted correspondence/verification or coherence/confirmation, many of the logical empiricists argued that true science has a central, characteristic method. While the account of this method differed among the philosophers, there are two characteristic tenets of logical empiricism that are central to the texts examined here: rational reconstruction, and the unity of science.

The method of rational reconstruction is often considered to go along with the "hypothetico-deductive" (H-D) or "deductive-nomological" (D-N) methods in philosophy of science (see, for example, Bechtel and Abrahamsen 2005, 421; Rosenberg 2013, 30ff.; Suppe 1977). As Rosenberg (2013, 30–31) puts it:

> In Hempel's original version the requirements on deductive nomological explanation were as follows:
> 1. The explanation must be a valid deductive argument.
> 2. The *explanans* must contain at least one general law actually needed in the deduction.
> 3. The *explanans* must be empirically testable.
> 4. The sentences in the explanans must be true.

These criteria distinguish scientific from other forms of explanation, according to Hempel. In the context of history and philosophy of science, it is significant that scientific theories generally are not first formulated as deductive nomological explanations of the phenomena. As Reichenbach observes, there is a difference between the way a scientist or mathematician actually arrives at a result, and the logical way that person reconstructs the result afterward. Reichenbach uses the term rational reconstruction, which he attributes to Rudolf Carnap, to describe this difference:

> Epistemology does not regard the processes of thinking in their actual occurrence; this task is entirely left to psychology. What epistemology

intends is to construct thinking processes in a way in which they ought to occur if they are to be ranged in a consistent system [. . .] Epistemology thus considers a logical substitute rather than thought-processes. For this logical substitute the term *rational reconstruction* has been introduced.

(Reichenbach 1938, 5–6)

In cases the D-N model takes to be characteristic, rational reconstruction shows how a result can be deduced from a theory, which may consist of a set of laws, axioms, definitions, and accepted background evidence (which may include facts, conventions, observations, and the like). The explicit statement of how such inferences are made yields an account of the "context of justification" of any given result:

> The way, for instance, in which a mathematician publishes a new demonstration, or a physicist his logical reasoning in the foundation of a new theory, would almost correspond to our concept of rational reconstruction; and the well-known difference between the thinker's way of finding this theorem and his way of presenting it before a public may illustrate the difference in question. I shall introduce the terms *context of discovery* and *context of justification* to mark this distinction. Then we have to say that epistemology is only occupied in constructing the context of justification.
>
> (*ibid.*, 6)

Defenders of the D-N method, again including Hempel, argue that the context of justification distinguishes scientific from non-scientific explanation. Information about the context of discovery, for instance, facts about the scientific community in which the discovery was made, the tacit or practical knowledge that community accepts, the material conditions in which a result is proven, and so forth, are germane to evaluation of science only if it is a material part of a rational reconstruction of a claim or claims to scientific knowledge.[5]

Rational reconstruction and the D-N method are taken sometimes to demarcate philosophy of science, and science itself, from history, sociology, and allied disciplines. This demarcation is, however, complex. Hempel concedes that *some* historical explanation can be lawlike, and thus can be rationally reconstructed. For Hempel, "The main function of general laws in the natural sciences is to connect events in patterns which are usually referred to as *explanation* and *prediction*" (Hempel 1942, 35). Hempel argues that history, as well as physics, can be considered an explanatory and even a predictive science:

The preceding considerations apply to *explanation in history* as well as in any other branch of empirical science. Historical explanation, too, aims at showing that the event in question was not "a matter of chance," but was to be expected in view of certain antecedent or simultaneous conditions. [. . .] If this view is correct, it would seem strange that while most historians do suggest explanations of historical events, many of them deny the possibility of resorting to any general laws in history.

(Hempel 1942, 39)

Hempel suggests that the reason historians deny this nomological character of history is their use of hermeneutic or interpretive methods, such as "historical narrative," in historical explanations, which may include techniques of "interpretation," sensitivity to "meaning," "empathy," or "feeling oneself into" the perspective of a historical actor (p. 44). Hempel argues that "the kind of 'understanding' thus conveyed must clearly be separated from scientific understanding" (p. 45).

Hempel argues explicitly in Hempel (1965) that only scientific explanation results in valid claims to scientific knowledge. Claims to have "understood" a historical episode, without an equivalent claim to have explained it scientifically using the D-N method, are erroneous. Several of the texts printed here (Rudwick, Shapin and Schaffer) reject this view of history. Recent work in philosophy subjects Hempel's skepticism about the validity of understanding to critical analysis (see de Regt, Leonelli, and Eigner 2009). Moreover, earlier historians and philosophers including Wilhelm Dilthey, Johann Droysen, and Johann Herder defended a hermeneutic view of history according to which the *methods* of the "human sciences" are independent from those of the natural sciences (see below). All three argued that historical events follow natural law, but they did not restrict claims to historical knowledge to the results of deductive-nomological explanation.

Hempel's discussion tacitly relies on some analogue to Reichenbach's distinction between discovery and justification, or Lakatos's account of rational reconstruction in the analysis of change over time. These arguments rule out hermeneutic analysis and the reconstruction of a historical narrative, for instance, arguing that these are reducible to empirical psychology or sociology, or are not valid forms of knowledge at all.

Hempel's approach is a later iteration of the logical empiricist response to discussions of the divisions between the sciences, which was to promote Ernst Mach's tenet of the "unity of science." A chapter of Mach's 1883 *The Science of Mechanics* called "The Economy of Science" explains Mach's "biological-economical" account of scientific knowledge, according to which scientific knowledge provides a stable and economical orientation to the environment (Mach 1960/1883). Mach challenged the distinction between the physical and the psychological, which arguably is one root of earlier discussions of the human and the natural sciences (see Pojman 2011, §4.4).

Along Machian lines, and, at first, along the lines set out in Carnap's *Logical Structure of the World* (1967/1928, usually called the "*Aufbau*"), the logical empiricists explored the project of showing how, in general, scientific knowledge is formulated. Carnap's account aims to show how scientific systems are "constituted" from observation and inference, a project Carnap refers to as "constitution theory."

There is a contemporary debate over how to read Carnap's *Aufbau*: two among the competing interpretations are W. v. O. Quine's empiricist reading, and Michael Friedman's and Alan Richardson's neo-Kantian reading (see Friedman 1987, 1999; Quine 1977/1969; Quine 1980/1951; Richardson 1998).

For our purposes here, it is enough to point out that Carnap is among the first to take a novel approach to the problem of the "unity of science." Previous debates, including the one between Windelband and Dilthey, had distinguished between the sciences based on their methodology. It is significant to Windelband's view, for instance, that history interprets specific events (is idiographic), while physics seeks general laws (is nomothetic). If scientific explanation is relative to a constitution system along the lines of the *Aufbau*, however, the methodological distinction is no longer relative to a science, but to a system. Moreover, for the early Carnap, the system of science progressively would become less and less differentiated, and thus would show increasing unity and coherence as time passed, as a larger number of facts could be accounted for by a more and more economical system.

In 1938, the first volume of the Vienna Circle's *International Encyclopedia of Unified Science* appeared. Over the following thirty years, twenty volumes appeared in this series, including Thomas Kuhn's *Structure of Scientific Revolutions* (1962). In a 1969 volume, Neurath explains his own theory of the unity of science, which differs somewhat from Carnap's (Neurath 1969). As Jordi Cat puts it:

Neurath [. . .] favored a less idealized and less reductive model of unity predicated on the complexity of empirical reality. He spoke of an "encyclopedia-model", instead of the classic ideal of the pyramidal, reductive "system-model". The encyclopedia-model took into account the presence within science of uneliminable and imprecise terms from ordinary language and the social sciences and emphasized a unity of language and the local exchanges of scientific tools. [. . .]

His view was not constrained by Carnap's ideals of conceptual precision, deductive systematicity and logical rigor. No unified science, like a boat at sea, would rest on firm foundations. This weaker model of unity emphasized empiricism and the normative unity of the natural and the human sciences.

(Cat 2013, §1.4)

Neurath's metaphor of science as a ship at sea, which would have to be repaired as it sailed, is a compelling statement of his "fallibilism" about science. According to Neurath, science must be overhauled as it is in progress, and no plank on the ship is immune from being replaced or re-designed. Fallibilism is Neurath's response to "foundationalism" in the philosophy of science, according to which certain elements of scientific theories are too central, too essential, to be jettisoned or amended.

The discussions of the unity of science informed later approaches to the "demarcation problem," that is, the question of how to distinguish science from non-science, from pseudoscience, and from metaphysics. Hempel's D-N approach is an early proposed solution to the demarcation problem. In the 1960s, Karl Popper (1902–1994) defended the claim that science is distinguished from non-science, or from pseudo-science, by the criterion of falsifiability: if conjectures, or hypotheses, made on the basis of a theory can be falsified by the evidence, then the claims are scientific (Popper 1963). Popper's falsifiability criterion would influence the approach taken by Imre Lakatos (1922–1974), among others, to the question of how to evaluate the progress through time of scientific theories, a problem that would come to the fore in in 1960s.

Scientific Change: Rational Reconstruction, Normal Science, and Scientific Revolutions

Thomas Kuhn's landmark *The Structure of Scientific Revolutions* appeared in 1962. In 1965, the International Colloquium in the Philosophy of Science was held in London, and Kuhn's work, along with the allied topic of scientific theory change, was the order of the day. Philosophers of science including Paul Feyerabend (1924–1994), Margaret Masterman (1910–1986), Alan Musgrave (1940–), Stephen Toulmin (1922–2009), Lakatos, and Popper responded to Kuhn's work in *Structure. Criticism and the Growth of Knowledge* (Lakatos and Musgrave 1970), the proceedings of the conference, has become a classic in the analysis of scientific change along with *Structure*; an excerpt from Lakatos's contribution is printed here.

In "What Are Scientific Revolutions," reprinted here, Kuhn restates his view of revolutionary change in science.[6]

> Revolutionary change is defined in part by its difference from normal change, and normal change is [. . .] the sort that results in growth, accretion, cumulative addition to what was known before ["normal science"—LP]. Scientific laws, for example, are usually products of this normal process: Boyle's law will illustrate what is involved. Its discoverers had previously possessed the concepts of gas pressure and volume as well as the instruments required to determine their magnitudes. The discovery that, for a given gas sample, the product of pressure and volume was a constant at constant temperature simply added to the knowledge of the way these antecedently understood variables behave. [. . .] Revolutionary changes are different and far more problematic. They involve discoveries that cannot be accommodated within the concepts in use before they were made. In order to make or to assimilate such a discovery one must alter the way one thinks about and describes some range of natural phenomena.

An alteration to "the way one thinks about and describes some range of natural phenomena" is an alteration to a paradigm, which Kuhn calls a paradigm shift. In *Structure*, Kuhn argues that paradigm shifts and revolutions in science likely were not rationally reconstructible, at least in their entirety. If a change is prompted by a discovery "that cannot be accommodated within the concepts in use," then new concepts and theoretical frameworks need to be developed. These concepts and frameworks are not direct inferences from evidence, but rather are ways of thinking about and ways of describing the phenomena. As such, they are not determined by direct inference from the new discovery that has prompted the paradigm shift. Instead, they are at least partly "volitional" or conventional. Adopting a new way of measuring space, for instance, in response to a challenging new discovery may solve the problem posed by the discovery. But the choice to adopt the new way of measurement is not uniquely determined by the discovery: another response to the challenging discovery, or even another class of responses, might solve the problem equally well.

In his contribution to *Criticism and the Growth of Knowledge*, Lakatos responded that scientific revolutions could be rationally reconstructed, just as scientific theories and discoveries were, but from a different perspective. Lakatos introduced the notion of a "scientific research programme," an entity that stretches over time. A research program contains a *hard core* (the elements essential to the theory), a protective belt of *auxiliary assumptions*, and a *positive and negative heuristic*.

The negative heuristic is the ability of researchers to block the falsification of a theory. Again, Popper (1963) distinguished scientific from non-scientific theories by arguing that the claims to knowledge of the former are each *falsifiable* by evidence or testing. Duhem had argued much earlier (see Part II-A of this book) that scientific claims are not falsifiable piecemeal, because what is tested in experiment is not a single claim or hypothesis. When positing hypotheses for confirmation or falsification, the framework of the scientific theory as a whole is also posited, along with a series of auxiliary hypotheses that make the testing of the hypothesis possible.

In "Physical Theory and Experiment," printed later in this book, Duhem asserts that "The physicist who carries out an experiment, or gives a report of one, implicitly recognizes the accuracy of a whole group of theories" (Duhem 1954/1914, 183). "The only thing the experiment teaches us is that among the propositions used to predict the phenomenon and to establish whether it would be produced, there is an error; but where this error lies is just what it does not tell us" (*ibid.*, 185). When an experiment constructed to test a hypothesis seems to show that the hypothesis is false, Duhem points out, it in fact only shows that entire system of the hypothesis, plus the theory's definitions and axioms, plus the auxiliary hypotheses (which may be theories themselves) that allow the researcher to test the theory has been shown to contain an error. According to Duhem, that particular experiment cannot identify which element of the system is at fault, since the entire system is tested at once.

Lakatos calls his "negative heuristic" the practice of scientific researchers to "block the modus tollens." The "modus tollens" is taken from Duhem's model of scientific testing: theory, plus auxiliaries, are the initial premises, and then observations are made to attempt to falsify the theory. The propositions of a theory are $\{T_1, T_2, \ldots, T_n\}$, and the auxiliary hypotheses are $\{A_1, A_2, \ldots, A_n\}$. The modus tollens may then be of a form analogous to

1. Hypothesis: If the theory $\{T_1, T_2, \ldots, T_n\}$ is true, then we will make an observation O_1 in an experiment.
2. Result: O_1 was not observed.

Therefore, the theory $\{T_1, T_2, \ldots, T_n\}$ is false.

Some even assert that a particular element of the theory can be isolated and falsified:

1. Hypothesis: If the claim T_1 is true, then we will make an observation O_1 in an experiment.
2. Result: O_1 was not observed.

Therefore, T_1 is false.

Duhem's point was that experimentation does not work this way in fact. Instead, the argument is, in practice:

1. Hypothesis: If the theory $\{T_1, T_2, \ldots, T_n\}$ is true, and the auxiliary hypotheses $\{A_1, A_2, \ldots, A_n\}$ hold, then we will make an observation O_1 in an experiment.
2. Result: O_1 was not observed.

Therefore, either one or more claims of the theory $\{T_1, T_2, \ldots, T_n\}$ is false, or one or more of the auxiliary hypotheses $\{A_1, A_2, \ldots, A_n\}$ does not hold, or both.

With the "negative heuristic," Lakatos is pointing out, with Duhem, that a scientist can modify the auxiliaries to block the modus tollens and protect the theory. However, scientists in a research programme do not protect all elements of the theory equally.

> The negative heuristic specifies the "hard core" of the programme which is "irrefutable" by the methodological decision of its protagonists; the positive heuristic consists of a partially articulated set of suggestions or hints on how to change, develop the "refutable variants" of the research-programme, how to modify, sophisticate, the "refutable" protective belt.

Lakatos accounts for the conventional or volitional aspects of a theory with his concept of the "hard core" plus the "negative heuristic" (Reichenbach and Poincaré also discuss these volitional aspects). This, he says, "rationalizes classical conventionalism" to an extent, because it makes clear that the "protective belt" around the hard core is considered above revision (immune from the modus tollens) by choice, not because it is actually essential to the theory. In other words, the choice to build a protective belt around part of the theory is not determined by the evidence or by properties of the theory, but by convention or volition.

The "positive heuristic of the program," on the other hand, "saves the scientist from becoming confused by the ocean of anomalies" (Lakatos and Musgrave 1970, 135). It consists of the theory, plus a number of idealized "models" which are employed in description, explanation, and prediction in "simulating reality." This simulation is what allows for the formulation of hypotheses to test the theory; this is how Lakatos interprets Popper's "supreme heuristic rule: 'devise conjectures which have more empirical content than their predecessors'" (Lakatos and Musgrave 1970; Popper 2002/1934, §11 and §70).

A Lakatosian research program is a Kuhnian paradigm plus pragmatic commitments: Lakatos appeals to the motivation of researchers in a program to save certain elements of a given theory rather than others. To Lakatos, this motivation can be given in Kuhn's own terms, in terms of the "puzzle-solving" ability of researchers working within a given paradigm, in "normal science." However, for Lakatos, the "puzzle-solving" ability is characterized, not just in terms of the power of a particular theory or even paradigm, but in terms of a progressive or degenerating "problemshift."

The point of disagreement between the early Kuhn and Lakatos, then, turns on which aspects of scientific change involve conventional choice or volition. To Kuhn, the change from one *paradigm* to another, a scientific revolution, is not always rationally reconstructible. Rather, there may be volitional or conventional aspects, or other non-logical moves, involved in abandoning a former theory and building a new theory.

To Lakatos, first, it is possible to assess even abandoned theories for their "heuristic power." The positive heuristic is not concerned with avoiding refutation, nor is it affected by refutation; rather, it is bolstered by successive verifications. "We may appraise research programmes, even after their 'elimination', for their *heuristic power:* how many new facts did they produce, how great was 'their capacity to explain their refutations in the course of their growth.'"

Second, insofar as a theory is eliminated, it is for a reason: it has become a degenerating, and not a progressive, research program, in terms of its ability to solve problems when confronted with new evidence. These problems are chosen, not in the face of anomalies, but as a result of trying to find new ways to provide the theory with increasing empirical content: "Which problems scientists working in powerful research programmes rationally choose, is determined by the positive heuristic of the programme rather than by psychologically worrying (or technologically urgent) anomalies" (Lakatos and Musgrave 1970, 137). This means, then, that if a given research program cannot solve a problem or problems specified by the positive heuristic, then a new program is needed *to solve that problem or problems*. The need for a solution to a specific problem or set of problems is a rational link across paradigms, a link that Kuhn denied, according to Lakatos.

A number of issues arise from this discussion. First, it is not clear whether Kuhn denied *any* rational link across paradigms, even in terms of problem-solving. Kuhn based his view on theoretical holism: paradigms are not accepted or rejected piecemeal, but as a whole. Kuhn denied that the choice of every specific element of a new paradigm is determined by rational means, not that there were not reasons to choose a new paradigm. Lakatos could respond that he was also objecting to Kuhn's account of anomalies building up and "causing" scientists to choose a new paradigm. For Lakatos, it is not anomalies per se, but the degenerating problem-solving ability of the old paradigm that prompts a revolution.

Second, a theme throughout *Criticism and the Growth of Knowledge* is the role of normal science in Kuhn's account. Lakatos, Masterman, Feyerabend, and others criticize Kuhn's account of normal science, as engaged only in "puzzle solving" that does not result in substantive scientific progress. One way to read Lakatos's contribution here is as an attempt to give a more robust characterization of the activity of normal science, and its relevance to the reasons for revolutionary change.

Third, a theme that is found in many of the texts of logical empiricism excerpted here, as well as in other texts from this tradition, is a careful distinction between epistemology and "psychology." Lakatos objects to the use of "psychological" or volitional accounts for scientific change, Reichenbach sets up the discovery-justification distinction in terms of "psychology" versus "epistemology," and Carnap's *Aufbau*, for instance, contains lengthy discussions distinguishing his methods from those of earlier "psychologists." A historian, sociologist, or philosopher of science looking for a research project in this tradition might find it productive to focus on a historical or sociological account of what was meant by "psychology" in nineteenth- and early twentieth-century Germany, and how this influenced Reichenbach's, Hempel's, Lakatos's, and Carnap's approach to the distinction between epistemology and psychology.

Fourth, much work is being done in Kuhn's account by a commitment to theoretical holism. This holism was defended earlier by Duhem, and throughout the twentieth century by Willard van Orman Quine. In the work excerpted for this volume, Kuhn's "What Are Scientific Revolutions?", Kuhn even argues for holism about revolutionary change in science:

> Revolutionary changes are somehow holistic. They cannot, that is, be made piecemeal, one step at a time, and they thus contrast with normal or cumulative changes like, for example, the discovery of Boyle's law. [. . .] An integrated picture of several aspects of nature has to be changed at the same time.
>
> (Kuhn 1987, 28–29)

Kuhn's last assertion, that the picture of nature *has* to be changed, is one statement of his thesis of the "incommensurability" of distinct paradigms. Kuhn and Paul Feyerabend are considered the authors of the view in philosophy of science that successive paradigms are "incommensurable." In *Structure*, as Oberheim and Hoyningen-Huene (2013, §2.1) remark,

> His investigations into the history of science revealed phenomena often now called 'Kuhn loss': Problems whose solution was vitally important to the older tradition may temporarily disappear, become obsolete or even unscientific. On the other hand, problems that had

not even existed, or whose solution had been considered trivial, may gain extraordinary significance in the new tradition.

In "Explanation, Reduction and Empiricism," Feyerabend (1962) argues that "formal accounts of reduction and explanation are impossible for general theories because they cannot accommodate the development of incommensurable concepts in the course of scientific advance" (Oberheim and Hoyningen-Huene 2013, §3.1).[7] Feyerabend's argument is against the "conceptual conservatism" of Thomas Nagel, and of Hempel, among others. He argues against the principle of "meaning invariance," that is, the view that concepts and kind terms remain invariant through changes in paradigms. Feyerabend marshals significant historical evidence in support of his arguments against claims of meaning invariance and instances of conceptual conservatism, including

> The dynamical characterization of impetus in the impetus theory of motion and the concept of force in the conceptual apparatus of Newtonian mechanics, the concepts of temperature and entropy in the transition from phenomenological thermodynamics to kinetic theory, and the concepts mass, length and time in the transition from Newtonian to relativistic mechanics.
>
> (Oberheim and Hoyningen-Huene 2013, §3.1)

In an earlier essay (1958), Feyerabend argues for a version of the incommensurability thesis, and

> challenged an implicit conceptual conservativism in logical positivism: The assumption that theoretical terms derive their meaning solely through their connection with experience, and that experience itself is a stable (or unchanging) foundation on which theoretical meaning can be based. Instead of such a bottom-up version of the relation of experience and theoretical knowledge, according to which experience determines the meanings of our theoretical terms, Feyerabend argued for a top-down version, according to which our theories determine the meaning we attach to our experiences.
>
> (Oberheim and Hoyningen-Huene 2013, §3.2)

This view is echoed by Kuhn's early and mature thought, in Kuhn's notion of "taxonomic incommensurability." In "What Are Scientific Revolutions?", Kuhn gives an account of the incommensurability between paradigms that is focused on "taxonomic categories":

the distinctive character of revolutionary change in language is that it alters not only the criteria by which terms attach to nature but also, massively, the set of objects or situations to which those terms attach. What had been paradigmatic examples of motion for Aristotle—acorn to oak or sickness to health—were not motions at all for Newton. [. . .] What characterizes revolutions is, thus, change in several of the taxonomic categories prerequisite to scientific descriptions and generalizations. That change, furthermore, is an adjustment not only of criteria relevant to categorization, but also of the way in which given objects and situations are distributed among preexisting categories. Since such redistribution always involves more than one category and since those categories are interdefined, this sort of alteration is necessarily holistic. That holism, furthermore, is rooted in the nature of language, for the criteria relevant to categorization are ipso facto the criteria that attach the names of those categories to the world. Language is a coinage with two faces, one looking outward to the world, the other inward to the world's reflection in the referential structure of the language.

(Kuhn 1987, 30)

Kuhn's and Feyerabend's arguments for "revolutionary change in language" have inspired much recent work. Oberheim and Hoyningen-Huene (2013) identify the Kuhn–Feyerabend incommensurability thesis as "a widely discussed, controversial idea that was instrumental in the historical turn in the philosophy of science and the establishment of the sociology of science as a professional discipline" (introduction, n.p.).

There has been a great deal of debate about the status of Feyerabend's and Kuhn's "incommensurability" thesis, and their related "taxonomic" claim that a change in paradigms changes "the way in which given objects and situations are distributed among preexisting categories." On this subject see, among others, Feyerabend 1993/1975, 1981, 1977, 1965, and 1962; Hoyningen-Huene 1993; Kuhn 2000, 1982, and 1977; Oberheim, Andersen, and Hoyningen-Huene 1996; Sankey 1993; and Wang 2002.

Another key theme for research following Kuhn's and Feyerabend's work was the status of "empiricism" in the philosophy of science. As can be seen in the discussion of Feyerabend, his work challenges the empiricist view that theories are built from the bottom up, beginning with observation, and that theoretical terms are defined by reference to observation terms. The distinction between theoretical and observation terms is found in the work of Rudolf Carnap and others in the logical empiricist tradition.

As Kuhn points out, the distinction between theoretical and observational terms did not die out as a result of his and Feyerabend's work. In particular, as he puts it in a footnote to "What Are Scientific Revolutions?":

the often elegant apparatus developed by logical empiricists for discussions of concept formation and of the definition of theoretical terms can be transferred as a whole to the historical approach and used to analyze the formation of new concepts and the definition of new terms, both of which usually take place in intimate association with the introduction of a new theory. A more systematic way of preserving an important part of the observational/theoretical distinction by embedding it in a developmental approach has been developed by ([Sneed 1971], pp. 1–64, 249–307). Wolfgang Stegmüller has clarified and extended Sneed's approach by positing a hierarchy of theoretical terms, each level introduced within a particular historical theory ([Stegmüller 1976], pp. 40–67, 196–231). The resulting picture of linguistic strata shows intriguing parallels to the one discussed by Michel Foucault in [Foucault 1972].

For more on Kuhn's relationship to Sneed and Stegmüller, see Kuhn 1976 and 2000, Stegmüller 1975, and Sneed 1983, among others. For more on Foucault's archaeology of scientific knowledge, see Gutting 1989 and 2005.

A more recent "developmental," historical approach to retaining the a priori elements of scientific theories, and to showing that such elements can be conserved and evaluated across paradigms, has been formulated by Michael Friedman in his analysis of the "dynamic" or "relativized" a priori. Friedman (2001) is the classic presentation of this account, based on a series of lectures given at Stanford University. According to Friedman, historical influences on the dynamic, relativized a priori include the philosophy of Rudolf Carnap, as well as the approaches of Immanuel Kant and the neo-Kantian Marburg School.

Friedman's arguments have inspired a new set of approaches to the philosophical analysis of paradigm shifts. His argument that one paradigm can be seen as a "limiting case" of its successor (as in the case of Newtonian and relativistic mechanics) has had a broad influence. Moreover, Friedman's analysis of the elements of scientific paradigms is a significant contribution to debates on the subject, for instance, his distinction between empirical laws, constitutive a priori principles, and overarching philosophical frameworks. For responses both sympathetic and critical, see, among others, Friedman 2000, 2001, and 2002; McArthur 2007; Richardson 2002; and Shaffer 2011. The essays in Domski and Dickson 2010, including the essays by Howard, Creath, Ryckman, di Salle, Koertge, Tanona, and Dickson, explore a number of topics inspired by Friedman's approach to the history and philosophy of science (see Patton 2011b for a brief, summarizing discussion of these essays).

Rival approaches to the history and philosophy of science were proposed around the same time that Kuhn wrote *Structure*. Kuhn himself mentions Foucault 1972 as a related, but distinct "archaeological" approach.

Michael Polanyi (1891–1976), a Hungarian physical chemist and philosopher of science, published seminal analyses of the scientific community in the 1950s and 1960s, including *The Tacit Dimension* (1966) and *Personal Knowledge* (1958). Both centered on the notion of "tacit knowledge," which Polanyi (1966) most famously expressed as "we know more than we can tell" (p. 4). Tacit knowledge is not usually put into words, but is nonetheless a form of knowledge and, Polanyi urges, some of the most significant knowledge we possess.[8] For instance, the "know-how" of the scientific community, including the ability to manipulate instruments in performing experiments, and to recognize, in practice, particular cases that fall under more general scientific categories, would qualify as tacit knowledge. Polanyi gives an analysis of cases from history in which tacit knowledge was the source of our ability to make progress in science.

Polanyi was a signal influence on a number of historians and philosophers of science. Nye 2011 is a detailed recent work that deals with Polanyi and his influence, along with others of his generation, on the "social turn" in philosophy of science. Nye cites J. D. Bernal (1901–1971), Ludwik Fleck (1896–1961, who had a profound influence on Feyerabend), Karl Mannheim (1893–1947), and Robert K. Merton (1910–2003) as members of Polanyi's generation who worked on allied topics and inaugurated the social approach to the history and philosophy of science. One might also mention the earlier thinker Edgar Zilsel (1891–1944). While he was a founder of the sociological approach to the philosophy of science, Zilsel also had interesting ties to logical empiricism. Hempel cites Zilsel in the essay printed here, and Zilsel published works in the *International Encyclopedia of Unified Science*.

Polanyi worked closely throughout his career with Marjorie Grene (1910–2009), one of the first modern philosophers of biology. Robert Brandon recalls that in the late 1970s, "I knew five philosophers of biology: Marjorie Grene, David Hull, Michael Ruse, Mary Williams and William Wimsatt" (Brandon 1996, xii–xiii; see also Griffiths 2011, §1). She was one of the first to formulate a philosophical position on the modern synthesis in biology.[9] Polanyi consulted with Grene when he was revising his 1950 lectures into the text that later became *Personal Knowledge*. Grene edited collections of Polanyi's essays, and Grene and Polanyi led a study group on the "Unity of Knowledge" over the course of five years (Burian and Ariew 2009).

Stephen Toulmin (1922–2009) contributed a skeptical perspective on scientific revolutions, including in his work with June Goodfield (1965, 1962, 1961), which centered precisely on historical cases that others considered to be "revolutions" (Nickles 2011, §2.1). Toulmin argues against Kuhn's distinction between normal and revolutionary science, at least, against the distinction as Kuhn draws it. Toulmin 1967 contains a revealing remark. Political historians, he says, have long used the term "revolution" to describe social and political upheavals. However,

> with experience, it has become clear to political historians that nothing is achieved by saying "and then there was a revolution," as though that exempted one from the need to give any historical analysis of a more explicit kind. To do only that is not to perform the historian's proper intellectual task, but to shirk it. (p. 84)

In the history of science as in political history, Toulmin urges, appealing to "revolutions" as historically ineffable, inexplicable events is to "shirk" the task of the historian.

In *The Road Since Structure*, Kuhn himself mentions Toulmin, Feyerabend, Polanyi, and Norwood Russell Hanson as thinkers who were working on topics similar to *Structure* at the time of its writing and afterward.[10] Nickles 2011 is an appreciation of work on scientific revolutions at the time and since.

History of Science and Philosophy of Science: Disciplines and Methods

Thomas Kuhn's essay on "The History of Science," printed below, contains an appreciation of work in the history of science up to the late 1960s, and, in fact, Kuhn also includes an analysis of work that would now be called the history and philosophy of science. The essay is worth reading for its appreciation of the discipline, but also for the copious references Kuhn gives to seminal works in the early tradition.

Kuhn's 1968 lecture "The Relations between the History and the Philosophy of Science," also printed here, is a partly a work of sociology on a small scale: a report of Kuhn's observations from a seminar he conducted repeatedly at Princeton in the 1960s. The seminar included students from the History of Science and the Philosophy program, and so it is helpful to provide some context about the education students were receiving.

Charles Coulston Gillispie established the program in History of Science at Princeton in 1960.[11] During the time Kuhn writes about here, Gillispie was a signal influence on the education of historians of science at Princeton. Other influential faculty included John Murdoch, and Seymour Mauskopf, a graduate student at the time, mentions making a pilgrimage to the Institute for Advanced Study to visit Alexandre Koyré.[12]

Princeton Philosophy faculty at the time included Margaret Wilson, David Lewis, Gilbert Harman, Carl Hempel (until 1973), James Ward Smith, Gregory Vlastos, Thomas Nagel, and others. The most significant in the context of philosophy of science are Hempel, Nagel, and Kuhn himself, though to say so is not to minimize the influence of the others on the students' philosophical education.

Kuhn argues that the methods, and final products, of history and philosophy are distinct, and thus he argues for separate training for historians and philosophers:

> The final product of most historical research is a narrative, a story, about particulars of the past. [. . .] Its success, however, depends not only on accuracy but also on structure. The historical narrative must render plausible and comprehensible the events it describes. [. . .] [H]istory is an explanatory enterprise; yet its explanatory functions are achieved with almost no recourse to explicit generalizations. [. . .] The philosopher, on the other hand, aims principally at explicit generalizations and at those with universal scope. He is no teller of stories, true or false. His goal is to discover and state what is true at all times and places rather than to impart understanding of what occurred at a particular time and place.
>
> (Kuhn 1977, reprinted here)

"To train a student simultaneously in both," Kuhn concludes, "would risk depriving him of any discipline at all." Kuhn contends that to attempt a training would be like trying to get someone to see the duck and the rabbit simultaneously when looking at Wittgenstein's and the Gestalt theorists' duck-rabbit.

Kuhn describes seminars he taught repeatedly at Princeton, consisting of students trained in history and students trained in philosophy. Kuhn claims to have found reason to group "the historians" and "the philosophers," not just in terms of which program they attended, but also in terms of their methods and conclusions in class work: "The Galileo or Descartes who appeared in the philosophers' papers was a better scientist or philosopher but a less plausible seventeenth-century figure than the figure presented by the historians" (Kuhn 1977).

Kuhn concludes that the history of science and philosophy of science can inform each other only under certain conditions:

> When speaking here of the history of science, I refer to that central part of the field that is concerned with the evolution of scientific ideas, methods, and techniques, not the increasingly significant portion that

emphasizes the social setting of science, particularly changing patterns of scientific education, institutionalization, and support, both moral and financial. [. . .]

[W]hen speaking of the philosophy of science, I have in mind [. . .] that central area that concerns itself with the scientific in general, asking, for example, about the structure of scientific theories, the status of theoretical entities, or the conditions under which scientists may properly claim to have produced sound knowledge. It is to this part of the philosophy of science, and very possibly to it alone, that the history of scientific ideas and techniques may claim relevance.

(Kuhn 1977)

Kuhn criticizes Hempel's so-called "covering law" or D-N model as a way of encompassing history within philosophy, remarking that the *practice* of putting together a historical narrative is usually what convinces students that this model is inadequate to describe historical explanation. Kuhn argues that this practice usually requires that students gain a holistic picture of the historical data, and that once this has happened, the student has a moment of recognition: "Now it makes sense, now I understand; what was for me previously a mere list of facts has fallen into a recognizable pattern." Kuhn's appreciation of Gestalt psychology is a clear influence on his view in this regard.

Works by Martin Rudwick (*The Great Devonian Controversy*) and Steven Shapin and Simon Schaffer (*Leviathan and the Air Pump*), both of which appeared in 1985, are printed here as examples of scholarship in a more sociological approach to the history of science.[13] Rudwick's explanation of his methods and theory contains clear references to the earlier work of Michael Polanyi, especially in Rudwick's reference to the tacit knowledge of the scientific community:

This picture of scientific work as skilled craftsmanship, practiced within a shared tradition that is maintained by a social collectivity, jarred not only against the fiercely held convictions of many philosophers but also against the routine assumptions of many historians of science. [. . .] In any case, it is clear that historians have not yet adjusted their modes of writing to take full account of the social dimension of scientific practice. [. . .] The work of individual scientists is often analyzed in admirable detail, and full weight may also be given to the ways in which that work was influenced by others. But few historians of science have set out deliberately to recover what such a network of individuals had in common, particularly what they held *tacitly* in common.

To do justice to this broader picture, Rudwick argues, historians must present each historical episode "in a style that does justice to its real complexity, muddle, and messiness."

Shapin and Schaffer 1985 is somewhat different in its approach, though it also contains an indictment of the logical empiricist approach. They begin by arguing that the history of science and the history of politics are, in fact, co-extensive, in three significant ways,

> There are three senses in which we want to say that the history of science occupies the same space as the history of politics. First, scientific practitioners have created, selected, and maintained a polity within which they operate and make their intellectual product; second, the intellectual product made within that polity has become an element in political activity in the state; third, there is a conditional relationship between the nature of the polity occupied by scientific intellectuals and the nature of the wider polity.

Shapin and Schaffer argue for including an account of material culture and of the community of investigators in analyses of scientific practice, as opposed to limiting that analysis to isolated thinkers or researchers. But they go further, to argue that the scientific community is in fact the human community. They argue against the restriction of scientific research to élite, highly trained subgroups within society, and for the democratization of science.

Figure 2.1 summarizes the points on the differing methodologies of history and philosophy made by Kuhn and by Rudwick. Kuhn argues that for a scholar to adopt both methodologies at the same time would turn her into a simultaneous duck-rabbit: no one can approach a text as a historian and as a philosopher at the same time. In Kuhn's view, a philosopher reads a historical text differently than a historian does, and these ways of reading are mutually inconsistent. Rudwick makes similar points about differences between historical and philosophical ways of reading a text.

This figure should make it clear that the methodological distinctions made here are very contentious. Moreover, as was emphasized in the overall introduction, contemporary programs in HOPOS, HPS, STS and the like are interdisciplinary, with scholars from distinct disciplines often sitting together on dissertation committees.

Scholars might ask whether these divisions, which may indeed be accurate descriptions in many cases, are necessary methodological divisions. Ruse 1999 is a history of the "Darwinian Revolution" that takes a philosophical approach and that is sensitive to the philosophical aspects of the events discussed. But the work also has many of the virtues of historical scholarship.

	Historians of science	Philosophers reconstructing the history of science
Kuhn	More accurate description of historical figures	Clearer, more consistent statements of the ideas and concepts of the historical figures
	Read for the details, Gestalt picture, and context	Read for arguments and concepts
	Amass the data first and then try to assemble a picture or narrative	Begin assembling a conceptual framework right away
	Look for the mistakes, infelicities, artefacts in the historical texts	May gloss over mistakes or infelicities in the process of trying to "reconstruct" a valid argument
Rudwick	Historical narrative	Argument and rational reconstruction
	Sensitive to the influence on science of the community of science, Polanyi's "tacit knowledge"	Focus on the individual thinker and her ideas
	Material culture	Intellectual culture

Figure 2.1

New Directions and Research Questions

In the beginning of this introduction, I presented the classic narrative of HOPOS as emerging from the articulation and critique of the characteristic arguments and approaches of logical empiricism. These arguments include the defense of rational reconstruction in history of science, the argument for the unity of scientific methods, and the primacy of the context of justification. All were called into question by early work in HOPOS, including the work of Feyerabend, Polanyi, Kuhn, Toulmin, Hanson, and others. I have reprinted below voices from the history and sociology of science who are sharply critical of the early philosophical approach.

Perhaps optimistically, I am of the view that the initial clash of methodologies between the logical empiricists and more socially and historically minded scholars no longer prevents productive collaboration; the Science Wars are a historical episode, and are being studied as such. Contemporary scholars do not have the same motivations as their counterparts in the history of the conflicts between the disciplines. We have the chance to build on what is increasingly becoming common ground.

Recent work on HOPOS has united philosophers from disparate traditions; Horwich 1993, with essays from Horwich, John Earman, Michael Friedman, Ernan McMullin, Jed Buchwald, Noel Swerdlow, John Heilbron, Nancy Cartwright, Ian Hacking, M. Norton Wise, Thomas Kuhn, and Carl Hempel, is a representative example.

The sociological approach to science has affinities with recent work in philosophy of science and in history of science. This includes work that focuses on the history and philosophy of experiment and theory: Buchwald 1995, 1994, 1989, and 1980; Chang 2012 and 2004; Darrigol 2012, 2005, and 2000; Franklin 2013, 2002, 1998, 1990, and 1986; Galison 1987 and 1997; Massimi 2004; Mayo 1996 and 1994; Mayo and Spanos 2010; Nickles 1980; Patton 2012 and 2011a; Pitt 2012 and 2001; Schickore 2007; Steinle 2002, 1997, and 1994; and others.

The social approach to philosophy of science informs the methods of feminist philosophy of science. Anderson 2012 is a frequently updated introductory source on this tradition, with a detailed bibliography of contemporary and classic work. For initial, detailed presentations of the field see Alcoff and Potter 1993; Anderson 1995; Antony and Witt 2002; and Fricker 1998; see Anderson 2012 for further resources.

The social approach also informs the growing tradition of work on social epistemology in its relationship to the history and philosophy of science (Fuller 1988; Goldman 1999; Kitcher 1990 and 1993; Kukla 2000; Latour and Woolgar 1986/1979; Longino 1990 and 2002; Searle 1995; Solomon 1992; Thagard 1997; and others). The work of Ian Hacking is distinctive for his seminal contributions to the social and philosophical analysis of experiment, and to the philosophical, social, and historical epistemology of science (Hacking 2004, 2000, 1983, among other works).

The philosophical approach to the influence of experiment, instrumentation, and the scientific community on scientific practice and on the development of scientific theories may differ from the sociological approach, at least in some aspects. However, surely the two traditions have much to say to each other. Franklin 1998 and Nickles, for instance, have cited the work of Harry Collins (the "experimenter's regress") in their analyses of experiment. Galison's work is perhaps in the middle of the two approaches, incorporating sociological and philosophical analyses. The work being done in the journals *Episteme* and *Social Epistemology* is interdisciplinary work, which incorporates philosophy, history, and sociology, though not always in the same project.

A philosophical approach inspired by Gaston Bachelard, Michel Foucault, and Georges Canguilhem (1904–1995), known as "historical epistemology," is gaining increasing traction.[14] A conference at the Max Planck Institute for the History of Science in Berlin, "What (Good) Is Historical Epistemology?" took place in 2008, organized by Uljana Feest and Thomas Sturm. The proceedings of the conference were published in volume 75, no. 3 of the journal *Erkenntnis*. The introduction to the proceedings, Feest and Sturm 2011, is to be recommended as a statement of what Feest and Sturm consider to be the most pressing research topics in the field, as of course are the essays in the volume. Classic works in the tradition include Bachelard 2002/1938; Canguilhem 1988/1977; Foucault 1972; Krüger 1973 and 2005; and the historical studies of Lecourt 1969 and 1975. More recent researchers in this field include, along with Feest and Sturm, Chimisso 2001; Daston 1994; Daston and Limbeck 2011; Hacking 2004 and 1999; Hyder 2003; Klein 2007; Kmita 1988; Renn 1996; Rheinberger 1997; Schmidgen, Schöttler; and Braunstein (forthcoming); Tiles 1987 and 1984; Tiles and Tiles 1993; and Wartofsky 1987; among others.

Notes

1. By this, he has been understood to mean that a "psychological" investigation of evidence or text can reveal facts not accessible to natural science. This strong psychological reading of Dilthey has been contested (Kluback 1956, Beiser 2012, Makkreel 1992, Patton forthcoming).
2. On Herder and Droysen see Beiser 2012 and Forster 2002 and 2007. For recent work on the erklären-verstehen distinction see the essays in Feest 2009.
3. In this text, I use the terms "logical empiricism" and "logical empiricists," because some members objected to the label "logical positivism." For a discussion, see Creath 2013.
4. The "Logisch-Philosophische Abhandlung" appeared in 1921, but the now-classic version of the *Tractatus* appeared in 1922 (Wittgenstein 1999/1922).
5. For a selection of recent work, and references to further reading, on the contexts of discovery and of justification, see Schickore and Steinle 2006.
6. I have not included an excerpt from *Structure* here, as it appears frequently in anthologies and often is assigned in courses. I have selected a more recent text that illuminates Kuhn's mature thought on scientific revolutions, which will complement material in *Structure* a professor may wish to assign.
7. Feyerabend had begun to develop his view of incommensurability ten years earlier, on his own account (Oberheim and Hoyningen-Huene 2013, §3.2).
8. Grene insists, in her inimitable style, that Polanyi's tacit knowledge, while not usually expressed, is not necessarily *inexpressible* (Cohen 2005).
9. Auxier and Hahn 2002 is an appreciation of Grene's work in philosophy.
10. Kuhn 2000, 312ff. I am grateful to Charles de Souza for pointing out this passage, and its significance, to me.
11. Buchwald 2012 is an appreciation of Gillispie's work and influence.
12. Buchwald 2012, 28; Mauskopf's essay is very illuminating as a chronicle of the department of history of science.
13. I am grateful to Matthew Wisnioski for recommending these texts as key reflections on methods and themes in the history and sociology of science.
14. As Feest and Sturm note, Canguilhem was ambivalent about the term as a description of his work.

References and Further Reading

Achinstein, Peter. 2004. *Science Rules: A Historical Introduction to Scientific Methods*. Baltimore: The Johns Hopkins University Press.

Agassi, Joseph. 1963. *Towards an Historiography of Science*, History and Theory, vol. 2. The Hague: Mouton.

Alcoff, Linda and Potter, Elizabeth (eds.). 1993. *Feminist Epistemologies*. New York: Routledge.

Anderson, Elizabeth. 2012. "Feminist Epistemology and Philosophy of Science," *The Stanford Encyclopedia of Philosophy*, Edward Zalta (ed.), URL = <http://plato.stanford.edu/archives/fall2012/entries/feminism-epistemology/>.

——. 1995. "Feminist Epistemology: An Interpretation and a Defense," *Hypatia* 10 (3): 50–84.

Antony, Louise, and Witt, Charlotte (eds.). 2002/1993. *A Mind of One's Own*. Boulder: Westview Press, second edition.

Aristotle. 1934/unknown. *Aristotle: The Physics. With an English Translation by Philip H. Wicksteed and Francis M. Cornford*. In two volumes. Loeb Classical Library. London: Heinemann. Cambridge, Mass.: Harvard University Press.

Auxier, Randall and Hahn, Lewis (eds.). 2002. *The Philosophy of Marjorie Grene*. Library of Living Philosophers, vol. 29. LaSalle, Ill.: Open Court.

Ayer, A. J. (ed.). 1959. *Logical Positivism*. New York: Dover.

Bachelard, Gaston. 2002/1938. *The Formation of the Scientific Mind*, trans. M. McAllester Jones. Manchester: Clinamen Press.

Bacon, Francis. 2000. *The New Organon*. Jardine Lisa and Silverthrone, Michael, eds. Cambridge: Cambridge University Press.

Banville, John. 2001. *The Revolutions Trilogy*. New York: Picador.

Bechtel, William and Abrahamsen, Adele. 2005. "Explanation: A Mechanistic Alternative," *Studies in History and Philosophy of the Biological and Biomedical Sciences* 36: 421–441.

Beiser, Frederick. 2012. *The German Historicist Tradition*. Oxford: Oxford University Press.

Bokulich, Alisa and Bokulich, Peter (eds). 2011. *Scientific Structuralism*. Boston Studies in the Philosophy of Science. New York: Springer.

Brandon, R. N. (ed.). 1996. *Concepts and Methods in Evolutionary Biology*. Cambridge: Cambridge University Press.

Brown, T. M. 1969. "The Electric Current in Early Nineteenth-Century French Physics," *Historical Studies in the Physical Sciences* 1.

Brunschvicg, Léon. 1912. *Les étapes de la philosophie mathématique*. Paris: Alcan.

——. 1922. *L'expérience humaine et la causalité physique*. Paris: Alcan.

——. 1929. *La valeur inductive de la relativité*. Paris: Alcan.

Buchwald, Jed (ed.). 2012. *A Master of Science History: Essays in Honor of Charles Coulston Gillispie*. Dordrecht: Springer.

——. (ed.). 1995. *Scientific Practice: Theories and Stories of Doing Physics*. Chicago: University of Chicago Press.

——. 1994. *The Creation of Scientific Effects: Heinrich Hertz and Electric Waves*. Chicago: University of Chicago Press.

——. 1989. *The Rise of the Wave Theory of Light*. Chicago: University of Chicago Press.

——. 1980. "Experimental Investigations of Double Refraction from Huygens to Malus," *Archive for History of the Exact Sciences* 21: 311–369.

Burian, Richard. 1977. "More than a Marriage of Convenience: On the Inextricability of History and Philosophy of Science," *Philosophy of Science* 44 (1): 1–42.

Burian, Richard and Ariew, Roger. 2009. "In Memoriam: Marjorie Glicksman Grene (1910–2009)," *Isis* 100: 856–859.

Burtt, Edwin. 1924. *The Metaphysical Foundations of Modern Physical Science: A Historical and Critical Essay*. London: Kegan Paul.

Butterfield, Herbert. 1957. *The Origins of Modern Science, 1300–1800*. Second revised edition. New York: Macmillan.

Canguilhem, Georges. 1988/1977. *Ideology and Rationality in the History of the Life Sciences*, trans. Arthur Goldhammer. Cambridge, Mass.: MIT Press.

Carnap, Rudolf. 1991/1955. "Logical Foundations of the Unity of Science," in *The Philosophy of Science*, ed. Boyd et al. Cambridge, Mass.: MIT Press. Reprinted from *International Encyclopedia of Unified Science*, vol. I (1938–1955), ed. Neurath, Carnap, and Morris. Chicago: University of Chicago Press.

——. 1950. "Empiricism, Semantics, and Ontology," *Revue Internationale de Philosophie* 4: 20–40.

——. 1967/1928. *The Logical Structure of the World: Pseudoproblems in Philosophy*, trans. Rolf A. George. Berkeley: University of California Press.

Carter, Jacoby Adeshei. 2012. "Alain LeRoy Locke," *The Stanford Encyclopedia of Philosophy*, Edward N. Zalta (ed.), URL = <http://plato.stanford.edu/archives/sum2012/entries/alain-locke/>.

Cassirer, Ernst. 1969. *The Problem of Knowledge: Philosophy, Science, and History Since Hegel*. New Haven, Conn.: Yale University Press.

Cat, Jordi. 2013. "The Unity of Science," *The Stanford Encyclopedia of Philosophy*, Edward N. Zalta (ed.), URL = <http://plato.stanford.edu/archives/sum2013/entries/scientific-unity/>.

Chalmers, Alan. 2009. *The Scientists' Atom and the Philosopher's Stone*. Dordrecht: Springer.

——. 1999. *What Is This Thing Called Science?* Third edition. Indianapolis, Ind.: Hackett Publishing Company.

Chang, Hasok. 2012. *Is Water H2O? Evidence, Pluralism and Realism*, Boston Studies in the Philosophy of Science. Dordrecht: Springer.

——. 2004. *Inventing Temperature: Measurement and Scientific Progress*. Oxford: Oxford University Press.

Chimisso, Cristina. 2001. *Gaston Bachelard*. New York: Routledge.

Cohen, Benjamin. 2005. "Marjorie Grene," *The Believer* 3 (2).

Comte, Auguste. 1975/1830–1842. *Cours de philosophie positive*. Paris: Hermann, 2 vols. First edition was translated and abridged by Harriet Martineau as *The Positive Philosophy of Auguste Comte*, London: J. Chapman, 1853.

Creath, Richard. 2013. "Logical Empiricism," *The Stanford Encyclopedia of Philosophy*, Edward N. Zalta (ed.), URL = <http://plato.stanford.edu/archives/spr2013/entries/logical-empiricism/>.

Curd, Martin and Cover, J. A. 1998. *Philosophy of Science: The Central Issues*. New York: W.W. Norton & Company.

Darrigol, Olivier. 2012. *A History of Optics from Greek Antiquity to the Nineteenth Century*. Oxford: Oxford University Press.

——. 2005. *Worlds of Flow: A History of Hydrodynamics from the Bernoullis to Prandtl*. Oxford: Oxford University Press.

——. 2000. *Electrodynamics from Ampère to Einstein*. Oxford: Oxford University Press.

Daston, Lorraine. 1994. "Historical epistemology," pp. 282–289, in J. Chandler, A. I. Davidson, and H. Harootunian (eds.). *Questions of Evidence*. Chicago: University of Chicago Press.

Daston, L., and Limbeck, E. (eds.). 2011. *Histories of Scientific Observation*. Chicago: University of Chicago Press.

de la Rive, Auguste. 1856. *Traité d'électricité théorique et appliquée*, vol. 2. Paris: J. B. Baillière.

De Regt, Henk, Leonelli, Sabina, and Eigner, Kai (eds.). 2009. *Scientific Understanding: Philosophical Perspectives*. Pittsburgh: University of Pittsburgh Press.

Dijksterhuis, Edward J. 1961/1950. *The Mechanization of the World Picture*. Oxford: Claren-don Press.

Dilthey, Wilhelm. 1991/1883. Wilhelm Dilthey: Selected Works Volume I: *Introduction to the Human Sciences*, trans. Rudolf Makkreel and Frithjof Rodi. Princeton, N.J.: Princeton University Press.

Domski, Mary and Dickson, Michael (eds.). 2010. *Discourse on a New Method: Reinvigorating the Marriage of History and Philosophy of Science*. Chicago: Open Court.

Duhem, Pierre. 1902. *L'évolution de la mécanique*. Paris: A. Hermann.

——. 1914. *La théorie physique: son objet et sa structure*, second ed. Paris: Chevalier et Rivière.

——. 1954/1914. *The Aim and Structure of Physical Theory*, trans. Phillip Wiener. Princeton, N.J.: Princeton University Press.

Feest, Uljana (ed.) 2009. *Historical Perspectives on Erklären and Verstehen*. Dordrecht: Springer.

Feest, Uljana and Sturm, Thomas. 2011. "What (Good) is Historical Epistemology? Editors' Introduction," *Erkenntnis* 75 (3): 285–302.

Feyerabend, Paul. 1993/1975. *Against Method: Outline of an Anarchistic Theory of Knowledge*. Third edition. New York: Verso. First edition was 1975, London: New Left Books.

——. 1981. *Realism, Rationalism and Scientific Method. Philosophical Papers*. Cambridge: Cambridge University Press.

——. 1977. "Changing Patterns of Reconstruction," *British Journal for the Philosophy of Science* 28: 351–382.

——. 1965. "On the 'Meaning' of Scientific Terms," *Journal of Philosophy* 62: 266–274.

——. 1962. "Explanation, Reduction and Empiricism," pp. 28–97 in H. Feigl and G. Maxwell (eds.), *Scientific Explanation, Space, and Time*. Minnesota Studies in the Philosophy of Science, Volume III. Minneapolis: University of Minneapolis Press.

——. 1958. "An Attempt at a Realistic Interpretation of Experience," *Proceedings of the Aristotelian Society* 58: 143–170.

Forster, Michael. 2007. "Hermeneutics," in *The Oxford Handbook of Continental Philosophy*, ed. Michael Rosen and Brian Leiter. Oxford: Oxford University Press.

——. 2002. "Herder's Philosophy of Language, Interpretation, and Translation," *Review of Metaphysics* 56 (2).

Foucault, Michel. 1972. *The Archeology of Knowledge*, trans. A. M. Sheridan Smith. New York: Pantheon.

Franklin, Allan. 2013. *Shifting Standards: Experiments in Particle Physics in the Twentieth Century*. Pittsburgh: University of Pittsburgh Press.

——. 2002. *Selectivity and Discord: Two Problems of Experiment*. Pittsburgh: University of Pittsburgh Press.

——. 1998. *Can That Be Right?: Essays on Experiment, Evidence, and Science*. Boston Studies in the Philosophy and History of Science. Dordrecht: Springer.

——. 1990. *Experiment, Right or Wrong*. Cambridge: Cambridge University Press.

——. 1986. *The Neglect of Experiment*. Cambridge: Cambridge University Press.

Fricker, Miranda. 1998. "Rational Authority and Social Power: Towards a Truly Social Epistemology," *Proceedings of the Aristotelian Society* 19 (2): 159–177.

Friedman, Michael. 2002. "Kant, Kuhn, and the Rationality of Science," *Philosophy of Science* 69 (2): 171–190.

——. 2001. *Dynamics of Reason*. Center for the Study of Language and Information, Stanford, Calif.: CSLI.

——. 2000. "Transcendental Philosophy and a Priori Knowledge: A Neo Kantian Perspective," in Paul Boghossian and Christopher Peacocke (eds.), *New Essays on the a Priori*. Oxford: Oxford University Press.

———. 1999. *Reconsidering Logical Positivism*. Cambridge: Cambridge University Press.

———. 1987. "Carnap's Aufbau Reconsidered," *Noûs* 21 (4): 521–545.

Fuller, Steven. 1988. *Social Epistemology*. Bloomington: Indiana University Press.

Galison, Peter. 1987. *How Experiments End*. Chicago: University of Chicago Press.

———. 1997. *Image and Logic*. Chicago: University of Chicago Press.

Godfrey-Smith, Peter. 2003. *Theory and Reality: An Introduction to the Philosophy of Science*. Chicago: University of Chicago Press.

Goldman, Alvin. 1999. *Knowledge in a Social World*. Oxford: Oxford University Press.

Griffiths, Paul. 2011. "Philosophy of Biology," *The Stanford Encyclopedia of Philosophy*, Edward Zalta (ed.), URL = <http://plato.stanford.edu/archives/sum2011/entries/biology-philosophy/>.

Gutting, Gary. 1989. *Michel Foucault's Archaeology of Scientific Reason*. Cambridge: Cambridge University Press

———. 2005. *Continental Philosophy of Science*. Oxford: Blackwell.

Hacking, Ian. 2004. *Historical Ontology*. Cambridge, Mass.: Harvard University Press.

———. 2000. *The Social Construction of What?* Cambridge, Mass.: Harvard University Press.

———. 1999. "Historical Meta-epistemology," pp. 53–77 in W. Carl and L. Daston (eds.), *Wahrheit und Geschichte*. Göttingen: Vandenhoeck & Ruprecht.

———. 1983. *Representing and Intervening: Introductory Topics in the Philosophy of Natural Science*. Cambridge: Cambridge University Press.

Hanson, Norwood Russell. 1958. *Patterns of Discovery*. Cambridge: Cambridge University Press.

Heidelberger, Michael and Stadler, Friedrich (eds). 2002. *History of Philosophy of Science: New Trends and Perspectives*. Vienna Circle Institute Yearbook 9/2001, Kluwer. New York: Springer.

Hempel, Carl. 1942. "The Function of General Laws in History," *Journal of Philosophy* 39: 35–48.

Hempel, Carl. 1965. *Aspects of Scientific Explanation and Other Essays in the Philosophy of Science*. New York: Free Press.

———. 1935. "On the Logical Positivists' Theory of Truth," *Analysis* 2 (4): 49–59.

Hesse, Mary. 1963. *Models and Analogies in Science*. London: Sheed & Ward.

Horwich, Paul (ed.) 1993. *World Changes: Thomas Kuhn and the Nature of Science*. Cambridge, Mass.: MIT Press.

Howard, Don. 2010. "'Let me briefly indicate why I do not find this standpoint natural.' Einstein, General Relativity, and the Contingent A Priori," in Domski and Dickson 2010.

Hoyningen-Huene, Paul. 1993. *Reconstructing Scientific Revolutions: Thomas S. Kuhn's Philosophy of Science*. Chicago: University of Chicago Press.

Hyder, David. 2003. "Foucault, Cavaillès, and Husserl on the historical epistemology of the sciences," *Perspectives on Science* 11: 107–129.

Kitcher, Philip. 1993. *The Advancement of Science*. New York: Oxford University Press.

———. 1990. "The Division of Cognitive Labor," *The Journal of Philosophy* 87: 5–22.

Klein, Ursula. 2007. *Materials in Eighteenth-Century Science: A Historical Ontology*. Cambridge: The MIT Press.

Kluback, William. 1956. *Wilhelm Dilthey's Philosophy of History*. New York: Columbia University Press.

Kmita, Jerzy. 1988. *Problems in Historical Epistemology*. Dordrecht: Reidel.

Kockelmans, Joseph J. (ed.) 1968. *Philosophy of Science: The Historical Background*. New York: Free Press.

Koyré, Alexandre. 1939. *Études galiléennes*. 3 vols. Paris: Hermann.

———. 1961. *La révolution astronomique*. Paris: Hermann.

———. 2005. *Why Does History Matter to Philosophy and the Sciences?* Thomas Sturm, Wolfgang Carl, and Lorraine Daston, eds. Berlin: De Gruyter.

Kragh, Helge. 1989. *An Introduction to the Historiography of Science.* Cambridge: Cambridge University Press.

Krüger, Lorenz. 1973. *Der Begriff des Empirismus. Erkenntnistheoretische Studien am Beispiel John Lockes.* Berlin: Walter De Gruyter.

Kuhn, Thomas. 2000. *The Road Since Structure: Philosophical Essays, 1970–1993.* James Conant and John Haugeland (eds.). Chicago: University of Chicago Press.

——. 1987. "What are Scientific Revolutions?" pp. 7 22 of *The Probabilistic Revolution,* volume I: Ideas in History, edited by Lorenz Kruger, Lorraine J. Daston, and Michael Heidelberger. Cambridge, Mass.: MIT Press.

——. 1982. "Commensurability, Comparability, Communicability," *PSA: Proceedings of the Biennial Meeting of the Philosophy of Science Association* 1982: 669–688.

——. 1978. *Black-Body Theory and the Quantum Discontinuity, 1894–1912.* Oxford and New York: Clarendon and Oxford University Presses.

——. 1977. *The Essential Tension.* Chicago: University of Chicago Press.

——. 1976. "Theory-Change as Structure-Change: Comments on the Sneed Formalism," *Erkenntnis* 10 (2): 179–199.

——. 1962. *The Structure of Scientific Revolutions.* Chicago: University of Chicago Press.

Kukla, Andre. 2000. *Social Construction and the Philosophy of Science.* London: Routledge.

Ladyman, James. 2002. *Understanding Philosophy of Science.* New York: Routledge.

Lakatos, Imre and Musgrave, Alan. 1970. *Criticism and the Growth of Knowledge,* Cambridge: Cambridge University Press.

Lange, Friedrich Albert. 1866. *Geschichte des Materialismus und Kritik seiner Bedeutung in der Gegenwart.* Iserlohn: J. Baedeker.

——.1877–1881. *The History of Materialism and Criticism of Its Importance,* second ed. trans. by Ernest Chester Thomas. London: Trübner & Company.

Latour, Bruno and Woolgar, Steve. 1986/1979. *Laboratory Life: The [Social] Construction of Scientific Facts.* Princeton, N.J.: Princeton University Press.

Laudan, Larry. 1978. *Progress and Its Problems.* Berkeley: University of California Press.

Lecourt, Dominique. 1969. *L'épistemologie historique de Gaston Bachelard.* Paris: Vrin.

——. 1975. *Marxism and epistemology: Bachelard, Canguilhem, and Foucault.* London: New Left Books.

Longino, Helen. 2002. *The Fate of Knowledge.* Princeton, N.J.: Princeton University Press.

——. 1990. *Science as Social Knowledge.* Princeton, N.J.: Princeton University Press.

Losee, John. 2001. *A Historical Introduction to the Philosophy of Science.* Fourth edition. Oxford: Oxford University Press.

Mach, Ernst. 1960/1883. *The Science of Mechanics: A Critical and Historical Account of Its Development,* trans. T. J. McCormack. La Salle, Ill.: Open Court. Translation of *Die Mechanik in ihrer Entwicklung Historisch-Kritisch Dargestellt.* Leipzig: F. A. Brockhaus.

Makkreel, Rudolf. 1992. *Dilthey: Philosopher of the Human Studies.* Princeton, N.J.: Princeton University Press.

Mandelbaum, Maurice. 1938. *The Problem of Historical Knowledge.* New York: Liveright.

Massimi, Michela. 2004. "Non-Defensible Middle Ground for Experimental Realism: Why We Are Justified to Believe in Colored Quarks," *Philosophy of Science* 71 (1): 36–60.

Mauskopf, Seymour H. and Schmaltz, Tad M. (eds.). 2011. *Integrating History and Philosophy of Science: Problems and Prospects.* Dordrecht: Springer Verlag.

Mayo, Deborah. 1996. *Error and the Growth of Experimental Knowledge.* Chicago: University of Chicago Press.

——. 1994. "The New Experimentalism, Topical Hypotheses, and Learning from Error," *PSA: Proceedings of the Biennial Meeting of the Philosophy of Science Association* 1994, 270–279.

Mayo, Deborah and Spanos, Aris (eds.). 2010. *Error and Inference: Recent Exchanges on Experi-*

mental Reasoning, Reliability, and the Objectivity and Rationality of Science. Cambridge: Cambridge University Press.

McArthur, Dan. 2007. "Laudan, Friedman and the Role of the A Priori in Science," *Journal of Philosophical Research* 32: 169–190.

McConnell, Donald. 1939. *Economic Behavior*. New York: Houghton Mifflin.

McGrew, Timothy, Alspector-Kelly, Marc, Allhoff, Fritz (eds.). 2009. *Philosophy of Science: An Historical Anthology*. Blackwell Philosophy Anthologies. Hoboken, N.J.: Wiley-Blackwell.

Meier, Anneliese. 1949–58. *Studien zur Naturphilosophie der Spätscholastik*. 5 vols. Rome: Editioni di "Storia e Letteratura".

Merton, Robert. 1967. *Science, Technology, and Society in Seventeenth Century England*. New York: Fertig.

Merz, John Theodore. 2000. *History of European Thought in the Nineteenth Century*. Bristol: Thoemmes Press.

Meyerson, Émile. 1964/1908. *Identity and Reality*. Second edition, London: Allen & Unwin.

Neugebauer, Otto. 1957. *The Exact Sciences in Antiquity*. Second edition. Providence, RI: Brown University Press.

Neurath, Otto. 1969. "Foundations of the Social Sciences," from *Foundations of the Unity of Science*, ed. Neurath, Carnap, and Morris. Chicago: University of Chicago Press.

Nickles, Thomas. 2011. "Scientific Revolutions," *The Stanford Encyclopedia of Philosophy* Edward Zalta (ed.) URL = <http://plato.stanford.edu/archives/spr2011/entries/scientific-revolutions/>.

———. (ed.). 2002. *Thomas Kuhn*. Cambridge: Cambridge University Press.

———. (ed.). 1980. *Scientific Discovery, Logic, and Rationality*, two volumes. Dordrecht: Springer.

Nye, Mary Jo. 2011. *Michael Polanyi and His Generation: Origins of the Social Construction of Science*. Chicago: University of Chicago Press.

Oberheim, Eric and Hoyningen-Huene, Paul. 2013. "The Incommensurability of Scientific Theories," *The Stanford Encyclopedia of Philosophy*, Edward Zalta (ed.), URL = <http://plato.stanford.edu/archives/spr2013/entries/incommensurability/>.

———. 1996. "On Incommensurability," *Studies in History and Philosophy of Science Part A* 27 (1): 131–141.

Patton, Lydia. Forthcoming. "Methodology of the Sciences," in Michael Forster and Kristin Gjesdal (eds.), *The Oxford Handbook of Nineteenth Century German Philosophy*. Oxford: Oxford University Press.

———. 2012. "Experiment and Theory Building," *Synthese* 184 (3): 235–246.

———. 2011a. "Reconsidering Experiments," *HOPOS* 1 (2): 209–226.

———. 2011b. "Review of *Discourse on a New Method: Reinvigorating the Marriage of History and Philosophy of Science*," *Notre Dame Philosophical Reviews*.

Pitt, Joseph. 2012. "Theory Change and Instrumentation," in Jan Friis, Stig Pedersen and Vincent Hendricks (eds.), *A Companion to the Philosophy of Technology*. Oxford: Wiley-Blackwell.

———. 2001. "The Dilemma of Case Studies: Toward a Heraclitian Philosophy of Science," *Perspectives on Science* 9 (4): 373–382.

Pojman, Paul. 2011. "Ernst Mach," *The Stanford Encyclopedia of Philosophy*, Edward N. Zalta (ed.), URL = <http://plato.stanford.edu/archives/win2011/entries/ernst-mach/>.

Polanyi, Michael. 1966. *The Tacit Dimension*. Chicago: University of Chicago Press.

———. 1958. *Personal Knowledge: Towards a Post-Critical Philosophy*. Chicago: University of Chicago Press.

Popper, Karl. 1963. *Conjectures and Refutations*. New York and London: Routledge.

———. 2002/1934. *The Logic of Scientific Discovery*. New York and London: Routledge.

Quine, W. v. O. 1977/1969. "Epistemology Naturalized," pp. 69–91 in *Ontological Relativity and Other Essays*. New York: Columbia University Press.

——. 1980/1951. "Two Dogmas of Empiricism," pp. 20–46 in *From a Logical Point of View*, second ed. Cambridge, Mass.: Harvard University Press.

Reichenbach, Hans. 1928. *Philosophie der Raum-Zeit-Lehre*. Berlin: De Gruyter. Translated into English 1958 as *The Philosophy of Space and Time*, trans. M. Reichenbach and J. Freund. New York: Dover Publications.

Reill, Peter. 1994. "Science and the Construction of the Cultural Sciences in Late Enlighten ment Germany," *History and Theory* 33 (3).

Renn, Jürgen. 1996. *Historical Epistemology and the Advancement of Science*. Berlin: Max Planck Institute for the History of Science Preprint Series, Preprint 36 (http://www. mpiwg-berlin.mpg.de/Preprints/P36.PDF).

Rheinberger, H.-J. 1997. *Toward a History of Epistemic Things: Synthesizing Proteins in the Test Tube*. Stanford, Calif.: Stanford University Press.

Richards, Robert. 2003. "Biology," in *From Natural Philosophy to the Sciences*, ed. David Cahan. Chicago: The University of Chicago Press.

Richardson, Alan. 2002. "Narrating the History of Reason Itself: Friedman, Kuhn, and a Constitutive A Priori for the Twenty-First Century," *Perspectives on Science* 10 (3): 253–274.

——. 1998. *Carnap's Construction of the World: The Aufbau and the Emergence of Logical Empiricism*. Cambridge: Cambridge University Press.

Rosenberg, Alex. 2013. *Philosophy of Science*. New York and London: Routledge.

Saatkamp, Herman, 2010, "George Santayana," *The Stanford Encyclopedia of Philosophy* Edward Zalta (ed.), URL = <http://plato.stanford.edu/archives/fall2010/entries/santayana/>.

Sankey, Howard. 1993. "Kuhn's Changing Concept of Incommensurability," *British Journal for the Philosophy of Science* 44 (4): 759–774.

Santayana, George. 1933. *Some Turns of Thought in Modern Philosophy: Five Essays*. New York: Scribner's; Cambridge: Cambridge University Press.

——. 1923. *Scepticism and Animal Faith*. New York: Scribner's; London: Constable.

——. 1905–1906. *The Life of Reason*, in five volumes. New York: Scribner's; London: Constable.

Sarton, George. 1927–48. *Introduction to the History of Science*. 3 vols. Baltimore: Williams & Wilkins.

Schickore, Jutta. 2007. *The Microscope and the Eye: A History of Reflections, 1740–1870*. Chicago: University of Chicago Press.

Schickore, Jutta and Steinle, Friedrich. 2006. *Revisiting Discovery and Justification: Historical and Philosophical Perspectives on the Context Distinction*. Dordrecht: Springer.

Schmidgen, H., Schöttler, P., and Braunstein, J.-F. (Eds.), *History and Epistemology: From Bachelard and Canguilhem to Today's History of Science*. Berlin: Max Planck Institute for the History of Science Preprint Series (forthcoming).

Searle, John. 1995. *The Construction of Social Reality*. New York: Free Press.

Shaffer, Michael. 2011. "The Constitutive A Priori and Epistemic Justification," in Michael Shaffer and Michael Veber (eds.), *What Place for the A Priori?* LaSalle, Ill.: Open Court.

Sneed, Joseph D. 1983. "Structuralism and Scientific Realism," *Erkenntnis* 19 (1–3): 345–370.

——. 1971. *The Logical Structure of Mathematical Physics*. Dordrecht: Reidel.

Snyder, Laura. 2012. *The Philosophical Breakfast Club*. New York: Broadway Books.

Sobel, Dava. 1999. *Galileo's Daughter*. New York: Walker & Company.

Solomon. Miriam, 1992. "Scientific Rationality and Human Reasoning," *Philosophy of Science* 59 (3): 439–454.

Stanford, Kyle. 2010. *Exceeding Our Grasp: Science, History, and the Problem of Unconceived Alternatives*. New York and Oxford: Oxford University Press.

Stegmüller, Wolfgang. 1976. *The Structure and Dynamics of Theories*. New York: Springer.

———. 1975. "Structures and Dynamics of Theories: Some Reflections on J. D. Sneed and T. S. Kuhn," *Erkenntnis* 9 (1): 75–100.

Steinle, Friedrich. 2002. "Experiments in History and Philosophy of Science," *Perspectives on Science* 10 (4): 408–432.

———. 1997. "Entering New Fields: Exploratory Uses of Experimentation," *Philosophy of Science* 64 (4): 65–74.

———. 1994. "Experiment, Speculation and Law: Faraday's Analysis of Arago's Wheel," *PSA: Proceedings of the Biennial Meeting of the Philosophy of Science Association* 1994: 293–303.

Sulloway, Frank. 1992. *Freud: Biologist of the Mind.* Cambridge, Mass.: Harvard University Press.

Suppe, Frederick. 1977. "The Search for Philosophical Understanding of Scientific Theories," pp. 3–241 in F. Suppe (ed.), *The Structure of Scientific Theories.* Urbana: University of Illinois Press.

Tannery, Paul. 1887. *Pour l'histoire de la science hellène.* Paris: F. Alcan.

Thagard, Paul. 1997. "Collaborative Knowledge," *Noûs* 31: 242–261.

Tiles, Mary. 1987. "Epistemological History: The Legacy of Bachelard and Canguilhem," pp. 141–156 in A. Phillips Griffiths (ed.), *Contemporary French Philosophy.* Cambridge: Cambridge University Press.

———. 1984. *Bachelard: Science and Objectivity.* Cambridge: Cambridge University Press.

Tiles, Mary and Tiles, James. 1993. *The Authority of Knowledge: An Introduction to Historical Epistemology.* Oxford: Blackwell.

Toulmin, Stephen. 1972. *Human Understanding, vol. 1: The Collective Use and Evolution of Concepts.* Oxford: Clarendon Press.

———. 1967. "Conceptual Revolutions in Science," *Synthese* 17 (1): 75–91.

———. 1961. *Foresight and Understanding.* Bloomington: University of Indiana Press.

———. 1953. *The Philosophy of Science: An Introduction.* London: Hutchinson.

Toulmin, Stephen and Goodfield, June. 1965. *The Discovery of Time.* New York: Harper.

———. 1962. *The Architecture of Matter.* New York: Harper.

———. 1961. *The Fabric of the Heavens: The Development of Astronomy and Dynamics.* New York: Harper.

Volta, Alessandro. 1800. "On the Electricity Excited by the Mere Contact of Conducting Substances of Different Kinds," *Philosophical Transactions* 90.

von Plato, Jan. 1994. *Creating Modern Probability: Its Mathematics, Physics and Philosophy in Historical Perspective.* Cambridge: Cambridge University Press.

Wang, Xinli. 2002. "Taxonomy, Truth-Value Gaps and Incommensurability: A Reconstruction of Kuhn's Taxonomic Interpretation of Incommensurability," *Studies in History and Philosophy of Science Part A* 33 (3): 465–485.

Wartofsky, Marx. 1987. "Epistemology Historicized," pp. 357–374 in A. Shimony and D. Nails (eds.), *Naturalistic Epistemology.* Dordrecht: Reidel.

Whewell, William. 1847. *The Philosophy of the Inductive Sciences, Founded Upon Their History.* Second edition, in two volumes. London: John W. Parker.

———. 1857. *History of the Inductive Sciences, from the Earliest to the Present Time.* Third edition, in two volumes, London: John W. Parker.

———. 1858a. *The History of Scientific Ideas,* in two volumes, London: John W. Parker.

———. 1858b. *Novum Organon Renovatum,* London: John W. Parker.

———. 1971 [1860]. *On the Philosophy of Discovery: Chapters Historical and Critical.* New York: Lenox Hill. Originally published 1860. London: John W. Parker.

Wiener, Philip and Noland, Aaron (eds.). 1957. *Roots of Scientific Thought.* Selections from the first 18 volumes of the *Journal of the History of Ideas.* New York: Basic Books.

Windelband, Wilhelm. 1980/1894. "History and Natural Science," trans. of "Geschichte und Naturwissenschaften" by Guy Oakes. *History and Theory* 19 (2).

Wittgenstein, Ludwig. 1999/1922. *Tractatus Logico-Philosophicus.* C. K. Ogden, prepared with assistance from G. E. Moore, Frank P. Ramsey, and Ludwig Wittgenstein. London: Routledge.

———. 1961/1922. *Tractatus Logico-Philosophicus,* trans. David Pears and Brian McGuinness. London: Routledge.

Zilsel, Edgar. 1941. "Physics and the Problem of Historico-Sociological Laws," *Philosophy of Science* 8.

———. 1942. "Problems of Empiricism," *International Encyclopedia of Unified Science,* Vol. II, 8.

CHAPTER 3

Reason in Science

GEORGE (JORGE) SANTAYANA

Reason in Science. Volume five of *The Life of Reason; or The Phases of Human Progress*, originally published by Charles Scribner's Sons, N.Y., in 1905. Unabridged republication by Dover, 1982. Excerpts from Chapter II, History.

Historical Research a Part of Physics

Historical investigation is the natural science of the past. The circumstance that its documents are usually literary may somewhat disguise the physical character and the physical principles of this science; but when a man wishes to discover what really happened at a given moment, even if the event were somebody's thought; he has to read his sources, not for what they say, but for what they imply. In other words, the witnesses cannot be allowed merely to speak for themselves, after the gossiping fashion familiar in Herodotus; their testimony has to be interpreted according to the laws of evidence. The past needs to be reconstructed out of reports, as in geology or archæology it needs to be reconstructed out of stratifications and ruins. A man's memory or the report in a newspaper is a fact justifying certain inferences about its probable causes according to laws which such phenomena betray in the present when they are closely scrutinised. This reconstruction is often very difficult, and sometimes all that can be established in the end is merely that the tradition before us is certainly false; somewhat as a perplexed geologist might venture on no conclusion except that the state of the earth's crust was once very different from what it is now.

Verification Here Indirect

A natural science dealing with the past labours under the disadvantage of not being able to appeal to experiment. The facts it terminates upon cannot be recovered, so that they may verify in sense the hypothesis that had inferred them. The hypothesis can be tested only by current events; it is then turned back upon the past, to give assurance of facts which themselves are hypothetical and remain hanging, as it were, to the loose end of the hypothesis itself. A hypothetical fact is a most dangerous creature, since it lives on the credit of a theory which in turn would be bankrupt if the fact should fail. Inferred past facts are more deceptive than facts prophesied, because while the risk of error in the inference is the same, there is no possibility of discovering that error; and the historian, while really as speculative as the prophet, can never be found out.

Most facts known to man, however, are reached by inference, and their reality may be wisely assumed so long as the principle by which they are inferred, when it is applied in the present, finds complete and constant verification. Presumptions involved in memory and tradition give the first hypothetical facts we count upon; the relations which these first facts betray supply the laws by which facts are to be concatenated; and these laws may then be used to pass from the first hypothetical facts to hypothetical facts of a second order, forming a background and congruous extension to those originally assumed. This expansion of discursive science can go on for ever, unless indeed the principles of inference employed in it involve some present existence, such as a skeleton in a given tomb, which direct experience fails to verify. Then the theory itself is disproved and the whole galaxy of hypothetical facts which clustered about it forfeit their credibility.

Futile Ideal to Survey All Facts

Historical investigation has for its aim to fix the order and character of events throughout past time in all places. The task is frankly superhuman, because no block of real existence, with its infinitesimal detail, can be recorded, nor if somehow recorded could it be dominated by the mind; and to carry on a survey of this social continuum *ad infinitum* would multiply the difficulty. The task might also be called infrahuman, because the sort of omniscience which such complete historical science would achieve would merely furnish materials for intelligence: it would be inferior to intelligence itself. There are many things which, as Aristotle says, it is better not to know than to know—namely, those things which do not count in controlling the mind's fortunes nor enter into its ideal expression. Such is the whole flux of immediate experience in other minds or in one's own past; and just as it is better to forget than to remember a nightmare or the by-gone sensations of sea-sickness, so it is better not to conceive the sensuous pulp of alien experience, something infinite in amount and insignificant in character.

An attempt to rehearse the inner life of everybody that has ever lived would be no rational endeavour. Instead of lifting the historian above the world and making him the most consummate of creatures, it would flatten his mind out into a passive after-image of diffuse existence, with all its horrible blindness, strain, and monotony. Reason is not come to repeat the universe but to fulfil it. Besides, a complete survey of events would perforce register all changes that have taken place in matter since time began, the fields of geology, astronomy, palæontology, and archæology being all, in a sense, included in history. Such learning would dissolve thought in a vertigo, if it had not already perished of boredom. Historical research is accordingly a servile science which may enter the Life of Reason to perform there some incidental service, but which ought to lapse as soon as that service is performed.

It Is Arbitrary

In fine, historical terms mark merely rhetorical unities, which have no dynamic cohesion, and there are no historical laws which are not at bottom physical, like the laws of habit—those expressions of Newton's first law of motion. An essayist may play with historical apperception as long as he will and always find something new to say, discovering the ideal nerve and issue of a movement in a different aspect of the facts. The truly proportionate, constant, efficacious relations between things will remain material. Physical causes traverse the moral units at which history stops, determining their force and duration, and the order, so irrelevant to intent, in which they succeed one another. Even the single man's life and character have subterranean sources; how should the outer expression and influence of that character have sources more superficial than its own? Yet we cannot trace mechanical necessity down to the more stable units composing a personal mechanism, and much less, therefore, to those composing a complex social evolution. We accordingly translate the necessity, obviously lurking under life's commonplace yet unaccountable shocks, into verbal principles, names for general impressive results, that play some rôle in our ideal philosophy. Each of these idols of the theatre is visible only on a single stage and to duly predisposed spectators. The next passion affected will throw a differently coloured calcium light on the same pageant, and there will be no end of rival evolutions and incompatible ideal principles crossing one another at every interesting event.

Such a manipulation of history, when made by persons who underestimate their imaginative powers, ends in asserting that events have directed themselves prophetically upon the interests which they arouse. Apart from the magic involved and the mockery of all science, there is a difficulty here which even a dramatic idealist ought to feel. The interests affected are themselves many and contrary. If history is to be understood teleologically, which of all the possible ends it might be pursuing shall we think really endowed with regressive influence and responsible for the movement that is going to realise it? Did Columbus, for instance, discover America so that George Washington might exist and that some day football and the Church of England may prevail throughout the world? Or was it (as has been seriously maintained) in order that the converted [inhabitants] of South America might console Saint Peter for the defection of the British and Germans? Or was America, as Hegel believed, ideally superfluous, the absolute having become self-conscious enough already in Prussia? Or shall we say that the real goal is at an infinite distance and unimaginable by us, and useless, therefore, for understanding anything?

In truth, whatever plausibility the providential view of a given occurrence may have is dependent on the curious limitation and selfishness of the observer's estimations. Sheep are providentially designed for men; but why not also for wolves, and men for worms and microbes? If the historian is willing to accept such a suggestion, and to become a blind worshipper of success, applauding every issue, however lamentable for humanity, and calling it admirable tragedy, he may seem for a while to save his theory by making it mystical; yet presently this last illusion will be dissipated when he loses his way in the maze and finds that all victors perish in their turn and everything, if you look far enough, falls back into the inexorable vortex. This is the sort of observation that the Indian sages made long ago; it is what renders their philosophy [. . .] such an irrefragable record of experience, such a superior, definitive perception of the flux. Beside it, our progresses of two centuries and our philosophies of history, embracing one-quarter of the earth for three thousand years, seem puerile vistas indeed. Shall all eternity and all existence be for the sake of what is happening here to-day, and to me? Shall we strive manfully to the top of this particular wave, on the ground that its foam is the culmination of all things for ever?

There is a sense, of course, in which definite political plans and moral aspirations may well be fulfilled by events. Our ancestors, sharing and anticipating our natures, may have had in many respects our actual interests in view, as we may have those of posterity. Such ideal co-operation extends far, where primary interests are concerned; it is rarer and more qualified where a fine and fragile organisation is required to support the common spiritual life. Even in these cases, the aim pursued and attained is not the force that operates, since the result achieved had many other conditions besides the worker's intent, and that intent itself had causes which it knew nothing of. Every "historical force" pompously appealed to breaks up on inspection into a cataract of miscellaneous natural processes and minute particular causes. It breaks into its mechanical constituents and proves to have been nothing but an *effet d'ensemble* produced on a mind whose habits and categories are essentially rhetorical.

CHAPTER 4

The Function of General Laws in History

CARL HEMPEL

This essay first appeared in *The Journal of Philosophy*, Vol. 39, No. 2 (Jan. 15, 1942), pp. 35–48.

1. It is a rather widely held opinion that history, in contradistinction to the so-called physical sciences, is concerned with the description of particular events of the past rather than with the search for general laws which might govern those events. As a characterization of the type of problem in which some historians are mainly interested, this view probably can not be denied; as a statement of the theoretical function of general laws in scientific historical research, it is certainly unacceptable. The following considerations are an attempt to substantiate this point by showing in some detail that general laws have quite analogous functions in history and in the natural sciences, that they form an indispensable instrument of historical research, and that they even constitute the common basis of various procedures which are often considered as characteristic of the social in contradistinction to the natural sciences.

By a general law, we shall here understand a statement of universal conditional form which is capable of being confirmed or disconfirmed by suitable empirical findings. The term "law" suggests the idea that the statement in question is actually well confirmed by the relevant evidence available; as this qualification is, in many cases, irrelevant for our purpose, we shall frequently use the term "hypothesis of universal form" or briefly "universal hypothesis" instead of "general law," and state the condition of satisfactory confirmation separately, if necessary. In the context of this paper, a universal hypothesis may be assumed to assert a regularity of the following type: In every case where an event of a specified kind C occurs at a certain place and time, an event of a specified kind E will occur at a place and time which is related in a specified manner to the place and time of the occurrence of the first event. (The symbols "C" and "E" have been chosen to suggest the terms "cause" and "effect," which are often, though by no means always, applied to events related by a law of the above kind.)

2.1 The main function of general laws in the natural sciences is to connect events in patterns which are usually referred to as *explanation* and *prediction*.

The explanation of the occurrence of an event of some specific kind E at a certain place and time consists, as it is usually expressed, in indicating the causes or determining factors of E. Now the assertion that a set of events— say, of the kinds $C_1, C_2, \ldots, C_n$—have caused the event to be explained, amounts to the statement that, according to certain general laws, a set of events of the kinds mentioned is regularly accompanied by an event of kind E. Thus, the scientific explanation of the event in question consists of

(1) a set of statements asserting the occurrence of certain events $C_1, \ldots C_n$ at certain times and places,
(2) a set of universal hypotheses, such that
   (a) the statements of both groups are reasonably well confirmed by empirical evidence,
   (b) from the two groups of statements the sentence asserting the occurrence of event E can be logically deduced.

In a physical explanation, group (1) would describe the initial and boundary conditions for the occurrence of the final event; generally, we shall say that group (1) states the *determining conditions* for the event to be explained, while group (2) contains the general laws on which the explanation is based; they imply the statement that whenever events of the kind described in the first group occur, an event of the kind to be explained will take place.

Illustration: Let the event to be explained consist in the cracking of an automobile radiator during a cold night. The sentences of group (1) may state the following initial and boundary conditions: The car was left in the street all night. Its radiator, which consists of iron, was completely filled with water, and the lid was screwed on tightly. The temperature during the night dropped from 39 °F. in the evening to 25 °F. in the morning; the air pressure was normal. The bursting pressure of the radiator material is so and so much.—Group (2) would contain empirical laws such as the following: Below 32 °F., under normal atmospheric pressure, water freezes. Below 39.2 °F., the pressure of a mass of water increases with decreasing temperature, if the volume remains constant or decreases; when the water freezes, the pressure again increases. Finally, this group would have to include a quantitative law concerning the change of pressure of water as a function of its temperature and volume.

From statements of these two kinds, the conclusion that the radiator cracked during the night can be deduced by logical reasoning; an explanation of the considered event has been established.

2.2 It is important to bear in mind that the symbols "E," "C," "C_1," "C_2," etc., which were used above, stand for kinds or properties of events, not for what is sometimes called individual events. For the object of description and explanation in every branch of empirical science is always the occurrence of an event of a certain *kind* (such as a drop in temperature by 14 °F., an eclipse of the moon, a cell-division, an earthquake, an increase in employment, a political assassination) at a given place and time, or in a given empirical object (such as the radiator of a certain car, the planetary system, a specified historical personality, etc.) at a certain time.

What is sometimes called the complete description of an individual event (such as the earthquake of San Francisco in 1906 or the assassination of Julius Caesar) would require a statement of all the properties exhibited by the spatial region or the individual object involved, for the period of time occupied by the event in question. Such a task can never be completely accomplished.

A fortiori, it is impossible to explain an individual event in the sense of accounting for *all* its characteristics by means of universal hypotheses, although the explanation of what happened at a specified place and time may gradually be made more and more specific and comprehensive.

But there is no difference, in this respect, between history and the natural sciences: both can give an account of their subject-matter only in terms of general concepts, and history can "grasp the unique individuality" of its objects of study no more and no less than can physics or chemistry.

3. The following points result more or less directly from the above study of scientific explanation and are of special importance for the questions here to be discussed.

3.1 A set of events can be said to have caused the event to be explained only if general laws can be indicated which connect "causes" and "effect" in the manner characterized above.

3.2 No matter whether the cause-effect terminology is used or not, a scientific explanation has been achieved only if empirical laws of the kind mentioned under (2) in 2.1 have been applied.[1]

3.3 The use of universal empirical hypotheses as explanatory principles distinguishes genuine from pseudo-explanation, such as, say, the attempt to account for certain features of organic behavior by reference to an entelechy, for whose functioning no laws are offered, or the explanation of the achievements of a given person in terms of his "mission in history," his "predestined fate," or similar notions. Accounts of this type are based on metaphors rather than laws; they convey pictorial and emotional appeals instead of insight into factual connections; they substitute vague analogies and intuitive "plausibility" for deduction from testable statements and are therefore unacceptable as scientific explanations.

Any explanation of scientific character is amenable to objective checks; these include

(a) an empirical test of the sentences which state the determining conditions;

(b) an empirical test of the universal hypotheses on which the explanation rests;

(c) an investigation of whether the explanation is logically conclusive in the sense that the sentence describing the event to be explained follows from the statements of groups (1) and (2).

4. The function of general laws in *scientific prediction* can now be stated very briefly. Quite generally, prediction in empirical science consists in deriving a statement about a certain future event (for example, the relative position of the planets to the sun, at a future date) from (1) statements describing certain known (past or present) conditions (for example, the positions and momenta of the planets at a past or present moment), and (2) suitable general laws (for example, the laws of celestial mechanics). Thus, the logical structure of a scientific prediction is the same as that of a scientific explanation, which has been described in 2.1. In particular, prediction no less than explanation throughout empirical science involves reference to universal empirical hypotheses.

The customary distinction between explanation and prediction rests mainly on a pragmatical difference between the two: While in the case of an explanation, the final event is known to have happened, and its determining conditions have to be sought, the situation is reversed in the case of a prediction: here, the initial conditions are given, and their "effect"—which, in the typical case, has not yet taken place—is to be determined.

In view of the structural equality of explanation and prediction, it may be said that an explanation as characterized in 2.1 is not complete unless it might as well have functioned as a prediction: If the final event can be derived from the initial conditions and universal hypotheses stated in the explanation, then it might as well have been predicted, before it actually happened, on the basis of a knowledge of the initial conditions and the general laws. Thus, e.g., those initial conditions and general laws which the astronomer would adduce in explanation of a certain eclipse of the sun are such that they might also have served as a sufficient basis for a forecast of the eclipse before it took place.

However, only rarely, if ever, are explanations stated so completely as to exhibit this predictive character (which the test referred to under (c) in 3.3 would serve to reveal). Quite commonly, the explanation offered for the occurrence of an event is incomplete. Thus, we may hear the explanation that a barn burnt down "because" a burning cigarette was dropped in the hay, or that a certain political movement has spectacular success "because" it takes advantage of widespread racial prejudices. Similarly, in the case of the broken radiator, the customary way of formulating an explanation would be restricted to pointing out that the car was left in the cold, and the radiator was filled with water.—In explanatory statements like these, the general laws which confer upon the stated conditions the character of "causes" or "determining factors" are completely omitted (sometimes, perhaps, as a "matter of course"), and, furthermore, the enumeration of the determining conditions of group (1) is incomplete; this is illustrated by the preceding examples, but even by the earlier analysis of the broken-radiator case: as a closer examination would reveal, even that much more detailed statement of determining conditions and universal hypotheses would require amplification in order to serve as a sufficient basis for the deduction of the conclusion that the radiator broke during the night.

In some instances, the incompleteness of a given explanation may be considered as inessential. Thus, e.g., we may feel that the explanation referred to in the last example could be made complete if we so desired; for we have reasons to assume that we know the kind of determining conditions and of general laws which are relevant in this context.

Very frequently, however, we encounter "explanations" whose incompleteness can not simply be dismissed as inessential. The methodological consequences of this situation will be discussed later (especially in 5.3 and 5.4).

5.1 The preceding considerations apply to *explanation in history* as well as in any other branch of empirical science. Historical explanation, too, aims at showing that the event in question was not "a matter of chance," but was to be expected in view of certain antecedent or simultaneous conditions.

The expectation referred to is not prophecy or divination, but rational scientific anticipation which rests on the assumption of general laws.

If this view is correct, it would seem strange that while most historians do suggest explanations of historical events, many of them deny the possibility of resorting to any general laws in history. It is, however, possible to account for this situation by a closer study of explanation in history, as may become clear in the course of the following analysis.

5.2 In some cases, the universal hypotheses underlying a historical explanation are rather explicitly stated, as is illustrated by the italicized passages in the following attempt to explain the tendency of government agencies to perpetuate themselves and to expand (italics the author's):

> As the activities of the government are enlarged, more people develop a vested interest in the continuation and expansion of governmental functions. *People who have jobs do not like to lose them; those who are habituated to certain skills do not welcome change; those who have become accustomed to the exercise of a certain kind of power do not like to relinquish their control*—if anything, *they want to develop greater power and correspondingly greater prestige.* . . .
>
> Thus, government offices and bureaus, once created, in turn institute drives, not only to fortify themselves against assault, but to enlarge the scope of their operations.[2]

Most explanations offered in history or sociology, however, fail to include an explicit statement of the general regularities they presuppose; and there seem to be at least two reasons which account for this:

First, the universal hypotheses in question frequently relate to individual or social psychology, which somehow is supposed to be familiar to everybody through his everyday experience; thus, they are tacitly taken for granted. This is a situation quite similar to that characterized in 4.

Second, it would often be very difficult to formulate the underlying assumptions explicitly with sufficient precision and at the same time in such a way that they are in agreement with all the relevant empirical evidence available. It is highly instructive, in examining the adequacy of a suggested explanation, to attempt a reconstruction of the universal hypotheses on which it rests. Particularly, such terms as "hence," "therefore," "consequently," "because," "naturally," "obviously," etc., are often indicative of the tacit presupposition of some general law: they are used to tie up the initial conditions with the event to be explained; but that the latter was "naturally" to be expected as "a consequence" of the stated conditions follows only if suitable general laws are presupposed. Consider, for example, the statement that the Dust Bowl farmers migrate to California "because" continual drought and sandstorms render their existence increasingly precarious, and because California seems to them to offer so much better living conditions. This explanation rests on some such universal hypothesis as that populations will tend to migrate to regions which offer better living conditions. But it would obviously be difficult accurately to state this hypothesis in the form of a general law which is reasonably well confirmed by all the relevant evidence available. Similarly, if a particular revolution is explained by reference to the growing discontent, on the part of a large part of the population, with certain prevailing conditions, it is clear that a general regularity is assumed in this explanation, but we are hardly in a position to state just what extent and what specific form the discontent has to assume, and what the environmental conditions have to be, to bring about a revolution. Analogous remarks apply to all historical explanations in terms of class struggle, economic or geographic conditions, vested interests of certain groups, tendency to conspicuous consumption, etc.: All of them rest on the assumption of universal hypotheses[3] which connect certain characteristics of individual or group life with others; but in many cases, the content of the hypotheses which are tacitly assumed in a given explanation can be reconstructed only quite approximately.

5.3 It might be argued that the phenomena covered by the type of explanation just mentioned are of a statistical character, and that therefore only probability hypotheses need to be assumed in their explanation, so that the question as to the "underlying general laws" would be based on a false premise. And indeed, it seems possible and justifiable to construe certain explanations offered in history as based on the assumption of probability hypotheses rather than of general "deterministic" laws, i.e., laws in the form of universal conditionals. This claim may be extended to many of the explanations offered in other fields of empirical science as well. Thus, e.g., if Tommy comes down with the measles two weeks after his brother, and if he has not been in the company of other persons having the measles, we accept the explanation that he caught the disease from his brother. Now, there is a general hypothesis underlying this explanation; but it can hardly be said to be a general law to the effect that any person who has not had the measles before will get them without fail if he stays in the company of somebody else who has the measles; that a contagion will occur can be asserted only with a high probability.

Many an explanation offered in history seems to admit of an analysis of this kind: if fully and explicitly formulated, it would state certain initial conditions, and certain probability hypotheses,[4] such that the occurrence of the event to be explained is made highly probable by the initial conditions in view of the probability hypotheses. But no matter whether explanations in history be construed as "causal" or as "probabilistic" in character, it remains true that in general the initial conditions and especially the universal hypotheses involved are not clearly indicated, and can not unambiguously be supplemented. (In the case of probability hypotheses, for example, the probability values involved will at best be known quite roughly.)

5.4 What the explanatory analyses of historical events offer is, then, in most cases not an explanation in one of the meanings developed above, but something that might be called an *explanation sketch*. Such a sketch consists of a more or less vague indication of the laws and initial conditions considered as relevant, and it needs "filling out" in order to turn into a full-fledged explanation. This filling-out requires further empirical research, for which the sketch suggests the direction. (Explanation sketches are common also outside of history; many explanations in psychoanalysis, for instance, illustrate this point.)

Obviously, an explanation sketch does not admit of an empirical test to the same extent as does a complete explanation; and yet, there is a difference between a scientifically acceptable explanation sketch and a pseudo-explanation (or a pseudo-explanation sketch). A scientifically acceptable explanation sketch needs to be filled out by more specific statements; but it points into the direction where these statements are to be found; and concrete research may tend to confirm or to infirm those indications; i.e., it may show that the kind of initial conditions suggested are actually relevant; or it may reveal that factors of a quite different nature have to be taken into account in order to arrive at a satisfactory explanation.—The filling-out process required by an explanation sketch will, in general, assume the form of a gradually increasing precision of the formulations involved; but at any stage of this process, those formulations will have some empirical import: it will be possible to indicate, at least roughly, what kind of evidence would be relevant in testing them, and what findings would tend to confirm them. In the case of non-empirical explanations or explanation sketches, on the other hand—say, by reference to the historical destination of a certain race, or to a principle of historical justice—the use of empirically meaningless terms makes it impossible even roughly to indicate the type of investigation that would have a bearing upon those formulations, and that might lead to evidence either confirming or infirming the suggested explanation.

5.5 In trying to appraise the soundness of a given explanation, one will first have to attempt to reconstruct as completely as possible the argument constituting the explanation or the explanation sketch. In particular, it is important to realize what the underlying explaining hypotheses are, and to judge of their scope and empirical foundation. A resuscitation of the assumptions buried under the gravestones "hence," "therefore," "because," and the like will often reveal that the explanation offered is poorly founded or downright unacceptable. In many cases, this procedure will bring to light the fallacy of claiming that a large number of details of an event have been explained when, even on a very liberal interpretation, only some broad characteristics of it have been accounted for. Thus, for example, the geographic or economic conditions under which a group lives may account for certain general features of, say, its art or its moral codes; but to grant this does not mean that the artistic achievements of the group or its system of morals has thus been explained in detail; for this would imply that from a description of the prevalent geographic or economic conditions alone, a detailed account of certain aspects of the cultural life of the group can be deduced by means of specifiable general laws.

A related error consists in singling out one of several important groups of factors which would have to be stated in the initial conditions, and then claiming that the phenomenon in question is "determined" by and thus can be explained in terms of that one group of factors.

Occasionally, the adherents of some particular school of explanation or interpretation in history will adduce, as evidence in favor of their approach, a successful historical prediction which was made by a representative of their school. But though the predictive success of a theory is certainly relevant evidence of its soundness, it is important to make sure that the successful prediction is in fact obtainable by means of the theory in question. It happens sometimes that the prediction is actually an ingenious guess which may have been influenced by the theoretical outlook of its author, but which can not be arrived at by means of his theory alone. Thus, an adherent of a quite metaphysical "theory" of history may have a sound feeling for historical developments and may be able to make correct predictions, which he will even couch in the terminology of his theory, though they could not have been attained by means of it. To guard against such pseudo-confirming cases would be one of the functions of test (*c*) in 3.3.

6. We have tried to show that in history no less than in any other branch of empirical inquiry, scientific explanation can be achieved only by means of suitable general hypotheses, or by theories, which are bodies of systematically related hypotheses. This thesis is clearly in contrast with the familiar view that genuine explanation in history is obtained by a method which characteristically distinguishes the social from the natural sciences, namely, *the method of empathetic understanding;* The historian, we are told, imagines himself in the place of the persons involved in the events which he wants to explain; he tries to realize as completely as possible the circumstances under which they acted, and the motives which influenced their actions; and by this imaginary self-identification with his heroes, he arrives at an understanding and thus at an adequate explanation of the events with which he is concerned.

This method of empathy is, no doubt, frequently applied by laymen and by experts in history. But it does not in itself constitute an explanation; it rather is essentially a heuristic device; its function is to suggest certain psychological hypotheses which might serve as explanatory principles in the case under consideration. Stated in crude terms, the idea underlying this function is the following: The historian tries to realize how he himself would act under the given conditions, and under the particular motivations of his heroes; he tentatively generalizes his findings into a general rule and uses the latter as an explanatory principle in accounting for the actions of the persons involved. Now, this procedure may sometimes prove heuristically helpful; but its use does not guarantee the soundness of the historical explanation to which it leads. The latter rather depends upon the factual correctness of the empirical generalizations which the method of understanding may have suggested.

Nor is the use of this method indispensable for historical explanation. A historian may, for example, be incapable of feeling himself into the rôle of a paranoiac historic personality, and yet he may well be able to explain certain of his actions; notably by reference to the principles of abnormal psychology. Thus, whether the historian is or is not in a position to identify himself with his historical hero, is irrelevant for the correctness of his explanation; what counts, is the soundness of the general hypotheses involved, no matter whether they were suggested by empathy or by a strictly behavioristic procedure. Much of the appeal of the "method of understanding" seems to be due to the fact that it tends to present the phenomena in question as somehow "plausible" or "natural" to us;[5] this is often done by means of attractively worded metaphors. But the kind of "understanding" thus conveyed must clearly be separated from scientific understanding. In history as anywhere else in empirical science, the explanation of a phenomenon consists in subsuming it under general empirical laws; and the criterion of its soundness is not whether it appeals to our imagination, whether it is presented in suggestive analogies, or is otherwise made to appear plausible—all this may occur in pseudo-explanations as well—but exclusively whether it rests on empirically well confirmed assumptions concerning initial conditions and general laws.

7.1 So far, we have discussed the importance of general laws for explanation and prediction, and for so-called understanding in history. Let us now survey more briefly some other procedures of historical research which involve the assumption of universal hypotheses.

Closely related to explanation and understanding is the so-called *interpretation of historical phenomena* in terms of some particular approach or theory. The interpretations which are actually offered in history consist either in subsuming the phenomena in question under a scientific explanation or explanation sketch; or in an attempt to subsume them under some general idea which is not amenable to any empirical test. In the former case, interpretation clearly is explanation by means of universal hypotheses; in the latter, it amounts to a pseudo-explanation which may have emotive appeal and evoke vivid pictorial associations, but which does not further our theoretical understanding of the phenomena under consideration.

7.2 Analogous remarks apply to the procedure of ascertaining the "*meaning*" of given historical events; its scientific import consists in determining what other events are relevantly connected with the event in question, be it as "causes," or as "effects"; and the statement of the relevant connections assumes, again, the form of explanations or explanation sketches which involve universal hypotheses; this will be seen more clearly in the subsequent section.

7.3 In the historical explanation of some social institutions great emphasis is laid upon an analysis of the *development* of the institution up to the stage under consideration. Critics of this approach have objected that a mere description of this kind is not a genuine explanation. This argument may be given a slightly different aspect in terms of the preceding reflections: A description of the development of an institution is obviously not simply a statement of *all* the events which temporally preceded it; only those events are meant to be included which are "*relevant*" to the formation of that institution. And whether an event is relevant to that development is not a question of the value attitude of the historian, but an objective question depending upon what is sometimes called a causal analysis of the rise of that institution.[6] Now, the causal analysis of an event consists in establishing an explanation for it, and since this requires reference to general hypotheses, so do assumptions about relevance, and, consequently, so does the adequate analysis of the historical development of an institution.

7.4 Similarly, the use of the notions of *determination* and of *dependence* in the empirical sciences, including history, involves reference to general laws.[7] Thus, e.g., we may say that the pressure of a gas depends upon its temperature and volume, or that temperature and volume determine the pressure, in virtue of Boyle's law. But unless the underlying laws are stated explicitly, the assertion of a relation of dependence or of determination between certain magnitudes or characteristics amounts at best to claiming that they are connected by some unspecified empirical law; and that is a very meager assertion indeed: If, for example, we know only that there is some empirical law connecting two metrical magnitudes (such as length and temperature of a metal bar), we can not even be sure that a change of one of the two will be accompanied by a change of the other (for the law may connect the same value of the "dependent" or "determined" magnitude with different values of the other), but only that with any specific value of one of the variables, there will always be associated one and the same value of the other; and this is obviously much less than most authors mean to assert when they speak of determination or dependence in historical analysis.

Therefore, the sweeping assertion that economic (or geographic, or any other kind of) conditions "determine" the development and change of all other aspects of human society, has explanatory value only in so far as it can be substantiated by explicit laws which state just what kind of change in human culture will regularly follow upon specific changes in the economic (geographic, etc.) conditions. Only the establishment of concrete laws can fill the general thesis with scientific content, make it amenable to empirical tests, and confer upon it an explanatory function. The elaboration of such laws with as much precision as possible seems clearly to be the direction in which progress in scientific explanation and understanding has to be sought.

8. The considerations developed in this paper are entirely neutral with respect to the problem of "*specifically historical laws*": neither do they presuppose a particular way of distinguishing historical from sociological and other laws, nor do they imply or deny the assumption that empirical laws can be found which are historical in some specific sense, and which are well confirmed by empirical evidence.

But it may be worth mentioning here that those universal hypotheses to which historians explicitly or tacitly refer in offering explanations, predictions, interpretations, judgments of relevance, etc., are taken from *various* fields of scientific research, in so far as they are not pre-scientific generalizations of everyday experiences. Many of the universal hypotheses underlying historical explanation, for instance, would commonly be classified as psychological, economical, sociological, and partly perhaps as historical laws; in addition, historical research has frequently to resort to general laws established in physics, chemistry, and biology. Thus, e.g., the explanation of the defeat of an army by reference to lack of food, adverse weather conditions, disease, and the like, is based on a—usually tacit—assumption of such laws. The use of tree rings in dating events in history rests on the application of certain biological regularities. Various methods of testing the authenticity of documents, paintings, coins, etc., make use of physical and chemical theories.

The last two examples illustrate another point which is relevant in this context: Even if a historian should propose to restrict his research to a "*pure description*" of the past, without any attempt at offering explanations, statements about relevance and determination, etc., he would continually have to make use of general laws. For the object of his studies would be the past—forever inaccessible to his direct examination. He would have to establish his knowledge by indirect methods: by the use of universal hypotheses which connect his present data with those past events. This fact has been obscured partly because some of the regularities involved are so familiar that they are not considered worth mentioning at all; and partly because of the habit of relegating the various hypotheses and theories which are used to ascertain knowledge about past events, to the "auxiliary sciences" of history. Quite probably, some of the historians who tend to minimize, if not to deny, the importance of general laws for history, are actuated by the feeling that only "genuinely historical laws" would be of interest for history. But once it is realized that the discovery of historical laws (in some specified sense of this very vague notion) would not make history methodologically autonomous and independent of the other branches of scientific research, it would seem that the problem of the existence of historical laws ought to lose some of its weight.

The remarks made in this section are but special illustrations of two broader principles of the theory of science: first, the separation of "pure description" and "hypothetical generalization and theory-construction" in empirical science is unwarranted; in the building of scientific knowledge the two are inseparably linked. And, second, it is similarly unwarranted and futile to attempt the demarcation of sharp boundary lines between the different fields of scientific research, and an autonomous development of each of the fields. The necessity, in historical inquiry, to make extensive use of universal hypotheses of which at least the overwhelming majority come from fields of research traditionally distinguished from history is just one of the aspects of what may be called the methodological unity of empirical science.

Notes

1. Maurice Mandelbaum, in his generally very clarifying analysis of relevance and causation in history ([Mandelbaum 1938], Chs. 7, 8) seems to hold that there is a difference between the "causal analysis" or "causal explanation" of an event and the establishment of scientific laws governing it in the sense stated above. He argues that "scientific laws can only be formulated on the basis of causal analysis," but that "they are not substitutes for full causal explanations" (l.c., p. 238). For the reasons outlined above, this distinction does not appear to be justified: every "causal explanation" is an "explanation by scientific laws"; for in no other way than by reference to empirical laws can the assertion of a causal connection between certain events be scientifically substantiated.
2. [McConnell 1939], pp. 894–895.
3. What is sometimes, misleadingly, called an explanation by means of a certain *concept* is, in empirical science, actually an explanation in terms of *universal hypotheses* containing that concept. "Explanations" involving concepts which do not function in empirically testable hypotheses—such as "entelechy" in biology, "historic destination of a race" or "self-unfolding of absolute reason" in history—are mere metaphors without cognitive content.

4. E[dgar] Zilsel, in a very stimulating paper on "Physics and the Problem of Historico-Sociological Laws" ([Zilsel 1941], pp. 567–579), suggests that all specifically historical laws are of a statistical character similar to that of the "macro-laws" in physics. The above remarks, however, are not restricted to specifically historical laws since explanation in history rests to a large extent on non-historical laws (cf. section 8 of this paper).

5. For a criticism of this kind of plausibility, cf. Zilsel, *l.c.*, pp. 577–578, and sections 7 and 8 in the same author's [Zilsel 1942].

6. See the detailed and clear exposition of this point in M. Mandelbaum's book; *l.c.*, Chs. 6–8.

7. According to Mandelbaum, history, in contradistinction to the physical sciences, consists "not in the formulation of laws of which the particular case is an instance, but in the description of the events in their actual determining relationships to each other; in seeing events as the products and producers of change" (*l.c.*, pp. 13–14). This is essentially a view whose untenability has been pointed out already by Hume; it is the belief that a careful examination of two specific events alone, without any reference to similar cases and to general regularities, can reveal that one of the events produces or determines the other. This thesis does not only run counter to the scientific meaning of the concept of determination which clearly rests on that of general law, but it even fails to provide any objective criteria which would be indicative of the intended relationship of determination or production. Thus, to speak of empirical determination independently of any reference to general laws means to use a metaphor without cognitive content.

CHAPTER 5

The Three Tasks of Epistemology

HANS REICHENBACH

This text was first published in Reichenbach's 1938 *Experience and Pre-diction*, Chapter I, "Meaning," §1, pp. 3–16.

Every theory of knowledge must start from knowledge as a given sociological fact. The system of knowledge as it has been built up by generations of thinkers, the methods of acquiring knowledge used in former times or used in our day, the aims of knowledge as they are expressed by the procedure of scientific inquiry, the language in which knowledge is expressed—all are given to us in the same way as any other sociological fact, such as social customs or religious habits or political institutions. The basis available for the philosopher does not differ from the basis of the sociologist or psychologist; this follows from the fact that, if knowledge were not incorporated in books and speeches and human actions, we never would know it. Knowledge, therefore, is a very concrete thing; and the examination into its properties means studying the features of a sociological phenomenon.

We shall call the first task of epistemology its *descriptive task*—the task of giving a description of knowledge as it really is. It follows, then, that epistemology in this respect forms a part of sociology. But it is only a special group of questions concerning the sociological phenomenon "knowledge" which constitutes the domain of epistemology. There are such questions as "What is the meaning of the concepts used in knowledge?" "What are the presuppositions contained in the method of science?" "How do we know whether a sentence is true, and do we know that at all?" and many others; and although, indeed, these questions concern the sociological phenomenon "science," they are of a very special type as compared with the form of questions occurring in general sociology.

What makes this difference? It is usually said that this is a difference of internal and external relations between those human utterances the whole of which is called "knowledge." Internal relations are such as belong to the content of knowledge, which must be realized if we want to understand knowledge, whereas external relations combine knowledge with utterances of another kind which do not concern the content of knowledge. Epistemology, then, is interested in internal relations only, whereas sociology, though it may partly consider internal relations, always blends them with external relations in which this science is also interested. A sociologist, for instance, might report that astronomers construct huge observatories containing telescopes in order to watch the stars, and in such a way the internal relation between telescopes and stars enters into a sociological description. The report on contemporary astronomy begun in the preceding sentence might be continued by the statement that astronomers are frequently musical men, or that they belong in general to the bourgeois class of society; if these relations do not interest epistemology, it is because they do not enter into the content of science—they are what we call external relations.

Although this distinction does not furnish a sharp line of demarcation, we may use it for a first indication of the design of our investigations. We may then say the descriptive task of epistemology concerns the internal structure of knowledge and not the external features which appear to an observer who takes no notice of its content.

We must add now a second distinction which concerns psychology. The internal structure of knowledge is the system of connections as it is followed in thinking. From such a definition we might be tempted to infer that epistemology is the giving of a description of thinking processes; but that would be entirely erroneous. There is a great difference between the system of logical interconnections of thought and the actual way in which thinking processes are performed. The psychological operations of thinking are rather vague and fluctuating processes; they almost never keep to the ways prescribed by logic and may even skip whole groups of operations which would be needed for a complete exposition of the subject in question. That is valid for thinking in daily life, as well as for the mental procedure of a man of science, who is confronted by the task of finding logical interconnections between divergent ideas about newly observed facts; the scientific genius has never felt bound to the narrow steps and prescribed courses of logical reasoning. It would be, therefore, a vain attempt to construct a theory of knowledge which is at the same time logically complete and in strict correspondence with the psychological processes of thought.

The only way to escape this difficulty is to distinguish carefully the task of epistemology from that of psychology. Epistemology does not regard the processes of thinking in their actual occurrence; this task is entirely left to psychology. What epistemology intends is to construct thinking processes in a way in which they ought to occur if they are to be ranged in a consistent system; or to construct justifiable sets of operations which can be intercalated between the starting-point and the issue of thought-processes, replacing the real intermediate links. Epistemology thus considers a logical substitute rather than real processes. For this logical substitute the term *rational reconstruction* has been introduced;[1] it seems an appropriate phrase to indicate the task of epistemology in its specific difference from the task of psychology. Many false objections and misunderstandings of modern epistemology have their source in not separating these two tasks; it will, therefore, never be a permissible objection to an epistemological construction that actual thinking does not conform to it.

In spite of its being performed on a fictive construction, we must retain the notion of the descriptive task of epistemology. The construction to be given is not arbitrary; it is bound to actual thinking by the postulate of correspondence. It is even, in a certain sense, a better way of thinking than actual thinking. In being set before the rational reconstruction, we have the feeling that only now do we understand what we think; and we admit that the rational reconstruction expresses what we mean, properly speaking. It is a remarkable psychological fact that there is such an advance toward understanding one's own thoughts, the very fact which formed the basis of the mäeutic of Socrates and which has remained since that time the basis of philosophical method; its adequate scientific expression is the principle of rational reconstruction.

If a more convenient determination of this concept of rational reconstruction is wanted, we might say that it corresponds to the form in which thinking processes are communicated to other persons instead of the form in which they are subjectively performed. The way, for instance, in which a mathematician publishes a new demonstration, or a physicist his logical reasoning in the foundation of a new theory, would almost correspond to our concept of rational reconstruction; and the well-known difference between the thinker's way of finding this theorem and his way of presenting it before a public may illustrate the difference in question. I shall introduce the terms *context of discovery* and *context of justification* to mark this distinction. Then we have to say that epistemology is only occupied in constructing the context of justification. But even the way of presenting scientific theories is only an approximation to what we mean by the context of justification.

Even in the written form scientific expositions do not always correspond to the exigencies of logic or suppress the traces of subjective motivation from which they started. If the presentation of the theory is subjected to an exact epistemological scrutiny, the verdict becomes still more unfavorable. For scientific language, being destined like the language of daily life for practical purposes, contains so many abbreviations and silently tolerated inexactitudes that a logician will never be fully content with the form of scientific publications. Our comparison, however, may at least indicate the way in which we want to have thinking replaced by justifiable operations; and it may also show that the rational reconstruction of knowledge belongs to the descriptive task of epistemology. It is bound to factual knowledge in the same way that the exposition of a theory is bound to the actual thoughts of its author.

In addition to its descriptive task, epistemology is concerned with another purpose which may be called its *critical task*. The system of knowledge is criticized; it is judged in respect of its validity and its reliability. This task is already partially performed in the rational reconstruction, for the fictive set of operations occurring here is chosen from the point of view of justifiability; we replace actual thinking by such operations as are justifiable, that is, as can be demonstrated as valid. But the tendency to remain in correspondence with actual thinking must be separated from the tendency to obtain valid thinking; and so we have to distinguish between the descriptive and the critical task. Both collaborate in the rational reconstruction. It may even happen that the description of knowledge leads to the result that certain chains of thoughts, or operations, cannot be justified; in other words, that even the rational reconstruction contains unjustifiable chains, or that it is not possible to intercalate a justifiable chain between the starting-point and the issue of actual thinking. This case shows that the descriptive task and the critical task are different; although description, as it is here meant, is not a copy of actual thinking but the construction of an equivalent, it is bound by the postulate of correspondence and may expose knowledge to criticism.

The critical task is what is frequently called *analysis of science;* and as the term "logic" expresses nothing else, at least if we take it in a sense corresponding to its use, we may speak here of the logic of science. The well-known problems of logic belong to this domain; the theory of the syllogism was built up to justify deductive thinking by reducing it to certain justifiable schemes of operation, and the modern theory of the tautological character of logical formulas is to be interpreted as a justification of deductive thinking as conceived in a more general form. The question of the synthetic a priori, which has played so important a role in the history of philosophy, also falls into this frame; and so does the problem of inductive reasoning, which has given rise to more than one "inquiry concerning human understanding." Analysis of science comprehends all the basic problems of traditional epistemology; it is, therefore, in the foreground of consideration when we speak of epistemology.

The inquiries of our book will belong, for the most part, to the same domain. Before entering upon them, however, we may mention a result of rather general character which has been furnished by previous investigations of this kind —a result concerning a distinction without which the process of scientific knowledge cannot be understood. Scientific method is not, in every step of its procedure, directed by the principle of validity; there are other steps which have the character of volitional decisions. It is this distinction which we must emphasize at the very beginning of epistemological investigations. That the idea of truth, or validity, has a directive influence in scientific thinking is obvious and has at all times been noticed by epistemologists. That there are certain elements of knowledge, however, which are not governed by the idea of truth, but which are due to volitional resolutions, and though highly influencing the makeup of the whole system of knowledge, do not touch its truth-character, is less known to philosophical investigators. The presentation of the volitional decisions contained in the system of knowledge constitutes, therefore, an integral part of the critical task of epistemology. To give an example of volitional decisions, we may point to the so-called *conventions*, e.g., the convention concerning the unit of length, the decimal system, etc. But not all conventions are so obvious, and it is sometimes a rather difficult problem to find out the points which mark conventions. The progress of epistemology has been frequently furthered by the discovery of the conventional character of certain elements taken, until that time, as having a truth-character; Helmholtz's discovery of the arbitrariness of the definition of spatial congruence, Einstein's discovery of the relativity of simultaneity, signify the recognition that what was deemed a statement is to be replaced by a decision. To find out all the points at which decisions are involved is one of the most important tasks of epistemology.

The conventions form a special class of decisions; they represent a choice between *equivalent* conceptions. The different systems of weights and measures constitute a good example of such an equivalence; they illustrate the fact that the decision in favor of a certain convention does not influence the content of knowledge. The examples chosen from the theory of space and time previously mentioned are likewise to be ranked among conventions. There are decisions of another character which do not lead to equivalent conceptions but to divergent systems; they may be called *volitional bifurcations*. Whereas a convention may be compared to a choice between different ways leading to the same place, the volitional bifurcation resembles a bifurcation of ways which will never meet again. There are some volitional bifurcations of an important character which stand at the very entrance of science: these are decisions concerning the aim of science. What is the purpose of scientific inquiry? This is, logically speaking, a question not of truth-character but of volitional decision, and the decision determined by the answer to this question belongs to the bifurcation type. If anyone tells us that he studies science for his pleasure and to fill his hours of leisure, we cannot raise the objection that this reasoning is "a false statement"—it is no statement at all but a decision, and everybody has the right to do what he wants. We may object that such a determination is opposed to the normal use of words and that what he calls the aim of science is generally called the aim of play—this would be a true statement. This statement belongs to the descriptive part of epistemology; we can show that in books and discourses the word "science" is always connected with "discovering truth," sometimes also with "foreseeing the future." But, logically speaking, this is a matter of volitional decision. It is obvious that this decision is not a convention because the two conceptions obtained by different postulates concerning the aims of science are not equivalent; it is a bifurcation. Or take a question as to the meaning of a certain concept—say, causality, or truth, or meaning itself. Logically speaking this is a question of a decision concerning the limitation of a concept, although, of course, the practice of science has already decided about this limitation in a rather precise way. In such a case, it must be carefully examined whether the decision in question is a convention or a bifurcation. The limitation of a concept may be of a conventional character, i.e., different limitations may lead to equivalent systems.

The character of being true or false belongs to statements only, not to decisions. We can, however, co-ordinate with a decision certain statements concerning it; and, above all, there are two types of statements which must be considered. The first one is a statement of the type we have already mentioned; it states which decision science uses in practice. It belongs to descriptive epistemology and is, therefore, of a sociological character. We may say that it states an *object fact*, i.e., a fact belonging to the sphere of the objects of knowledge,[2] a sociological fact being of this type. It is, of course, the same type of fact with which natural science deals. The second statement concerns the fact that, logically speaking, there is a decision and not a statement; this kind of fact may be called a *logical fact*. There is no contradiction in speaking here of a fact concerning a decision; although a decision is not a fact, its character of being a decision is a fact and may be expressed in a statement. That becomes obvious by the cognitional character of such a statement; the statement may be right or wrong, and in some cases the wrong statement has been maintained for centuries, whereas the right statement was discovered only recently. The given examples of Helmholtz's and

Einstein's theories of space and time may illustrate this. But the kind of fact maintained here does not belong to the sphere of the objects of science, and so we call it a logical fact. It will be one of our tasks to analyze these logical facts and to determine their logical status; but for the present we shall use the term "logical fact" without further explanation.

The difference between statements and decisions marks a point at which the distinction between the descriptive task and the critical task of epistemology proves of utmost importance. Logical analysis shows us that within the system of science there are certain points regarding which no question as to truth can be raised, but where a decision is to be made; descriptive epistemology tells us what decision is actually in use. Many misunderstandings and false pretensions of epistemology have their origin here. We know the claims of Kantianism, and Neo-Kantianism, to maintain Euclidean geometry as the only possible basis of physics; modern epistemology showed that the problem as it is formulated in Kantianism is falsely constructed, as it involves a decision which Kant did not see. We know the controversies about the "meaning of meaning"; their passionate character is due to the conviction that there is an absolute meaning of meaning which we must discover, whereas the question can only be put with respect to the concept of meaning corresponding to the use of science, or presupposed in certain connections. But we do not want to anticipate the discussion of this problem, and our later treatment of it will contain a more detailed explanation of our distinction between statements and decisions.

The concept of decision leads to a third task with which we must charge epistemology. There are many places where the decisions of science cannot be determined precisely, the words or methods used being too vague; and there are others in which two or even more different decisions are in use, intermingling and interfering within the same context and confusing logical investigations. The concept of meaning may serve as an example; simpler examples occur in the theory of measurement. The concrete task of scientific investigation may put aside the exigencies of logical analysis; the man of science does not always regard the demands of the philosopher. It happens, therefore, that the decisions presupposed by positive science are not clarified. In such a case, it will be the task of epistemology to suggest a proposal concerning a decision; and we shall speak, therefore, of the *advisory task* of epistemology as its third task. This function of epistemology may turn out to be of great practical value; but it must be kept clearly in mind that what is to be given here is a proposal and not a determination of a truth-character. We may point out the advantages of our proposed decision, and we may use it in our own expositions of related subjects; but never can we demand agreement to our proposal in the sense that we can demand it for statements which we have proven to be true.

There is, however, a question regarding facts which is to be considered in connection with the proposal of a decision. The system of knowledge is interconnected in such a way that some decisions are bound together; one decision, then, involves another, and, though we are free in choosing the first one, we are no longer free with respect to those following. We shall call the group of decisions involved by one decision its *entailed decisions*. To give a simple example: the decision for the English system of measures leads to the impossibility of adding measure numbers according to the technical rules of the decimal system; so the renunciation of these rules would be an entailed decision. Or a more complicated example: the decision expressed in the acceptance of Euclidean geometry in physics may lead to the occurrence of strange forces, "universal forces," which distort all bodies to the same extent, and may lead to even greater inconveniences concerning the continuous character of causality.[3] The discovery of interconnections of this kind is an important task of epistemology, the relations between different decisions being frequently hidden by the complexity of the subject; it is only by adding the group of entailed decisions that a proposal respecting a new decision becomes complete.

The discovery of entailed decisions belongs to the critical task of epistemology, the relation between decisions being of the kind called a logical fact. We may therefore reduce the advisory task of epistemology to its critical task by using the following systematic procedure: we renounce making a proposal but instead construe a list of all possible decisions, each one accompanied by its entailed decisions. So we leave the choice to our reader after showing him all factual connections to which he is bound. It is a kind of logical signpost which we erect; for each path we give its direction together with all connected directions and leave the decision as to his route to the wanderer in the forest of knowledge. And perhaps the wanderer will be more thankful for such a signpost than he would be for suggestive advice directing him into a certain path. Within the frame of the modern philosophy of science there is a movement bearing the name of *conventionalism;* it tries to show that most of the epistemological questions contain no questions of truth-character but are to be settled by arbitrary decisions. This tendency, and above all, in its founder Poincaré, had historical merits, as it led philosophy to stress the volitional elements of the system of knowledge which had been previously neglected. In its further development, however, the tendency has largely trespassed beyond its proper boundaries by highly exaggerating the part occupied by decisions in knowledge. The relations between different decisions were overlooked, and the task of reducing arbitrariness to a minimum by showing the logical interconnections between the arbitrary decisions was forgotten. The concept of entailed decisions, therefore, may be regarded as a dam erected against extreme conventionalism; it allows us to separate the arbitrary part of the system of knowledge from its substantial content, to distinguish the subjective and the objective part of science. The relations between decisions do not depend on our choice but are prescribed by the rules of logic, or by the laws of nature.

It even turns out that the exposition of entailed decisions settles many quarrels about the choice of decisions. Certain basic decisions enjoy an almost universal assent; and, if we succeed in showing that one of the contended decisions is entailed by such a basic decision, the acceptance of the first decision will be secured. Basic decisions of such a kind are, for instance, the principle that things of the same kind shall receive the same names, or the principle that science is to furnish methods for foreseeing the future as well as possible (a demand which will be accepted even if science is also charged with other tasks). I will not say that these basic decisions must be assumed and retained in every development of science; what I want to say is only that these decisions are actually maintained by most people and that many quarrels about decisions are caused only by not seeing the implication which leads from the basic decisions to the decision in question.

The objective part of knowledge, however, may be freed from volitional elements by the method of reduction transforming the advisory task of epistemology into the critical task. We may state the connection in the form of an implication: If you choose this decision, then you are obliged to agree to this statement, or to this other decision. This implication, taken as a whole, is free from volitional elements; it is the form in which the objective part of knowledge finds its expression.

Notes

1. The term *rationale Nachkonstruktion* was used by Carnap in *Der logische Aufbau der Welt* (Berlin and Leipzig, 1928).
2. The term "objective fact" taken in the original sense of the word "objective" would express the same point; but we avoid it, as the word "objective" suggests an opposition to "subjective," an opposition which we do not intend.
3. Cf. the author's [Reichenbach 1928], § 12.

CHAPTER 6

What Are Scientific Revolutions?

THOMAS KUHN

Pages 7–22 of *The Probabilistic Revolution*, volume I: Ideas in History, edited by Lorenz Kruger, Lorraine J. Daston, and Michael Heidelberger (Cambridge, MA: MIT Press, 1987). As Conant and Haugeland note in their introduction to Chapter One of Kuhn 2000, "under the title 'From Revolutions to Salient Features,' the paper was read to the third annual conference of the Cognitive Science Society in August 1981" (James Conant and John Haugeland, p. 13).

IT IS NOW ALMOST TWENTY YEARS since I first distinguished what I took to be two types of scientific development, normal and revolutionary.[1] Most successful scientific research results in change of the first sort, and its nature is well captured by a standard image: normal science is what produces the bricks that scientific research is forever adding to the growing stockpile of scientific knowledge. That cumulative conception of scientific development is familiar, and it has guided the elaboration of a considerable methodological literature. Both it and its methodological by-products apply to a great deal of significant scientific work. But scientific development also displays a noncumulative mode, and the episodes that exhibit it provide unique clues to a central aspect of scientific knowledge. Returning to a long-standing concern, I shall therefore here attempt to isolate several such clues, first by describing three examples of revolutionary change and then by briefly discussing three characteristics which they all share. Doubtless revolutionary changes share other characteristics as well, but these three provide a

71

sufficient basis for the more theoretical analyses on which I am currently engaged, and on which I shall be drawing somewhat cryptically when concluding this paper.

Before turning to a first extended example, let me try—for those not previously familiar with my vocabulary—to suggest what it is an example of. Revolutionary change is defined in part by its difference from normal change, and normal change is, as already indicated, the sort that results in growth, accretion, cumulative addition to what was known before. Scientific laws, for example, are usually products of this normal process: Boyle's law will illustrate what is involved. Its discoverers had previously possessed the concepts of gas pressure and volume as well as the instruments required to determine their magnitudes. The discovery that, for a given gas sample, the product of pressure and volume was a constant at constant temperature simply added to the knowledge of the way these antecedently understood[2] variables behave. The overwhelming majority of scientific advance is of this normal cumulative sort, but I shall not multiply examples.

Revolutionary changes are different and far more problematic. They involve discoveries that cannot be accommodated within the concepts in use before they were made. In order to make or to assimilate such a discovery one must alter the way one thinks about and describes some range of natural phenomena. The discovery (in cases like these "invention" may be a better word) of Newton's second law of motion is of this sort. The concepts of force and mass deployed in that law differed from those in use before the law was introduced, and the law itself was essential to their definition. A second, fuller, but more simplistic example is provided by the transition from Ptolemaic to Copernican astronomy. Before it occurred, the sun and moon were planets, the earth was not. After it, the earth was a planet, like Mars and Jupiter; the sun was a star; and the moon was a new sort of body, a satellite. Changes of that sort were not simply corrections of individual mistakes embedded in the Ptolemaic system. Like the transition to Newton's laws of motion, they involved not only changes in laws of nature but also changes in the criteria by which some terms in those laws attached to nature. These criteria, furthermore, were in part dependent upon the theory with which they were introduced.

When referential changes of this sort accompany change of law or theory, scientific development cannot be quite cumulative. One cannot get from the old to the new simply by an addition to what was already known. Nor can one quite describe the new in the vocabulary of the old or vice versa. Consider the compound sentence, "In the Ptolemaic system planets revolve about the earth; in the Copernican they revolve about the sun." Strictly construed, that sentence is incoherent. The first occurrence of the term 'planet' is Ptolemaic, the second Copernican, and the two attach to nature differently. For no univocal reading of the term 'planet' is the compound sentence true.

No example so schematic can more than hint at what is involved in revolutionary change. I therefore turn at once to some fuller examples, beginning with the one that, a generation ago, introduced me to revolutionary change, the transition from Aristotelian to Newtonian physics. Only a small part of it, centering on problems of motion and mechanics, can be considered here, and even about it I shall be schematic. In addition, my account will invert historical order and describe, not what Aristotelian natural philosophers required to reach Newtonian concepts, but what I, raised a Newtonian, required to reach those of Aristotelian natural philosophy. The route I traveled backward with the aid of written texts was, I shall simply assert, nearly enough the same one that earlier scientists had traveled forward with no text but nature to guide them.

I first read some of Aristotle's physical writings in the summer of 1947, at which time I was a graduate student of physics trying to prepare a case study on the development of mechanics for a course in science for nonscientists. Not surprisingly, I approached Aristotle's texts with the Newtonian mechanics I had previously read clearly in mind. The question I hoped to answer was how much mechanics Aristotle had known, how much he had left for people like Galileo and Newton to discover. Given that formulation, I rapidly discovered that Aristotle had known almost no mechanics at all. Everything was left for his successors, mostly those of the sixteenth and seventeenth centuries. That conclusion was standard, and it might in principle have been right. But I found it bothersome because, as I was reading him, Aristotle appeared not only ignorant of mechanics, but a dreadfully bad physical scientist as well. About motion, in particular, his writings seemed to me full of egregious errors, both of logic and of observation.

These conclusions were unlikely. Aristotle, after all, had been the much admired codifier of ancient logic. For almost two millennia after his death, his work played the same role in logic that Euclid's played in geometry. In addition, Aristotle had often proved an extraordinarily acute naturalistic observer. In biology, especially, his descriptive writings provided models that were central in the sixteenth and seventeenth centuries to the emergence of the modern biological tradition. How could his characteristic talents have deserted him so systematically when he turned to the study of motion and mechanics? Equally, if his talents had so deserted him, why had his writings in physics been taken so seriously for so many centuries after his death? Those questions troubled me. I could easily believe that Aristotle had stumbled, but not that, on entering physics, he had totally collapsed. Might not the fault be mine rather than Aristotle's, I asked myself. Perhaps his words had not always meant to him and his contemporaries quite what they meant to me and mine.

Feeling that way, I continued to puzzle over the text, and my suspicions ultimately proved well-founded. I was sitting at my desk with the text of Aristotle's *Physics* open in front of me and with a four-colored pencil in my hand. Looking up, I gazed abstractedly out the window of my room—the visual image is one I still retain. Suddenly the fragments in my head sorted themselves out in a new way, and fell into place together. My jaw dropped, for all at once Aristotle seemed a very good physicist indeed, but of a sort I'd never dreamed possible. Now I could understand why he had said what he'd said, and what his authority had been. Statements that had previously seemed egregious mistakes, now seemed at worst near misses within a powerful and generally successful tradition. That sort of experience—the pieces suddenly sorting themselves out and coming together in a new way—is the first general characteristic of revolutionary change that I shall be singling out after further consideration of examples. Though scientific revolutions leave much piecemeal mopping up to do, the central change cannot be experienced piecemeal, one step at a time. Instead, it involves some relatively sudden and unstructured transformation in which some part of the flux of experience sorts itself out differently and displays patterns that were not visible before.

To make all this more concrete let me now illustrate some of what was involved in my discovery of a way of reading Aristotelian physics, one that made the texts make sense. A first illustration will be familiar to many. When the term 'motion' occurs in Aristotelian physics, it refers to change in general, not just to the change of position of a physical body. Change of position, the exclusive subject of mechanics for Galileo and Newton, is one of a number of subcategories of motion for Aristotle. Others include growth (the transformation of an acorn to an oak), alterations of intensity (the heating of an iron bar), and a number of more general qualitative changes (the transition from sickness to health). As a result, though Aristotle recognizes that the various subcategories are not alike in *all* respects, the basic characteristics relevant to the recognition and analysis of motion must apply to changes of all sorts. In some sense that is not merely metaphorical; all varieties of change are seen as like each other, as constituting a single natural family.[3]

A second aspect of Aristotle's physics—harder to recognize and even more important—is the centrality of qualities to its conceptual structure. By that I do not mean simply that it aims to explain quality and change of quality, for other sorts of physics have done that. Rather I have in mind that Aristotelian physics inverts the ontological hierarchy of matter and quality that has been standard since the middle of the seventeenth century. In Newtonian physics a body is constituted of particles of matter, and its qualities are a consequence of the way those particles are arranged, move, and interact. In Aristotle's physics, on the other hand, matter is very nearly dispensable. It is a neutral substrate, present wherever a body could be—which means wherever there's space or place. A particular body, a substance, exists in whatever place this neutral substrate, a sort of sponge, is sufficiently impregnated with qualities like heat, wetness, color, and so on to give it individual identity. Change occurs by changing qualities, not matter, by removing some qualities from some given matter and replacing them with others. There are even some implicit conservation laws that the qualities must apparently obey.[4]

Aristotle's physics displays other similarly general aspects, some of great importance. But I shall work toward the points that concern me from these two, picking up one other well-known one in passing. What I want now to begin to suggest is that, as one recognizes these and other aspects of Aristotle's viewpoint, they begin to fit together, to lend each other mutual support, and thus to make a sort of sense collectively that they individually lack. In my original experience of breaking into Aristotle's text, the new pieces I have been describing and the sense of their coherent fit actually emerged together.

Begin from the notion of a qualitative physics that has just been sketched. When one analyzes a particular object by specifying the qualities that have been imposed on omnipresent neutral matter, one of the qualities that must be specified is the object's position, or, in Aristotle's terminology, its place. Position is thus, like wetness or hotness, a quality of the object, one that changes as the object moves or is moved. Local motion (motion *tout court* in Newton's sense) is therefore change-of-quality or change-of-state for Aristotle, rather than being itself a state as it is for Newton. But it is precisely seeing motion as change-of-quality that permits its assimilation to all other sorts of change—acorn to oak or sickness to health, for examples. That assimilation is the aspect of Aristotle's physics from which I began, and I could equally well have traveled the route in the other direction. The conception of motion-as-change and the conception of a qualitative physics prove deeply interdependent, almost equivalent notions, and that is a first example of the fitting or the locking together of parts.

If that much is clear, however, then another aspect of Aristotle's physics—one that regularly seems ridiculous in isolation—begins to make sense as well. Most changes of quality, especially in the organic realm, are asymmetric, at least when left to themselves. An acorn naturally develops into an oak, not vice versa. A sick man often grows healthy by himself, but an external agent is needed, or believed to be needed, to make him sick. One set of qualities, one end point of change, represents a body's natural state, the one that it realizes voluntarily and thereafter rests. The same asymmetry should be characteristic of local motion, change of position, and indeed it is. The quality that a stone or other heavy body strives to realize is position at the center of the universe; the natural position of fire is at the periphery. That is why stones fall toward the center until blocked by an obstacle and why fire flies to the heavens. They are realizing their natural properties just as the acorn does through its growth. Another initially strange part of Aristotelian doctrine begins to fall into place.

One could continue for some time in this manner, locking individual bits of Aristotelian physics into place in the whole. But I shall instead conclude this first example with a last illustration, Aristotle's doctrine about the vacuum or void. It displays with particular clarity the way in which a number of theses that appear arbitrary in isolation lend each other mutual authority and support. Aristotle states that a void is impossible: his underlying position is that the notion itself is incoherent. By now it should be apparent how that might be so. If position is a quality, and if qualities cannot exist separate from matter, then there must be matter wherever there's position, wherever body might be. But that is to say that there must be matter everywhere in space: the void, space without matter, acquires the status of, say, a square circle.[5]

That argument has force, but its premise seems arbitrary. Aristotle need not, one supposes, have conceived position as a quality. Perhaps, but we have already noted that that conception underlies his view of motion as change-of-state, and other aspects of his physics depend on it as well. If there could be a void, then the Aristotelian universe or cosmos could not be finite. It is just because matter and space are coextensive that space can end where matter ends, at the outermost sphere beyond which there is nothing at all, neither space nor matter. That doctrine, too, may seem dispensable. But expanding the stellar sphere to infinity would make problems for astronomy, since that sphere's rotations carry the stars about the earth. Another, more central, difficulty arises earlier. In an infinite universe there is no center—any point is as much the center as any other—and there is thus no natural position at which stones and other heavy bodies realize their natural quality. Or, to put the point in another way, one that Aristotle actually uses, in a void a body could not be aware of the location of its natural place. It is just by being in contact with all positions in the universe through a chain of intervening matter that a body is able to find its way to the place where its natural qualities are fully realized. The presence of matter is what provides space with structure.[6] Thus, both Aristotle's theory of natural local motion and ancient geocentric astronomy are threatened by an attack on Aristotle's doctrine of the void. There is no way to "correct" Aristotle's views about the void without reconstructing much of the rest of his physics.

Those remarks, though both simplified and incomplete, should sufficiently illustrate the way in which Aristotelian physics cuts up and describes the phenomenal world. Also, and more important, they should indicate how the pieces of that description lock together to form an integral whole, one that had to be broken and reformed on the road to Newtonian mechanics. Rather than extend them further, I shall therefore proceed at once to a second example, returning to the beginning of the nineteenth century for the purpose. The year 1800 is notable, among other things, for Volta's discovery of the electric battery. That discovery was announced in a letter to Sir Joseph Banks, President of the Royal Society.[7] It was intended for publication and was accompanied by the illustration reproduced here as figure 1. For a modern audience there is something odd about it, though the oddity is seldom noticed, even by historians. Looking at any one of the so-called "piles" (of coins) in the lower two-thirds of the diagram, one sees, reading upward from the bottom right, a piece of zinc, Z, then a piece of silver, A, then a piece of wet blotting paper, then a second piece of zinc, and so on. The cycle zinc, silver, wet blotting paper is repeated an integral number of times, eight in Volta's original illustration. Now suppose that, instead of having all this spelled out, you had been asked simply to look at the diagram, then to put it aside and reproduce it from memory. Almost certainly, those of you who know even the most elementary physics would have drawn zinc (or silver), followed by wet blotting paper, followed by silver (or zinc). In a battery, as we all know, the liquid belongs between the two different metals.

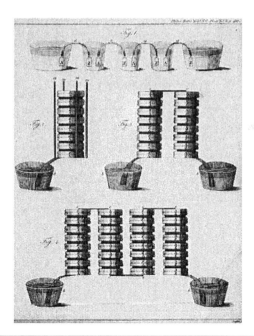

Figure 1

If one recognizes this difficulty and puzzles over it with the aid of Volta's texts, one is likely to realize suddenly that for Volta and his followers, the unit cell consists of the two pieces of metal in contact. The source of power is the metallic interface, the bimetallic junction that Volta had previously found to be the source of an electrical tension, what we would call a voltage. The role of the liquid then is simply to connect one unit cell to the next without generating a contact potential, which would neutralize the initial effect. Pursuing Volta's text still further, one realizes that he is assimilating his new discovery to electrostatics. The bimetallic junction is a condenser or Leyden jar, but one that charges itself. The pile of coins is, then, a linked assemblage or "battery" of charged Leyden jars, and that is where, by specialization from the group to its members, the term 'battery' comes from in its application to electricity. For confirmation, look at the top part of Volta's diagram, which illustrates an arrangement he called "the crown of cups." This time the resemblance to diagrams in elementary modem textbooks is striking, but there is again an oddity. Why do the cups at the two ends of the diagram contain only one piece of metal? Why does Volta include two half-cells? The answer is the same as before. For Volta the cups are not cells but simply containers for the liquids that connect cells. The cells themselves are the bimetallic horseshoe strips. The apparently unoccupied positions in the outermost cups are what we would think of as binding posts. In Volta's diagram there are no half-cells.

As in the previous example, the consequences of this way of looking at the battery are widespread. For example, as shown in figure 2, the transition from Volta's viewpoint to the modern one reverses the direction of current flow. A modern cell diagram (figure 2, bottom) can be derived from Volta's (top left) by a process like turning the latter inside out (top right). In that process what was previously current flow internal to the cell becomes the external current and vice versa. In the Voltaic diagram the external current flow is from black metal to white, so that the black is positive. In the modern diagram both the direction of flow and the polarity are reversed. Far more important conceptually is the change in the current source effected by the transition. For Volta the metallic interface was the essential element of the cell and necessarily the source of the current the cell produced. When the cell was turned inside out, the liquid and its two interfaces with the metals provided its essentials, and the source of the current became the chemical effects at these interfaces. When both viewpoints were briefly in the field at once, the first was known as the contact theory, the second as the chemical theory of the battery.

Those are only the most obvious consequences of the electrostatic view of the battery, and some of the others were even more immediately important. For example, Volta's viewpoint suppressed the conceptual role of the external circuit. What we would think of as an external circuit is simply a discharge path like the short circuit to ground that discharges a Leyden jar.

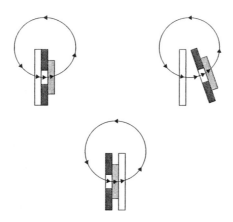

Figure 2 (redrawn)

As a result, early battery diagrams do not show an external circuit unless some special effect, like electrolysis or heating a wire, is occurring there, and then, very often the battery is not shown. Not until the 1840s do modern cell diagrams begin to appear regularly in books on electricity. When they do, either the external circuit or explicit points for its attachment appears with them.[8] Examples are shown in figures 3 and 4.

Finally, the electrostatic view of the battery leads to a concept of electrical resistance very different from the one now standard. There is an electrostatic concept of resistance, or there was in this period. For an insulating material of given cross section, resistance was measured by the shortest length the material could have without breaking down or leaking—ceasing to insulate—when subjected to a given voltage. For a conducting material of given cross section, it was measured by the shortest length the material could have without melting when connected across a given voltage. It is possible to measure resistance conceived in this way, but the results are not compatible with Ohm's law. To get those results one must conceive the battery and circuit on a more hydrodynamic model. Resistance must become like the frictional resistance to the flow of water in pipes. The assimilation of Ohm's law required a noncumulative change of that sort, and that is part of what made his law so difficult for many people to accept. It has for some time provided a standard example of an important discovery that was initially rejected or ignored.

At this point I end my second example and proceed at once to a third, this one both more modern and more technical than its predecessors. Substantively, it is controversial, involving a new version, not yet everywhere accepted, of the origins of the quantum theory.[9] Its subject is Max Planck's work on the so-called black-body problem, and its structure may

Figure 3

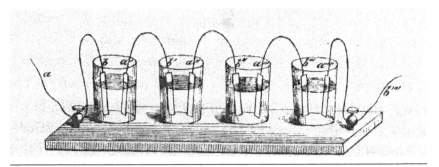

Figure 4

usefully be anticipated as follows. Planck first solved the black-body problem in 1900 using a classical method developed by the Austrian physicist Ludwig Boltzmann. Six years later a small but crucial error was found in his derivation, and one of its central elements had to be reconceived. When that was done Planck's solution did work, but it then also broke radically with tradition. Ultimately that break spread through and caused the reconstruction of a good deal of physics.

Begin with Boltzmann, who had considered the behavior of a gas, conceived as a collection of many tiny molecules, moving rapidly about within a container, and colliding frequently both with each other and with the container's walls. From previous works of others, Boltzmann knew the average velocity of the molecules (more precisely, the average of the square of their velocity). But, many of the molecules were, of course, moving much more slowly than the average, others much faster. Boltzmann wanted to know what proportion of them were moving at, say, ½ the average velocity, what proportion at ⅓ the average, and so on. Neither that question nor the answer he found to it was new. But Boltzmann reached the answer by a new route, from probability theory, and that route was fundamental for Planck, since whose work it has been standard.

Only one aspect of Boltzmann's method is of present concern. He considered the total kinetic energy E of the molecules. Then, to permit the introduction of probability theory, he mentally subdivided that energy into little cells or elements of size ε, as in figure 5. Next, he imagined distributing the molecules at random among those cells, drawing numbered slips from an urn to specify the assignment of each molecule and then excluding all distributions with total energy different from E. For example, if the first molecule were assigned to the last cell (energy E), then the only acceptable distribution would be the one that assigned all other molecules to the first cell (energy o). Clearly, that particular distribution is a most improbable one. It is far more likely that most molecules will have appreciable energy, and by probability theory one can discover the most probable distribution of all. Boltzmann showed how to do so, and his result was the same as the one he and others had previously gotten by more problematic means.

That way of solving the problem was invented in 1877, and twenty-three years later, at the end of 1900, Max Planck applied it to an apparently rather different problem, black-body radiation. Physically the problem is to explain the way in which the color of a heated body changes with temperature. Think, for example, of the radiation from an iron bar, which, as the temperature increases, first gives off heat (infrared radiation), then glows dull red, and then gradually becomes a brilliant white. To analyze that situation Planck imagined a container or cavity filled with radiation, that is, with light, heat, radio waves, and so on. In addition, he supposed that the cavity contained a lot of what he called "resonators" (think of them as tiny

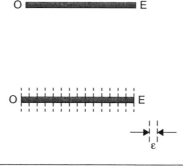

Figure 5 (redrawn)

electrical tuning forks, each sensitive to radiation at one frequency, not at others). These resonators absorb energy from the radiation, and Planck's question was: How does the energy picked up by each resonator depend on its frequency? What is the frequency distribution of the energy over the resonators?

Conceived in that way, Planck's problem was very close to Boltzmann's, and Planck applied Boltzmann's probabilistic techniques to it. Roughly speaking, he used probability theory to find the proportion of resonators that fell in each of the various cells, just as Boltzmann had found the proportion of molecules. His answer fit experimental results better than any other then or since known, but there turned out to be one unexpected difference between his problem and Boltzmann's. For Boltzmann's, the cell size ε could have many different values without changing the result. Though permissible values were bounded, could not be too large or too small, an infinity of satisfactory values was available in between. Planck's problem proved different: other aspects of physics determined ε, the cell size. It could have only a single value given by the famous formula $\varepsilon = h\nu$, where ν is the resonator frequency and h is the universal constant subsequently known by Planck's name. Planck was, of course, puzzled about the reason for the restriction on cell size, though he had a strong hunch about it, one he attempted to develop. But, excepting that residual puzzle, he had solved his problem, and his approach remained very close to Boltzmann's. In particular—the presently crucial point—in both solutions the division of the total energy E into cells of size ε was a mental division made for statistical purposes. The molecules and resonators could lie anywhere along the line and were governed by all the standard laws of classical physics.

The rest of this story is very quickly told. The work just described was done at the end of 1900. Six years later, in the middle of 1906, two other physicists argued that Planck's result could not be gained in Planck's way. One small but absolutely crucial alteration of the argument was required. The resonators could not be permitted to lie anywhere on the continuous energy line but only at the divisions between cells. A resonator might, that is, have energy 0, ε, 2ε, 3ε, . . ., and so on, but not $(\frac{1}{3})\varepsilon$, $(\frac{4}{5})\varepsilon$, etc. When a resonator changed energy it did not do so continuously but by discontinuous jumps of size ε or a multiple of ε.

After those alterations, Planck's argument was both radically different and very much the same. Mathematically it was virtually unchanged, with the result that it has been standard for years to read Planck's 1900 paper as presenting the subsequent modern argument. But physically, the entities to which the derivation refers are very different. In particular, the element ε has gone from a mental division of the total energy to a separable physical energy atom, of which each resonator may have 0, 1, 2, 3, or some other number. Figure 6 tries to capture that change in a way that suggests its resemblance to the inside-out battery of my last example. Once again the transformation is subtle, difficult to see. But also once again, the change is consequential. Already the resonator has been transformed from a familiar sort of entity governed by standard classical laws to a strange creature the very existence of which is incompatible with traditional ways of doing physics. As most of you know, changes of the same sort continued for another twenty years as similar nonclassical phenomena were found in other parts of the field.

Those later changes, I shall not attempt to follow, but instead conclude this example, my last, by pointing to one other sort of change that occurred near its start. In discussing the earlier examples, I pointed out that revolutions were accompanied by changes in the way in which terms like 'motion' or 'cell' attached to nature. In this example there was actually a change in the words themselves, one that highlights those features of the physical situation that the revolution had made prominent. When Planck around 1909 was at last persuaded that discontinuity had come to stay he switched to a vocabulary that has been standard since. Previously he had ordinarily referred to the cell size ε as the energy "element." Now, in 1909, he began regularly to speak instead of the energy "quantum"; for 'quantum', as used in German physics, was a separable element, an atomlike entity that could exist by itself. While ε had been merely the size of a mental subdivision, it had not been a quantum but an element. Also in 1909, Planck abandoned the acoustic analogy. The entities he had introduced as "resonators" now became "oscillators," the latter a neutral term that refers to any entity that simply vibrates regularly back and forth. By contrast, 'resonator' refers in the first instance to an acoustic entity or, by extension, to a vibrator that

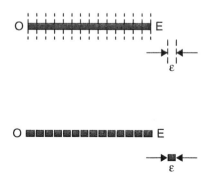

Figure 6 (redrawn)

responds gradually to stimulation, swelling and diminishing with the applied stimulus. For one who believed that energy changes discontinuously, 'resonator' was not an appropriate term, and Planck gave it up in and after 1909.

That vocabulary change concludes my third example. Rather than give others, I shall conclude this discussion by asking what characteristics of revolutionary change are displayed by the examples at hand. Answers will fall under three headings, and I shall be relatively brief about each. The extended discussion they require, I am not quite ready to provide.

A first set of shared characteristics was mentioned near the start of this paper. Revolutionary changes are somehow holistic. They cannot, that is, be made piecemeal, one step at a time, and they thus contrast with normal or cumulative changes like, for example, the discovery of Boyle's law. In normal change, one simply revises or adds a single generalization, all others remaining the same. In revolutionary change one must either live with incoherence or else revise a number of inter-related generalizations together. If these same changes were introduced one at a time, there would be no intermediate resting place. Only the initial and final sets of generalizations provide a coherent account of nature. Even in my last example, the most nearly cumulative of the three, one cannot simply change the description of the energy element ε. One must also change one's notion of what it is to be a resonator, for resonators, in any normal sense of the term, cannot behave as these do. Simultaneously, to permit the new behavior, one must change, or try to, laws of mechanics and of electromagnetic theory. Again, in the second example, one cannot simply change one's mind about the order of elements in a battery cell. The direction of the current, the role of the external circuit, the concept of electrical resistance, and so on, must also be changed. Or still again, in the case of Aristotelian physics, one cannot simply discover that a vacuum is possible or that motion is a state, not a change-of-state. An integrated picture of several aspects of nature has to be changed at the same time.

A second characteristic of these examples is closely related. It is the one I have in the past described as meaning change and which I have here been describing, somewhat more specifically, as change in the way words and phrases attach to nature, change in the way their referents are determined. Even that version is, however, somewhat too general. As recent studies of reference have emphasized, anything one knows about the referents of a term may be of use in attaching that term to nature. A newly discovered property of electricity, of radiation, or of the effects of force on motion may thereafter be called upon (usually with others) to determine the presence of electricity, radiation, or force and thus to pick out the referents of the corresponding term. Such discoveries need not be and usually are not revolutionary. Normal science, too, alters the way in which terms attach to nature. What characterizes revolutions is not, therefore, simply change in the way referents are determined, but change of a still more restricted sort.

How best to characterize that restricted sort of change is among the problems that currently occupy me, and I have no full solution. But roughly speaking, the distinctive character of revolutionary change in language is that it alters not only the criteria by which terms attach to nature but also, massively, the set of objects or situations to which those terms attach. What had been paradigmatic examples of motion for Aristotle—acorn to oak or sickness to health—were not motions at all for Newton. In the transition, a natural family ceased to be natural; its members were redistributed among preexisting sets; and only one of them continued to bear the old name. Or again, what had been the unit cell of Volta's battery was no longer the referent of any term forty years after his invention was made. Though Volta's successors still dealt with metals, liquids, and the flow of charge, the units of their analyses were different and differently interrelated.

What characterizes revolutions is, thus, change in several of the taxonomic categories prerequisite to scientific descriptions and generalizations. That change, furthermore, is an adjustment not only of criteria relevant to categorization, but also of the way in which given objects and situations are distributed among preexisting categories. Since such redistribution always involves more than one category and since those categories are interdefined, this sort of alteration is necessarily holistic. That holism, furthermore, is rooted in the nature of language, for the criteria relevant to categorization are ipso facto the criteria that attach the names of those categories to the world. Language is a coinage with two faces, one looking outward to the world, the other inward to the world's reflection in the referential structure of the language.

Look now at the last of the three characteristics shared by my three examples. It has been the most difficult of the three for me to see, but now seems the most obvious and probably the most consequential. Even more than the others, it should repay further exploration. All of my examples have involved a central change of model, metaphor, or analogy—a change in one's sense of what is similar to what, and of what is different. Sometimes, as in the Aristotle example, the similarity is internal to the subject matter. Thus, for Aristotelians, motion was a special case of change, so that the falling stone was *like* the growing oak, or *like* the person recovering from illness. That is the pattern of similarities that constitutes these phenomena a natural family, that places them in the same taxonomic category, and that had to be replaced in the development of Newtonian physics. Elsewhere the similarity is external. Thus, Planck's resonators were *like* Boltzmann's molecules, or Volta's battery cells were *like* Leyden jars, and resistance was *like* electrostatic leakage. In these cases, too, the old pattern of similarities had to be discarded and replaced before or during the process of change.

All these cases display interrelated features familiar to students of metaphor. In each case two objects or situations are juxtaposed and said to be the same or similar. (An even slightly more extended discussion would have also to consider examples of dissimilarity, for they, too, are often important in establishing a taxonomy.) Furthermore, whatever their origin—a separate issue with which I am not presently concerned—the primary function of all these juxtapositions is to transmit and maintain a taxonomy. The juxtaposed items are exhibited to a previously uninitiated audience by someone who can already recognize their similarity, and who urges that audience to learn to do the same. If the exhibit succeeds, the new initiates emerge with an acquired list of features salient to the required similarity relation—with a feature-space, that is, within which the previously juxtaposed items are durably clustered together as examples of the same thing and are simultaneously separated from objects or situations with which they might otherwise have been confused. Thus, the education of an Aristotelian associates the flight of an arrow with a falling stone and both with the growth of an oak and the return to health. All are thereafter changes-of-state; their end points and the elapsed time of transition are their salient features. Seen in that way, motion cannot be relative and must be in a category distinct from rest, which is a state. Similarly, on that view, an infinite motion, because it lacks an end point, becomes a contradiction in terms.

The metaphor-like juxtapositions that change at times of scientific revolution are thus central to the process by which scientific and other language is acquired. Only after that acquisition or learning process has passed a certain point can the practice of science even begin. Scientific practice always involves the production and the explanation of generalizations about nature, those activities presuppose a language with some minimal richness, and the acquisition of such a language brings knowledge of nature with it. When the exhibit of examples is part of the process of learning terms like 'motion', 'cell', or 'energy element', what is acquired is knowledge of language and of the world together. On the one hand, the student learns what these terms mean, what features are relevant to attaching them to nature, what things cannot be said of them on pain of self-contradiction, and so on. On the other hand, the student learns what categories of things populate the world, what their salient features are, and something about the behavior that is and is not permitted to them. In much of language learning these two sorts of knowledge—knowledge of words and knowledge of nature—are acquired together, not really two sorts of knowledge at all, but two faces of the single coinage that a language provides.

The reappearance of the double-faced character of scientific language provides an appropriate terminus for this paper. If I am right, the central characteristic of scientific revolutions is that they alter the knowledge of nature that is intrinsic to the language itself and that is thus prior to anything quite describable as description or generalization, scientific or everyday. To make the void or an infinite linear motion part of science required observation reports that could only be formulated by altering the language with which nature was described. Until those changes had occurred, language itself resisted the invention and introduction of the sought-after new theories. The same resistance by language is, I take it, the reason for Planck's switch from 'element' and 'resonator' to 'quantum' and 'oscillator'. Violation or distortion of a previously unproblematic scientific language is the touchstone for revolutionary change.

Notes

1. Kuhn 1962.
2. The phrase 'antecedently understood' was introduced by C. G. Hempel, who shows that it will serve many of the same purposes as 'observational' in discussions involving the distinction between observational and theoretical terms (cf., particularly, [Hempel 1965], pp. 208ff.). I borrow the phrase because the notion of an antecedently understood term is intrinsically developmental or historical, and its use within logical empiricism points to important areas of overlap between that traditional approach to philosophy of science and the more recent historical approach. In particular, the often elegant apparatus developed by logical empiricists for discussions of concept formation and of the definition of theoretical terms can be transferred as a whole to the historical approach and used to analyze the formation of new concepts and the definition of new terms, both of which usually take place in intimate association with the introduction of a new theory. A more systematic way of preserving an important part of the observational/theoretical distinction by embedding it in a developmental approach has been developed by [Sneed 1971], pp. 1–64, 249–307). Wolfgang Stegmüller has clarified and extended Sneed's approach by positing a hierarchy of theoretical terms, each level introduced within a particular historical theory ([Stegmüller 1976], pp. 40–67, 196–231). The resulting picture of linguistic strata shows intriguing parallels to the one discussed by Michel Foucault in [Foucault 1972].
3. For all of this see Aristotle's *Physics*, book V, chapters 1–2 (224a21–226b16). Note that Aristotle does have a concept of change that is broader than that of motion. Motion is change of substance, change from something to something (225a1). But change also includes coming to be and passing away, i.e., change from nothing to something and from something to nothing (225a34–225b9), and these are not motions.
4. Compare Aristotle's *Physics*, book I, and especially his *On Generation and Corruption*, book II, chapters 1–4.
5. There is an ingredient missing from my sketch of this argument: Aristotle's doctrine of place, developed in the *Physics*, book IV, just before his discussion of the vacuum. Place, for Aristotle, is always the place of body or, more precisely, the interior surface of the containing or surrounding body (212a2–7). Turning to his next topic, Aristotle says, "Since the void (if there is any) must be conceived as place in which there might be body but is not, it is clear that, so conceived, the void cannot exist at all, either as inseparable or separable" (214a16–20). (I quote from the Loeb Classical Library translation by Philip H.

Wickstead and Francis M. Cornford, a version that, on this difficult aspect of the *Physics*, seems to me clearer than most, both in text and commentary.) That it is not merely a mistake to substitute 'position' for 'place' in a sketch of the argument is indicated by the last part of the next paragraph of my text.

6. For this and closely related arguments see Aristotle, *Physics*, book IV, chapter 8 (especially 214b27–215a24).
7. [Volta 1800]: 403–31. On this subject, see [Brown 1969]: 61–103.
8. The illustrations are from [de la Rive 1856], pp. 600, 656. Structurally similar but schematic diagrams appear in Faraday's experimental researches from the early 1830s. My choice of the 1840s as the period when such diagrams became standard results from a casual survey of electricity texts lying ready to hand. A more systematic study would, in any case, have had to distinguish between British, French, and German responses to the chemical theory of the battery.
9. For the full version with supporting evidence see my [Kuhn 1978].

CHAPTER 7

Falsification and the Methodology of Scientific Research Programmes

IMRE LAKATOS[1]

Pp. 132–138 from *Criticism and the Growth of Knowledge,* ed. Imre Lakatos and Alan Musgrave. Cambridge: Cambridge University Press, 1970.

A Methodology of Scientific Research Programmes

I have discussed the problem of objective appraisal of scientific growth in terms of progressive and degenerating problemshifts in series of scientific theories. The most important such series in the growth of science are characterized by a certain *continuity* which connects their members. This continuity evolves from a genuine research programme adumbrated at the start. The programme consists of methodological rules: some tell us what paths of research to avoid (*negative heuristic*), and others what paths to pursue (*positive heuristic*).[2]

Even science as a whole can be regarded as a huge research programme with Popper's supreme heuristic rule: 'devise conjectures which have more empirical content than their predecessors.' Such methodological rules may be formulated, as Popper pointed out, as metaphysical principles.[3] For instance, the *universal* anti-conventionalist rule against exception-barring may be stated as the metaphysical principle: 'Nature does not allow exceptions'. This is why Watkins called such rules 'influential metaphysics'.[4]

But what I have primarily in mind is not science as a whole, but rather *particular* research programmes, such as the one known as 'Cartesian metaphysics'. Cartesian metaphysics, that is, the mechanistic theory of the universe—according to which the universe is a huge clockwork (and system of vortices) with push as the only cause of motion—functioned as a powerful heuristic principle. It discouraged work on scientific theories—like [the 'essentialist' version of] Newton's theory of action at a distance—which were inconsistent with it (*negative heuristic*). On the other hand, it encouraged work on auxiliary hypotheses which might have saved it from apparent counterevidence—like Keplerian ellipses (*positive heuristic*).[5]

(a) Negative heuristic: the 'hard core' of the programme

All scientific research programmes may be characterized by their 'hard core'. The negative heuristic of the programme forbids us to direct the *modus tollens* at this 'hard core'. Instead, we must use our ingenuity to articulate or even invent 'auxiliary hypotheses', which form a *protective belt* around this core, and we must redirect the *modus tollens* to *these*. It is this protective belt of auxiliary hypotheses which has to bear the brunt of tests and get adjusted and re-adjusted, or even completely replaced, to defend the thus-hardened core. A research programme is successful if all this leads to a progressive problemshift; unsuccessful if it leads to a degenerating problemshift.

The classical example of a successful research programme is Newton's gravitational theory: possibly the most successful research programme ever. When it was first produced, it was submerged in an ocean of 'anomalies' (or, if you wish, 'counterexamples'[6]), and opposed by the observational theories supporting these anomalies. But Newtonians turned, with brilliant tenacity and ingenuity, one counter-instance after another into corroborating instances, primarily by overthrowing the original observational theories in the light of which this 'contrary evidence' was established. In the process they themselves produced new counter-examples which they again resolved. They 'turned each new difficulty into a new victory of their programme'.[7]

In Newton's programme the negative heuristic bids us to divert the *modus tollens* from Newton's three laws of dynamics and his law of gravitation. This 'core' is 'irrefutable' by the methodological decision of its protagonists: anomalies must lead to changes only in the 'protective' belt of auxiliary, 'observational' hypothesis and initial conditions.[8]

I have given a contrived micro-example of a progressive Newtonian problemshift.[9] If we analyse it, it turns out that each successive link in this exercise predicts some new fact; each step represents an increase in empirical content: the example constitutes a *consistently progressive theoretical shift*. Also, each prediction is in the end verified; although on three subsequent occasions they may have seemed momentarily to be 'refuted'.[10] While 'theoretical progress' (in the sense here described) may be verified immediately,[11] 'empirical progress' cannot, and in a research programme we may be frustrated by a long series of 'refutations' before ingenious and lucky content-increasing auxiliary hypotheses turn a chain of defeats—*with hindsight*—into a resounding success story, either by revising some false 'facts' or by adding novel auxiliary hypotheses. We may then say that we must require that each step of a research programme be consistently content-increasing: that each step constitute a *consistently progressive theoretical problemshift*. All we need in addition to this is that at least every now and then the increase in content should be seen to be retrospectively corroborated: the programme as a whole should also display an *intermittently progressive empirical shift*. We do not demand that each step produce *immediately* an *observed* new fact. Our term '*intermittently*' gives sufficient *rational* scope for dogmatic adherence to a programme in face of *prima facie* 'refutations'.

The idea of 'negative heuristic' of a scientific research programme rationalizes classical conventionalism to a considerable extent. We may rationally decide not to allow 'refutations' to transmit falsity to the hard core as long as the corroborated empirical content of the protecting belt of auxiliary hypotheses increases. But our approach differs from Poincaré's justificationist conventionalism in the sense that, unlike Poincaré's, we maintain that if and when the programme ceases to anticipate novel facts, its hard core might have to be abandoned: that is, *our* hard core, unlike Poincaré's, may crumble under certain conditions. In this sense we side with Duhem who thought that such a possibility must be allowed for;[12] but for Duhem the reason for such crumbling is purely *aesthetic*,[13] while for us it is mainly *logical and empirical*.

(b) Positive heuristic: the construction of the 'protective belt' and the relative autonomy of theoretical science

Research programmes, besides their negative heuristic, are also characterized by their positive heuristic.

Even the most rapidly and consistently progressive research programmes can digest their 'counter-evidence' only piecemeal: anomalies are never completely exhausted. But it should not be thought that yet unexplained anomalies—'puzzles' as Kuhn might call them—are taken in random order, and the protective belt built up in an eclectic fashion, without any preconceived order. The order is usually decided in the theoretician's cabinet, independently of the *known* anomalies. Few theoretical scientists engaged in a research programme pay undue attention to 'refutations'. They have a long-term research policy which anticipates these refutations. This research policy, or order of research, is set out—in more or less detail—in the *positive heuristic* of the research programme. The negative heuristic specifies the 'hard core' of the programme which is 'irrefutable' by the methodological decision of its protagonists; the positive heuristic consists of a partially articulated set of suggestions or hints on how to change, develop the 'refutable variants' of the research-programme, how to modify, sophisticate, the 'refutable' protective belt.

The positive heuristic of the programme saves the scientist from becoming confused by the ocean of anomalies. The positive heuristic sets out a programme which lists a chain of ever more complicated *models* simulating reality: the scientist's attention is riveted on building his models following instructions which are laid down in the positive part of his programme. He ignores the *actual* counterexamples, the available '*data*'.[14] Newton first worked out his programme for a planetary system with a fixed point-like sun and one single point-like planet. It was in this model that he derived his inverse square law for Kepler's ellipse. But this model was forbidden by Newton's own third law of dynamics, therefore the model had to be replaced by one in which both sun and planet revolved round their common centre of gravity. This change was not motivated by any observation (the data did not suggest an 'anomaly' here) but by a theoretical difficulty in developing the programme. Then he worked out the programme for more planets as if there were only heliocentric but no interplanetary forces. Then he worked out the case where the sun and planets were not mass-points but mass-*balls*. Again, for this change he did not *need* the observation of an anomaly; infinite density was forbidden by an (inarticulated) touchstone theory, therefore planets *had* to be extended. This change involved considerable mathematical difficulties, held up Newton's work—and delayed the publication of the *Principia* by more than a decade. Having solved this 'puzzle', he started work on *spinning balls* and their wobbles. Then he admitted interplanetary forces and started work on *perturbations*. At this point he started to look more anxiously at the facts. Many of them were beautifully explained (qualitatively) by this model, many were not. It was then that he started to work on *bulging* planets, rather than round planets, etc.

Newton despised people who, like Hooke, stumbled on a first naive model but did not have the tenacity and ability to develop it into a research programme, and who thought that a first version, a mere aside, constituted a 'discovery'. He held up publication until his programme had achieved a remarkable progressive shift.[15]

Most, if not all, Newtonian 'puzzles', leading to a series of new variants superseding each other, were forseeable at the time of Newton's first naive model and no doubt Newton and his colleagues *did* forsee them: Newton must have been fully aware of the blatant falsity of his first variants. Nothing shows the existence of a positive heuristic of a research programme clearer than this fact: this is why one speaks of 'models' in research programmes. A '*model*' is a set of initial conditions (possibly together with some of the observational theories) which one knows is *bound* to be replaced during the further development of the programme, and one even knows, more or less, how. This shows once more how irrelevant 'refutations' of any specific variant are in a research programme: their existence is fully expected, the positive heuristic is there as the strategy both for predicting (producing) and digesting them. Indeed, if the positive heuristic is clearly spelt out, the difficulties of the programme are mathematical rather than empirical.[16]

One may formulate the 'positive heuristic' of a research programme as a 'metaphysical' principle. For instance one may formulate Newton's programme like this: 'the planets are essentially gravitating spinning-tops of roughly spherical shape'. This idea was never *rigidly* maintained: the planets are not *just* gravitational, they have also, for example, electromagnetic characteristics which may influence their motion. Positive heuristic is thus in general more flexible than negative heuristic. Moreover, it occasionally happens that when a research programme gets into a degenerating phase, a little revolution or a *creative shift* in its positive heuristic may push it forward again.[17] It is better therefore to separate the 'hard core' from the more flexible metaphysical principles expressing the positive heuristic.

Our considerations show that the positive heuristic forges ahead with almost complete disregard of 'refutations': it may seem that it is the '*verifications*'[18] rather than the refutations which provide the contact points with reality. Although one must point out that any 'verification' of the $n+1$-th version of the programme is a refutation of the n-th version, we cannot deny that *some* defeats of the subsequent versions are always foreseen: it is the 'verifications' which keep the programme going, recalcitrant instances notwithstanding.

We may appraise research programmes, even after their 'elimination', for their *heuristic power:* how many new facts did they produce, how great was 'their capacity to explain their refutations in the course of their growth'?[19]

(We may also appraise them for the stimulus they gave to mathematics. The real difficulties for the theoretical scientist arise rather from the *mathematical difficulties* of the programme than from anomalies. The greatness of the Newtonian programme comes partly from the development—by Newtonians—of classical infinitesimal analysis which was a crucial precondition of its success.)

Notes

1. This paper is a considerably improved version of my [1968*b*] and a crude version of my [1973]. Some parts of the former are here reproduced without change with the permission of the Editor of the *Proceedings of the Aristotelian Society*. In the preparation of the

new version I received much help from Tad Beckman, Colin Howson, Clive Kilmister, Larry Laudan, Eliot Leader, Alan Musgrave, Michael Sukale, John Watkins and John Worrall.

2. One may point out that the negative and positive heuristic gives a rough (implicit) definition of the 'conceptual framework' (and consequently of the language). The recognition that the history of science is the history of research programmes rather than of theories may therefore be seen as a partial vindication of the view that the history of science is the history of conceptual frameworks or of scientific languages.

3. Popper [1934], sections 11 and 70. I use 'metaphysical' as a technical term of naive falsificationism: a contingent proposition is 'metaphysical' if it has no 'potential falsifiers'

4. Watkins [1958]. Watkins cautions that 'the logical gap between statements and prescriptions in the metaphysical-methodological field is illustrated by the fact that a person may reject a [metaphysical] doctrine in its fact-stating form while subscribing to the prescriptive version of it' (*Ibid.*, pp. 356–7).

5. For this Cartesian research programme, cf. Popper [1958] and Watkins [1958], pp. 350–1.

6. For the clarification of the concepts of 'counterexample' and 'anomaly' cf. *above*, p. 110, and especially *below*, p. 159, text to footnote 1.

7. Laplace [1796], livre IV, chapter ii.

8. The actual hard core of a programme does not actually emerge fully armed like Athene from the head of Zeus. It develops slowly, by a long, preliminary process of trial and error. In this paper this process is not discussed.

9. Cf. *above*, pp. 100–1.

10. The 'refutation' was each time successfully diverted to 'hidden lemmas'; that is, to lemmas emerging, as it were, from the *ceteris paribus* clause.

11. But cf. *below*, pp. 155–7.

12. Cf. *above*, p. 105.

13. *Ibid.*

14. If a scientist (or mathematician) has a positive heuristic, he refuses to be drawn into observation. He will 'lie down on his couch, shut his eyes and forget about the data'. (Cf. my [1963–4], especially pp. 300ff., where there is a detailed case study of such a programme.) Occasionally, of course, he will ask Nature a shrewd question: he will then be encouraged by Nature's YES, but not discouraged by its NO.

15. Reichenbach, following Cajori, gives a different explanation of what delayed Newton in the publication of his *Principia:* 'To his disappointment he found that the observational results disagreed with his calculations. Rather than set any theory, however beautiful, before the facts, Newton put the manuscript of his theory into his drawer. Some twenty years later, after new measurements of the circumference of the earth had been made by a French expedition, Newton saw that the figures on which he had based his test were false and that the improved figures agreed with his theoretical calculation. It was only after this test that he published his law … The story of Newton is one of the most striking illustrations of the method of modern science' (Reichenbach [1951], pp. 101–2). Feyerabend criticizes Reichenbach's account (Feyerabend [1965], p. 229), but does not give an alternative *rationale*.

16. For this point cf. Truesdell [1960].

17. Soddy's contribution to Prout's programme or Pauli's to Bohr's (old quantum theory) programme are typical examples of such creative shifts.

18. A 'verification' is a corroboration of excess content in the expanding programme. But, of course, a 'verification' does not *verify* a programme: it shows only its heuristic power.

19. Cf. my [1963–4], pp. 324–30. Unfortunately in 1963–4 I had not yet made a clear terminological distinction between theories and research programmes, and this impaired my exposition of a research programme in informal, quasi-empirical mathematics.

CHAPTER 8

The Relations between the History and the Philosophy of Science

THOMAS KUHN

Isenberg Lecture, delivered at Michigan State University, 1 March 1968; revised October 1976. Published pp. 3–20 in *The Essential Tension*, Chicago: University of Chicago Press, 1977.

The subject on which I have been asked to speak today is the relations between the history and the philosophy of science. For me, more than for most, it has deep personal as well as intellectual significance. I stand before you as a practicing historian of science. Most of my students mean to be historians, not philosophers. I am a member of the American Historical, not the American Philosophical, Association. But for almost ten years after I first encountered philosophy as a college freshman, it was my primary avocational interest, and I often considered making it my vocation, displacing theoretical physics, the only field in which I can claim to have been properly trained. Throughout those years, which lasted until around 1948, it never occurred to me that history or history of science could hold the slightest interest. To me then, as to most scientists and philosophers still, the historian was a man who collects and verifies facts about the past and who later arranges them in chronological order. Clearly the production of chronicles could have little appeal to someone whose fundamental concerns were with deductive inference and fundamental theory.

I shall later ask why the image of the historian as chronicler has such special appeal to both philosophers and scientists. Its continued and selective attraction is not due either to coincidence or to the nature of history, and it may therefore prove especially revealing. But my present point is still autobiographical. What drew me belatedly from physics and philosophy to history was the discovery that science, when encountered in historical source materials, seemed a very different enterprise from the one implicit in science pedagogy and explicit in standard philosophical accounts of scientific method. History might, I realized with astonishment, be relevant to the philosopher of science and perhaps also to the epistemologist in ways that transcended its classic role as a source of examples for previously occupied positions. It might, that is, prove to be a particularly consequential source of problems and of insights. Therefore, though I became a historian, my deepest interests remained philosophical, and in recent years those interests have become increasingly explicit in my published work. To an extent, then, I do both history and philosophy of science. Of course I therefore think about the relation between them, but I also live it, which is not the same thing. That duality of my involvement will inevitably be reflected in the way I approach today's topic. From this point my talk will divide into two quite different, though closely related parts. The first is a report, often quite personal, of the difficulties to be encountered in any attempt to draw the two fields closer together. The second, which deals with problems more explicitly intellectual, argues that the *rapprochement* is fully worth the quite special effort it requires.

Few members of this audience will need to be told that, at least in the United States, the history and the philosophy of science are separate and distinct disciplines. Let me, from the very start, develop reasons for insisting that they be kept that way. Though a new sort of dialogue between these fields is badly needed, it must be inter- not intra-disciplinary. Those of you aware of my involvement with Princeton University's Program in History and Philosophy of Science may find odd my insistence that there is no such field. At Princeton, however, the historians and the philosophers of science pursue different, though overlapping, courses of study, take different general examinations, and receive their degrees from different departments, either history or philosophy. What is particularly admirable in that design is that it provides an institutional basis for a dialogue between fields without subverting the disciplinary basis of either.

Subversion is not, I think, too strong a term for the likely result of an attempt to make the two fields into one. They differ in a number of their central constitutive characteristics, of which the most general and apparent is their goals. The final product of most historical research is a narrative, a story, about particulars of the past. In part it is a description of what occurred (philosophers and scientists often say, a *mere* description). Its success, however, depends not only on accuracy but also on structure. The historical narrative must render plausible and comprehensible the events it describes. In a sense to which I shall later return, history is an explanatory enterprise; yet its explanatory functions are achieved with almost no recourse to explicit generalizations. (I may point out here, for later exploitation, that when philosophers discuss the role of covering laws in history, they characteristically draw their examples from the work of economists and sociologists, not of historians. In the writings of the latter, lawlike generalizations are extraordinarily hard to find.) The philosopher, on the other hand, aims principally at explicit generalizations and at those with universal scope. He is no teller of stories, true or false. His goal is to discover and state what is true at all times and places rather than to impart understanding of what occurred at a particular time and place.

Each of you will want to articulate and to qualify those crass generalizations, and some of you will recognize that they raise deep problems of discrimination. But few will feel that distinctions of this sort are entirely empty, and I therefore turn from them to their consequences. It is these that make the distinction of aims important. To say that history of science and philosophy of science have different goals is to suggest that no one can practice them both at the same time. But it does not suggest that there are also great difficulties about practicing them alternately, working from time to time on historical problems and attacking philosophical issues in between. Since I obviously aim at a pattern of that sort myself, I am committed to the belief that it can be achieved. But it is nonetheless important to recognize that each switch is a personal wrench, the abandonment of one discipline for another with which it is not quite compatible. To train a student simultaneously in both would risk depriving him of any discipline at all. [. . .] Nor, I think, is a compromise possible, for it presents problems of the same sort as a compromise between the duck and the rabbit of the well-known Gestalt diagram. Though most people can readily see the duck and the rabbit alternately, no amount of ocular exercise and strain will educe a duck-rabbit.

That view of the relation between enterprises is not at all the one I had at the time of my conversion to history twenty years ago. Rather it derives from much subsequent experience, sometimes painful, as a teacher and writer. In the former role I have, for example, repeatedly taught graduate seminars in which prospective historians and philosophers read and discussed the same classic works of science and philosophy. Both groups were conscientious and both completed the assignments with care, yet it was often difficult to believe that both had been engaged with the same texts. Undoubtedly the two had looked at the same signs, but they had been trained (programmed, if you will) to process them differently. Inevitably, it was the processed signs—for example their reading notes or their memory of the text—rather than the signs themselves that provided the basis for their reports, paraphrases, and contributions to discussion.

Subtle analytic distinctions that had entirely escaped the historians would often be central when the philosophers reported on their reading. The resulting confrontations were invariably educational for the historians, but the fault was not always theirs. Sometimes the distinctions dwelt upon by the philosophers were not to be found at all in the original text. They were products of the subsequent development of science or philosophy, and their introduction during the philosophers' processing of signs altered the argument. Or again, listening to the historians' paraphrase of a position, the philosophers would often point out gaps and inconsistencies that the historians had failed to see. But the philosophers could then sometimes be shocked by the discovery that the paraphrase was accurate, that the gaps were there in the original. Without quite knowing they were doing so, the philosophers had improved the argument while reading it, knowing what its subsequent form must be. Even with the text open before them it was regularly difficult and sometimes impossible to persuade them that the gap was really there, that the author had not seen the logic of the argument quite as they did. But if the philosophers could be brought to see that much, they could usually see something more important as well—that what they took to be gaps had in fact been introduced by analytic distinctions they had themselves supplied, that the original argument, if no longer viable philosophy, was sound in its own terms. At this point the whole text might begin to look different to them. Both the extent of the transformation and the pedagogic difficulty in deliberately bringing it about are reminiscent of the Gestalt switch.

Equally impressive, as evidence of different processing, was the range of textual material noticed and reported by the two groups. The historians always ranged more widely. Important parts of their reconstructions might, for example, be built upon passages in which the author had introduced a metaphor designed, he said, "to aid the reader." Or again, having noticed an apparent error or inconsistency in the text, the historian might spend some time explaining how a brilliant man could have slipped in this way. What aspect of the author's thought, the historian would ask, can be discovered by noting that an inconsistency obvious to us was invisible to him and was perhaps no inconsistency at all? For the philosophers, trained to construct an argument, not to reconstruct historical thought, both metaphors and errors were irrelevant and were sometimes not noticed at all. Their concern, which they pursued with a subtlety, skill, and persistence seldom found among the historians, was the explicit philosophical generalization and the arguments that could be educed in its defense. As a result, the papers they submitted at the end of the term were regularly shorter and usually far more coherent than those produced by the historians. But the latter, though often analytically clumsy, usually came far closer to reproducing the major conceptual ingredients in the thought of the men the two groups had studied together. The Galileo or Descartes who appeared in the philosophers papers was a better scientist or philosopher but a less plausible seventeenth-century figure than the figure presented by the historians.

I have no quarrel with either of these modes of reading and reporting. Both are essential components as well as central products of professional training. But the professions are different, and they quite properly put different first things first. For the philosophers in my seminars the priority tasks were, first, to isolate the central elements of a philosophical position and, then, to criticize and develop them. Those students were, if you will, honing their wits against the developed opinions of their greatest predecessors. Many of them would continue to do so in their later professional life. The historians, on the other hand, were concerned with the viable and the general only in the forms that had, in fact, guided the men they studied. Their first concern was to discover what each one had thought, how he had come to think it, and what the consequences had been for him, his contemporaries, and his successors. Both groups thought of themselves as attempting to grasp the essentials of a past philosophical position, but their ways of doing the job were conditioned by the primary values of their separate disciplines, and their results were often correspondingly distinct. Only if the philosophers were converted to history or the historians to philosophy did additional work produce significant convergence. [. . .]

These are only first steps in a quasi-sociological account of history and philosophy as knowledge-producing enterprises. They should, however, be sufficient to suggest why, admiring both, I suspect that an attempt to make them one would be subversive. Those whom I have convinced or those who, for one another reason, have needed no convincing will, however, have a different question. Given the deep and consequential differences between the two enterprises, what can they have to say to each other? Why have I insisted that an increasingly active dialogue between them is an urgent desideratum? To that question, particularly to one part of it, the remainder of my remarks this evening are directed.

Any answer must divide into two far-from-symmetrical parts, of which the first here requires no more than cursory summary. Historians of science need philosophy for reasons that are, at once, apparent and well known. For them it is a basic tool, like knowledge of science. Until the end of the seventeenth century, much of science was philosophy. After the disciplines separated, they continued to interact in often consequential ways. A successful attack on many of the problems central to the history of science is impossible for the man who does not command the thought of the main philosophic schools of the periods and areas he studies. [. . .] Scientists are not often philosophers, but they do deal in ideas, and the analysis of ideas has long been the philosopher's province. The men who did most to establish the flourishing contemporary tradition in the history of science—I think particularly of A. O. Lovejoy and, above all, Alexandre Koyré—were philosophers before they turned to the history of scientific ideas. From them my colleagues and I learned to recognize the structure and coherence of idea systems other than our own. [. . .]

Those remarks will suggest what I had in mind in saying that the problem of the relations between history and philosophy of science divides into two parts, which are far from symmetrical. Though I do not think current philosophy of science has much relevance for the historian of science, I deeply believe that much writing on philosophy of science would be improved if history played a larger background role in its preparation. Before attempting to justify that belief, I must, however, introduce a few badly needed limitations. When speaking here of the history of science, I refer to that central part of the field that is concerned with the evolution of scientific ideas, methods, and techniques, not the increasingly significant portion that emphasizes the social setting of science, particularly changing patterns of scientific education, institutionalization, and support, both moral and financial. The philosophical import of the latter sort of work seems to me far more problematic than that of the former, and its consideration would, in any case, require a separate lecture. By the same token, when speaking of the philosophy of science, I have in mind neither those portions that shade over into applied logic nor, at least not with much assurance, those parts that are addressed to the implications of particular current theories for such longstanding philosophical problems as causation or space and time. Rather I am thinking of that central area that concerns itself with the scientific in general, asking, for example, about the structure of scientific theories, the status of theoretical entities, or the conditions under which scientists may properly claim to have produced sound knowledge. It is to this part of the philosophy of science, and very possibly to it alone, that the history of scientific ideas and techniques may claim relevance.

[. . .] I do want to suggest what it is about history that makes it a possible source for a rational reconstruction of science different from that now current. For that purpose, furthermore, I must first insist that history is not itself the enterprise much contemporary philosophy takes it to be. I must, that is, argue briefly the case for what Louis Mink has perceptively called "the autonomy of historical understanding."

No one, I think, still believes that history is mere chronicle, a collection of facts arranged in the order of their occurrence. It is, most would concede, an explanatory enterprise, one that induces understanding, and it must thus display not only facts but also connections between them. No historian has, however, yet produced a plausible account of the nature of these connections, and philosophers have recently filled the resulting void with what is known as the "covering law model." My concern with it is as an articulated version of a widely diffused image of history, one that makes the discipline seem uninteresting to those who seek lawlike generalizations, philosophers, scientists, and social scientists in particular.

According to proponents of the covering law model, a historical narrative is explanatory to the extent that the events it describes are governed by laws of nature and society to which the historian has conscious or unconscious access. Given the conditions that obtained at the point in time when the narrative opens, and given also a knowledge of the covering laws, one should be able to predict, perhaps with the aid of additional boundary conditions inserted along the way, the future course of some central parts of the narrative. It is these parts, and only these, that the historian may be said to have explained. If the laws permit only rough predictions, one speaks of having provided an "explanation sketch" rather than an explanation. If they permit no prediction at all, the narrative has provided no explanation.

Clearly the covering law model has been drawn from a theory of explanation in the natural sciences and applied to history. I suggest that, whatever its merits in the fields for which it was first developed, it is an almost total misfit in this application. Very likely there are or will be laws of social behavior capable of application to history. As they come into being, historians sooner or later use them. But laws of that sort are primarily the business of the social sciences, and except in economics very few are yet in hand. I have already pointed out that philosophers turn generally to writings by social scientists for the laws they attribute to historians. I now add that, when they do draw examples from historical writing, the laws they educe are at once obvious and dubious: for example, "Hungry men tend to riot." Probably, if the words "tend to" are heavily underscored, the law is valid. But does it follow that an account of starvation in eighteenth-century France is less essential to a narrative dealing with the first decade of the century, when there were no riots, than to one dealing with the last, when riots did occur?

Surely the plausibility of a historical narrative does not depend upon the power of a few scattered and doubtful laws like this one. If it did, then history would explain virtually nothing at all. With few exceptions, the facts that fill the pages of its narratives would be mere window dressing, facts for the sake of facts, unconnected to each other or to any larger goal. Even the few facts actually connected by law would become uninteresting, for precisely to the extent that they were "covered," they would add nothing to what everyone already knew, I am not claiming, let me be clear, that the historian has access to no laws and generalizations, nor that he should make no use of them when they are at hand. But I do claim that, however much laws may add substance to an historical narrative, they are not essential to its explanatory force. That is carried, in the first instance, by the facts the historian presents and the manner in which he juxtaposes them.

During my days as a philosophically inclined physicist, my view of history resembled that of the covering law theorists, and the philosophers in my seminars usually begin by viewing it in a similar way. What changed my mind and often changes theirs is the experience of putting together a historical narrative. That experience is vital, for the difference between learning history and doing it is far larger than that in most other creative fields, philosophy certainly included. From it I conclude, among other things, that an ability to predict the future is no part of the historian's arsenal. He is neither a social scientist nor a seer. It is no mere accident that he knows the end of his narrative as well as the start before he begins to write. History cannot be written without that information. Though I have no alternate philosophy of history or of historical explanation to offer here. I can at least outline a better image of the historian's task and suggest why its performance might produce a sort of understanding.

The historian at work is not, I think, unlike the child presented with one of those picture puzzles of which the pieces are square; but the historian is given many extra pieces in the box. He has or can get the data, not all of them (what would that be?) but a very considerable collection. His job is to select from them a set that can be juxtaposed to provide the elements of what, in the child's case, would be a picture of recognizable objects plausibly juxtaposed and of what, for the historian and his reader, is a plausible narrative involving recognizable motives and behaviors. Like the child with the puzzle, the historian at work is governed by rules that may not be violated. There may be no empty spaces in the middle either of the puzzle or of the narrative. Nor may there be any discontinuities. If the puzzle displays a pastoral scene, the legs of a man may not be joined to the body of a sheep. In the narrative a tyrannical monarch may not be transformed by sleep alone to a benevolent despot. For the historian there are additional rules that do not apply to the child. Nothing in the narrative may, for example, do violence to the facts the historian has elected to omit from his story. That story must, in addition, conform to any laws of nature and society the historian knows. Violation of rules like these is ground for rejecting either the assembled puzzle or the historian's narrative.

Such rules, however, only limit but do not determine the outcome of either the child's or the historian's task. In both cases the basic criterion for having done the job right is the primitive recognition that the pieces fit to form a familiar, if previously unseen, product. The child has seen pictures, the historian behavior patterns, similar to these before. That recognition of similarity is, I believe, prior to any answers to the question, similar with respect to what? Though it can be rationally understood and perhaps even modelled on a computer (I once attempted something of the sort myself), the similarity relation does not lend itself to lawlike reformulation. It is global, not reducible to a unique set of prior criteria more primitive than the similarity relation itself. One may not replace it with a statement of the form. "*A* is similar to *B*, if and only if the two share the characteristics *c, d, e,* and *f*." I have elsewhere argued that the cognitive content of the physical sciences is in part dependent on the same primitive similarity relation between concrete examples, or paradigms, of successful scientific work, that scientists model one problem solution on another without at all knowing what characteristics of the original must be preserved to legitimate the process. Here I am suggesting that in history that obscure global relationship carries virtually the entire burden of connecting fact. If history is explanatory, that is not because its narratives are covered by general laws. Rather it is because the reader who says, "Now I know what happened," is simultaneously saying, "Now it makes sense; now I understand; what was for me previously a mere list of facts has fallen into a recognizable pattern." I urge that the experience he reports be taken seriously.

What has just been said is, of course, the early stage of a program for philosophical contemplation and research, not yet the solution of a problem. If many of you differ with me about its likely outcome, that is not because you are more aware than I of its incompleteness and difficulty, but because you are less convinced that the occasion demands so radical a break with tradition. That point, however, I shall not argue here. The object of the digression from which I now return has been to identify my convictions, not to defend them. What has troubled me about the covering law model is that it makes of the historian a social scientist *manqué*, the gap being filled by assorted factual details. It makes it hard to recognize that he has another and a profound discipline of his own, that there is an autonomy (and integrity) of historical understanding. If that claim now seems even remotely plausible, it prepares the way for my principal conclusion. When the historian of science emerges from the contemplation of sources and the construction of narrative, he may have a right to claim acquaintance with essentials. If he then says, "I cannot construct a viable narrative without giving a central place to aspects of science that philosophers ignore, nor can I find a trace of elements they consider essential," then he deserves an audience. What he is claiming is that the enterprise reconstructed by the philosopher is not, as to certain of its essentials, science.

What sort of lessons might the philosopher learn by taking the historian's narrative constructions more seriously? I shall close this lecture with a single global example, referring you to my earlier work for other illustrations, many of them dependent on the examination of individual cases. The overwhelming majority of historical work is concerned with process, with development over time. In principle, development and change need not play a similar role in philosophy, but in practice, I now want to urge, the philosopher's view of even static science, and thus of such questions as theory structure and theory confirmation, would be fruitfully altered if they did.

Consider, for example, the relation between empirical laws and theories, both of which I shall, for purposes of this brief conclusion, construe quite broadly. Despite real difficulties, which I have elsewhere perhaps overemphasized, empirical laws fit the received tradition in philosophy of science relatively well. They can, of course, be confronted directly with observation or experiment. More to my present point, when they first emerge, they fill an apparent gap, supplying information that was previously lacking. As science develops, they may be refined, but the original versions remain approximations to their successors, and their force is therefore either obvious or readily recaptured. Laws, in short, to the extent that they are purely empirical, enter science as net additions to knowledge and are never thereafter entirely displaced. They may cease to be of interest and therefore remain uncited, but that is another matter, important difficulties do, I repeat, confront the elaboration of this position, for it is no longer clear just what it would be for a law to be purely empirical. Nevertheless, as an admitted idealization, this standard account of empirical laws fits the historian's experience quite well.

With respect to theories the situation is different. The tradition introduces them as collections or sets of law. Though it concedes that individual members of a set can be confronted with experience only through the deductive consequences of the set as a whole, it thereafter assimilates theories to laws as closely as possible. That assimilation does not fit the historian's experience at all well. When he looks at a given period in the past he can find gaps in knowledge later to be filled by empirical laws. The ancients knew that air was compressible but were ignorant of the regularity that quantitatively relates its volume and pressure; if asked, they would presumably have conceded the lack. But the historian seldom or never finds similar gaps to be filled by later theory. In its day, Aristotelian physics covered the accessible and imaginable world as completely as Newtonian physics later would. To introduce the latter, the former had to be literally displaced. After that occurred, furthermore, efforts to recapture Aristotelian theory presented difficulties of a very different nature from those required to recapture an empirical law. Theories, as the historian knows them, cannot be decomposed into constituent elements for purposes of direct comparison either with nature or with each other. That is not to say that they cannot be analytically decomposed at all, but rather that the lawlike parts produced by analysis cannot, unlike empirical laws, function individually in such comparisons.

A central tenet of Aristotle's physics was, for example, the impossibility of a void. Suppose that a modern physicist had told him that an arbitrarily close approximation to a void could now be produced in the laboratory. Probably Aristotle would have responded that a container emptied of air and other gases was not in his sense a void. That response would suggest that the impossibility of a void was not, in his physics, a merely empirical matter. Suppose now instead that Aristotle had conceded the physicist's point and announced that a void could, after all, exist in nature. Then he would have required a whole new physics, for his concept of the finite cosmos, of place within it, and of natural motion stand or fall together with his concept of the void. In that sense, too, the lawlike statement "there are no voids in nature" did not function within Aristotelian physics quite as a law. It could not, that is, be eliminated and replaced by an improved version, leaving the rest of the structure standing.

For the historian, therefore, or at least for this one, theories are in certain essential respects holistic. So far as he can tell, they have always existed (though not always in forms one would comfortably describe as scientific), and they then always cover the entire range of conceivable natural phenomena (though often without much precision). In these respects they are clearly unlike laws, and there are inevitably corresponding differences in the ways they develop and are evaluated. About these latter processes we know very little, and we shall not learn more until we learn properly to reconstruct selected theories of the past. As of today, the people taught to do that job are historians, not philosophers. Doubtless the latter could learn, but in the process, as I have suggested, they would likely become historians too. I would of course welcome them, but would be saddened if they lost sight of their problems in the transition, a risk that I take to be real. To avoid it I urge that history and philosophy of science continue as separate disciplines. What is needed is less likely to be produced by marriage than by active discourse.

CHAPTER 9

The History of Science

THOMAS KUHN

Pages 105–126 in *The Essential Tension*, Chicago: University of Chicago Press, 1977. Reprinted from *International Encyclopedia of the Social Sciences*, vol, 14 (New York: Crowell Collier and Macmillan, 1968), pp. 74–83. © 1968 by Crowell Collier and Macmillan.

As an independent professional discipline, the history of science is a new field still emerging from a long and varied prehistory. Only since 1950, and initially only in the United States, has the majority of even its youngest practitioners been trained for, or committed to, a full-time scholarly career in the field. From their predecessors, most of whom were historians only by avocation and thus derived their goals and values principally from some other field, this younger generation inherits a constellation of sometimes irreconcilable objectives. The resulting tensions, though they have relaxed with the increasing maturation of the profession, are still perceptible, particularly in the varied primary audiences to which the literature of the history of science continues to be addressed. Under the circumstances any brief report on development and current state is inevitably more personal and prognostic than for a longer-established profession.

Development of the Field

Until very recently most of those who wrote the history of science were practicing scientists, sometimes eminent ones. Usually history was for them a by-product of pedagogy. They saw in it, besides intrinsic appeal, a means to elucidate the concepts of their specialty, to establish its tradition, and to attract students. The historical section with which so many technical treatises and monographs still open is contemporary illustration of what was for many centuries the primary form and exclusive source for the history of science. That traditional genre appeared in classical antiquity both in historical sections of technical treatises and in a few independent histories of the most developed ancient sciences, astronomy and mathematics. Similar works—together with a growing body of heroic biography—had a continuous history from the Renaissance through the eighteenth century, when their production was much stimulated by the Enlightenment's vision of science as at once the source and the exemplar of progress. From the last fifty years of that period come the earliest historical studies that are sometimes still used as such, among them the historical narratives embedded in the technical works of Lagrange (mathematics) as well as the imposing separate treatises by Montucla (mathematics and physical science), Priestley (electricity and optics), and Delambre (astronomy). In the nineteenth and early twentieth centuries, though alternative approaches had begun to develop, scientists continued to produce both occasional biographies and magistral histories of their own specialties, for example, Kopp (chemistry), Poggendorff (physics), Sachs (botany), Zittel and Geikie (geology), and Klein (mathematics).

A second main historiographic tradition, occasionally indistinguishable from the first, was more explicitly philosophical in its objectives. Early in the seventeenth century Francis Bacon proclaimed the utility of histories of learning to those who would discover the nature and proper use of human reason. Condorcet and Comte are only the most famous of the philosophically inclined writers who, following Bacon's lead, attempted to base normative descriptions of true rationality on historical surveys of Western scientific thought. Before the nineteenth century this tradition remained predominantly programmatic, producing little significant historical research. But then, particularly in the writings of Whewell, Mach, and Duhem, philosophical concerns became a primary motive for creative activity in the history of science, and they have remained important since.

Both of these historiographic traditions, particularly when controlled by the textual-critical techniques of nineteenth-century German political history, produced occasional monuments of scholarship, which the contemporary historian ignores at his peril. But they simultaneously reinforced a concept of the field that has today been largely rejected by the nascent profession. The objective of these older histories of science was to clarify and deepen an understanding of *contemporary* scientific methods or concepts by displaying their evolution. Committed to such goals, the historian

characteristically chose a single established science or branch of science—
one whose status as sound knowledge could scarcely be doubted—and
described when, where, and how the elements that in his day constituted
its subject matter and presumptive method had come into being. Observations, laws, or theories which contemporary science had set aside as error or
irrelevancy were seldom considered unless they pointed a methodological
moral or explained a prolonged period of apparent sterility. Similar selective principles governed discussion of factors external to science. Religion,
seen as a hindrance, and technology, seen as an occasional prerequisite
to advance in instrumentation, were almost the only such factors which
received attention. The outcome of this approach has recently been brilliantly parodied by the philosopher Joseph Agassi.

Until the early nineteenth century, of course, characteristics very much
like these typified most historical writing. The romantics' passion for distant times and places had to combine with the scholarly standards of bibli-
cal criticism before even general historians could be brought to recognize
the interest and integrity of value systems other than their own. (The nineteenth century is, for example, the period when the Middle Ages were first
observed to have a history.) That transformation of sensibility which most
contemporary historians would suppose essential to their field was not,
however, at once reflected in the history of science. Though they agreed
about nothing else, both the romantic and the scientist-historian continued to view the development of science as a quasi-mechanical march of
the intellect, the successive surrender of nature's secrets to sound methods
skillfully deployed. Only in this century have historians of science gradually
learned to see their subject matter as something different from a chronology of accumulating positive achievement in a technical specialty defined
by hindsight. A number of factors contributed to this change.

Probably the most important was the influence, beginning in the late
nineteenth century, of the history of philosophy. In that field only the most
partisan could feel confident of his ability to distinguish positive knowledge
from error and superstition. Dealing with ideas that had since lost their
appeal, the historian could scarcely escape the force of an injunction which
Bertrand Russell later phrased succinctly: "In studying a philosopher, the
right attitude is neither reverence nor contempt, but first a kind of hypothetical sympathy, until it is possible to know what it feels like to believe
in his theories." That attitude toward past thinkers came to the history of
science from philosophy. Partly it was learned from men like Lange and
Cassirer who dealt historically with people or ideas that were also important
for scientific development. (Burtt's *Metaphysical Foundations of Modern
Physical Science* and Lovejoy's *Great Chain of Being* were, in this respect,
especially influential.) And partly it was learned from a small group of

neo-Kantian epistemologists, particularly Brunschvicg and Meyerson, whose search for quasi-absolute categories of thought in older scientific ideas produced brilliant genetic analyses of concepts which the main tradition in the history of science had misunderstood or dismissed.

These lessons were reinforced by another decisive event in the emergence of the contemporary profession. Almost a century after the Middle Ages had become important to the general historian, Pierre Duhem's search for the sources of modern science disclosed a tradition of medieval physical thought which, in contrast to Aristotle's physics, could not be denied an essential role in the transformation of physical theory that occurred in the seventeenth century. Too many of the elements of Galileo's physics and method were to be found there. But it was not possible, either, to assimilate it quite to Galileo's physics or to Newton's, leaving the structure of the so-called Scientific Revolution unchanged but extending it greatly in time. The essential novelties of seventeenth-century science would be understood only if medieval science were explored first on its own terms and then as the base from which the "new science" sprang. More than any other, that challenge has shaped the modern historiography of science. The writings which it has evoked since 1920, particularly those of E. J. Dijksterhuis, Anneliese Maier, and especially Alexandre Koyré, are the models which many contemporaries aim to emulate. In addition, the discover of medieval science and its Renaissance role has disclosed an area in which the history of science can and must be integrated with more traditional types of history. That task has barely begun, but the pioneering synthesis by Butterfield and the special studies by Panofsky and Frances Yates mark a path which will surely be broadened and followed.

A third factor in the formation of the modern historiography of science has been a repeated insistence that the student of scientific development concern himself with positive knowledge as a whole and that general histories of science replace histories of special sciences. Traceable as a program to Bacon, and more particularly to Comte, that demand scarcely influenced scholarly performance before the beginning of this century, when it was forcefully reiterated by the universally venerated Paul Tannery and then put to practice in the monumental researches of George Sarton. Subsequent experience has suggested that the sciences are not, in fact, all of a piece and that even the superhuman erudition required for a general history of science could scarcely tailor their joint evolution to a coherent narrative. But the attempt has been crucial, for it has highlighted the impossibility of attributing to the past the divisions of knowledge embodied in contemporary science curricula. Today, as historians increasingly turn back to the detailed investigation of indvidual branches of science, they study fields which actually existed in the periods that concern them, and they do so with an awareness of the state of other sciences at the time.

Still more recently, one other set of influences has begun to shape contemporary work in the history of science. Its result is an increased concern, deriving partly from general history and partly from German sociology and Marxist historiography, with the role of nonintellectual, particularly institutional and socioeconomic, factors in scientific development. Unlike the ones discussed above, however, these influences and the works responsive to them have to date scarcely been assimilated by the emerging profession. For all its novelties, the new historiography is still directed predominantly to the evolution of scientific ideas and of the tools (mathematical, observational, and experimental) through which these interact with each other and with nature. Its best practitioners have, like Koyré, usually minimized the importance of nonintellectual aspects of culture to the historical developments they consider. A few have acted as though the obtrusion of economic or institutional considerations into the history of science would be a denial of the integrity of science itself. As a result, there seems at times to be two distinct sorts of history of science, occasionally appearing between the same covers but rarely making firm or fruitful contact. The still dominant form, often called the "internal approach," is concerned with the substance of science as knowledge. Its newer rival, often called the "external approach," is concerned with the activities of scientists as a social group within a larger culture. Putting the two together is perhaps the greatest challenge now faced by the profession, and there are increasing signs of a response. Nevertheless, any survey of the field's present state must unfortunately still treat the two as virtually separate enterprises.

Internal History

What are the maxims of the new internal historiography? Insofar as possible (it is never entirely so, nor could history be written if it were), the historian should set aside the science that he knows. His science should be learned from the textbooks and journals of the period he studies, and he should master these and the indigenous traditions they display before grappling with innovators whose discoveries or inventions changed the direction of scientific advance. Dealing with innovators, the historian should try to think as they did. Recognizing that scientists are often famous for results they did not intend, he should ask what problems his subject worked at and how these became problems for him. Recognizing that a historic discovery is rarely quite the one attributed to its author in later textbooks (pedagogic goals inevitably transform a narrative), the historian should ask what his subject thought he had discovered and what he took the basis of that discovery to be. And in this process of reconstruction the historian should pay particular attention to his subject's apparent errors, not for their own sake but because they reveal far more of the mind at work than do the passages in which a scientist seems to record a result or an argument that modern science still retains.

For at least thirty years the attitudes which these maxims are designed to display have increasingly guided the best interpretive scholarship in the history of science, and it is with scholarship of that sort that this article is predominantly concerned. (There are other types, of course, though the distinction is not sharp, and much of the most worthwhile effort of historians of science is devoted to them. But this is not the place to consider work like that of, say, Needham, Neugebauer, and Thorndike, whose indispensable contribution has been to establish and make accessible texts and traditions previously known only through myth.) Nevertheless, the subject matter is immense; there have been few professional historians of science (in 1950 scarcely more than half a dozen in the United States); and their choice of topic has been far from random. [. . .]

External History

Attempts to set science in a cultural context which might enhance understanding both of its development and of its effects have taken three characteristic forms, of which the oldest is the study of scientific institutions. Bishop Sprat prepared his pioneering history of the Royal Society of London almost before that organization had received its first charter, and there have since been innumerable in-house histories of individual scientific societies. These books are, however, useful principally as source materials for the historian, and only in this century have students of scientific development started to make use of them. Simultaneously they have begun seriously to examine the other types of institutions, particularly educational, which may promote or inhibit scientific advance. As elsewhere in the history of science, most of the literature on institutions deals with the seventeenth century. The best of it is scattered through periodicals (the once standard book-length accounts are regrettably out-of-date) from which it can be retrieved, together with much else concerning the history of science, through the annual "Critical Bibliography" of the journal *Isis* and through the quarterly *Bulletin signalétique* of the Centre National de la Recherche Scientifique, Paris. Guerlac's classic study on the professionalization of French chemistry (Schofield's history of the Lunar Society, and a recent collaborative volume (Taton) on scientific education in France are among the very few works on eighteenth-century scientific institutions. For the nineteenth, only Cardwell's study of England, Dupree's of the United States, and Vucinich's of Russia begin to replace the fragmentary but immensely suggestive remarks scattered, often in footnotes, through the first volume of Merz's *History of European Thought in the Nineteenth Century*.

Intellectual historians have frequently considered the impact of science on various aspects of Western thought, particularly during the seventeenth and eighteenth centuries. For the period since 1700, however, these studies are peculiarly unsatisfying insofar as they aim to demonstrate the influence, and not merely the prestige, of science. The name of a Bacon, a Newton, or a Darwin is a potent symbol: there are many reasons to invoke it besides recording a substantive debt. And the recognition of isolated conceptual parallels, for example, between the forces that keep a planet in its orbit and the system of checks and balances in the U.S. constitution, more often demonstrates interpretive ingenuity than the influence of science on other areas of life. No doubt scientific concepts, particularly those of broad scope, do help to change extra-scientific ideas. But the analysis of their role in producing this kind of change demands immersion in the literature of science. The older historiography of science does not, by its nature, supply what is needed, and the new historiography is too recent and its products too fragmentary to have had much effect. Though the gap seems small, there is no chasm that more needs bridging than that between the historian of ideas and the historian of science. Fortunately there are a few works to point the way. Among the more recent are Nicolson's pioneering studies of science in seventeenth- and eighteenth-century literature, Westfall's discussion of natural religion, Gillispie's chapter on science in the Enlightenment, and Roger's monumental survey of the role of the life sciences in eighteenth-century French thought.

The concern with institutions and that with ideas merge naturally in a third approach to scientific development. This is the study of science in a geographical area too small to permit concentration on the evolution of any particular technical specialty but sufficiently homogeneous to enhance an understanding of the social role and setting of science. Of all the types of external history, this is the newest and most revealing, for it calls forth the widest range of historical and sociological experience and skill. The small but rapidly growing literature on science in America (Dupree, Hindle, Shryock) is a prominent example of this approach, and there is promise that current studies of science in the French Revolution may yield similar illumination. Merz, Lilley, and Ben-David point to aspects of the nineteenth century on which much similar effort must be expended. The topic which has, however, evoked the greatest activity and attention is the development of science in seventeenth-century England. Because it has become the center of vociferous debate both about the origin of modern science and about the nature of the history of science, this literature is an appropriate focus for separate discussion. Here it stands for a type of research: the problems it presents will provide perspective on the relations between the internal and external approaches to the history of science.

The Merton Thesis

The most visible issue in the debate about seventeenth-century science has been the so-called Merton thesis, really two overlapping theses with distinguishable sources. Both aim ultimately to account for the special productiveness of seventeenth-century science by correlating its novel goals and values—summarized in the program of Bacon and his followers—with other aspects of contemporary society. The first, which owes something to Marxist historiography, emphasizes the extent to which the Baconians hoped to learn from the practical arts and in turn to make science useful. Repeatedly they studied the techniques of contemporary craftsmen—glassmakers, metallurgists, mariners, and the like—and many also devoted at least a portion of their attention to pressing practical problems of the day, for example, those of navigation, land drainage, and deforestation. The new problems, data, and methods fostered by these novel concerns are, Merton supposes, a principal reason for the substantive transformation experienced by a number of sciences during the seventeenth century. The second thesis points to the same novelties of the period but looks to Puritanism as their primary stimulant. (There need be no conflict. Max Weber, whose pioneering suggestion Merton was investigating, had argued that Puritanism helped to legitimize a concern with technology and the useful arts.) The values of settled Puritan communities—for example, an emphasis upon justification through works and on direct communion with God through nature—are said to have fostered both the concern with science and the empirical, instrumental, and utilitarian tone which characterized it during the seventeenth century.

Both of these theses have since been extended and also attacked with vehemence, but no consensus has emerged. (An important confrontation, centering on papers by Hall and de Santillana, appears in the symposium of the Institute for the History of Science edited by Clagett; Zilsel's pioneering paper on William Gilbert can be found in the collection of relevant articles from the *Journal of the History of Ideas* edited by Wiener and Noland. Most of the rest of the literature, which is voluminous, can be traced through the footnotes in a recently published controversy over the work of Christopher Hill.) In this literature the most persistent criticisms are those directed to Merton's definition and application of the label "Puritan," and it now seems clear that no term so narrowly doctrinal in its implications will serve. Difficulties of this sort can surely be eliminated, however, for the Baconian ideology was neither restricted to scientists nor uniformly spread through all classes and areas of Europe. Merton's label may be inadequate, but there is no doubt that the phenomenon he describes did exist. The more significant arguments against his position are the residual ones which derive from the recent transformation in the history of science. Merton's image of the Scientific Revolution, though longstanding, was rapidly being discredited as he wrote, particularly in the role it attributed to the Baconian movement.

Participants in the older historiographic tradition did sometimes declare that science as they conceived it owed nothing to economic values or religious doctrine. Nevertheless, Merton's emphases on the importance of manual work, experimentation, and the direct confrontation with nature were familiar and congenial to them. The new generation of historians, in contrast, claims to have shown that the radical sixteenth- and seventeenth-century revisions of astronomy, mathematics, mechanics, and even optics owed very little to new instruments, experiments, or observations. Galileo's primary method, they argue, was the traditional thought experiment of scholastic science brought to a new perfection. Bacon's naive and ambitious program was an impotent delusion from the start. The attempts to be useful failed consistently; the mountains of data provided by new instruments were of little assistance in the transformation of existing scientific theory. If cultural novelties are required to explain why men like Galileo, Descartes, and Newton were suddenly able to see well-known phenomena in a new way, those novelties are predominantly intellectual and include Renaissance Neoplatonism, the revival of ancient atomism, and the rediscovery of Archimedes. Such intellectual currents were, however, at least as prevalent and productive in Roman Catholic Italy and France as in Puritan circles in Britain or Holland. And nowhere in Europe, where these currents were stronger among courtiers than among craftsmen, do they display a significant debt to technology. If Merton were right, the new image of the Scientific Revolution would apparently be wrong.

In their more detailed and careful versions, which include essential qualification, these arguments are entirely convincing, up to a point. The men who transformed scientific theory during the seventeenth century sometimes talked like Baconians, but it has yet to be shown that the ideology which a number of them embraced had a major effect, substantive or methodological, on their central contributions to science. Those contributions are best understood as the result of the internal evolution of a cluster of fields which, during the sixteenth and seventeenth centuries, were pursued with renewed vigor and in a new intellectual milieu. That point, however, can be relevant only to the revision of the Merton thesis, not to its rejection. One aspect of the ferment which historians have regularly labeled "the Scientific Revolution" was a radical programmatic movement centering in England and the Low Countries, though it was also visible for a time in Italy and France. That movement, which even the present form of Merton's argument does make more comprehensible, drastically altered the appeal, the locus, and the nature of much scientific research during the seventeenth century, and the changes have been permanent. Very likely, as contemporary historians argue, none of these novel features played a large role in transforming scientific concepts during the seventeenth century, but historians must learn to deal with them nonetheless. Perhaps the following suggestions, whose more general import will be considered in the next section, may prove helpful.

Omitting the biological sciences, for which close ties to medical crafts and institutions dictate a more complex developmental pattern, the main branches of science transformed during the sixteenth and seventeenth centuries were astronomy, mathematics, mechanics, and optics. It is their development which makes the Scientific Revolution seem a revolution in concepts. Significantly, however, this cluster of fields consists exclusively of classical sciences. Highly developed in antiquity, they found a place in the medieval university curriculum where several of them were significantly further developed. Their seventeenth-century metamorphosis, in which university-based men continued to play a significant role, can reasonably be portrayed as primarily an extension of an ancient and medieval tradition developing in a new conceptual environment. Only occasionally need one have recourse to the Baconian programmatic movement when explaining the transformation of these fields.

By the seventeenth century, however, these were not the only areas of intense scientific activity, and the others—among them the study of electricity and magnetism, of chemistry, and of thermal phenomena—display a different pattern. As sciences, as fields to be scrutinized systematically for an increased understanding of nature, they were all novelties during the Scientific Revolution. Their main roots were not in the learned university tradition but often in the established crafts, and they were all critically dependent both on the new program of experimentation and on the new instrumentation which craftsmen often helped to introduce. Except occasionally in medical schools, they rarely found a place in universities before the nineteenth century, and they were meanwhile pursued by amateurs loosely clustered around the new scientific societies that were the institutional manifestation of the Scientific Revolution. Obviously these are the fields, together with the new mode of practice they represent, which a revised Merton thesis may help us understand. Unlike that in the classical sciences, research in these fields added little to man's understanding of nature during the seventeenth century, a fact which has made them easy to ignore when evaluating Merton's viewpoint. But the achievements of the late eighteenth and of the nineteenth centuries will not be comprehensible until they are taken fully into account. The Baconian program, if initially barren of conceptual fruits, nevertheless inaugurated a number of the major modern sciences.

Internal and External History

Because they underscore distinctions between the earlier and later stages of a science's evolution, these remarks about the Merton thesis illustrate aspects of scientific development recently discussed in a more general way by Kuhn. Early in the development of a new field, he suggests, social needs and values are a major determinant of the problems on which its practitioners concentrate. Also during this period, the concepts they deploy in solving problems are extensively conditioned by contemporary common sense, by a prevailing philosophical tradition, or by the most prestigious contemporary sciences. The new fields which emerged in the seventeenth century and a number of the modern social sciences provide examples. Kuhn argues, however, that the later evolution of a technical specialty is significantly different in ways at least foreshadowed by the development of the classical sciences during the Scientific Revolution. The practitioners of a mature science are men trained in a sophisticated body of traditional theory and of instrumental, mathematical, and verbal technique. As a result, they constitute a special subculture, one whose members are the exclusive audience for, and judges of, each other's work. The problems on which such specialists work are no longer presented by the external society but by an internal challenge to increase the scope and precision of the fit between existing theory and nature. And the concepts used to resolve these problems are normally close relatives of those supplied by prior training for the specialty. In short, compared with other professional and creative pursuits, the practitioners of a mature science are effectively insulated from the cultural milieu in which they live their extraprofessional lives.

That quite special, though still incomplete, insulation is the presumptive reason why the internal approach to the history of science, conceived as autonomous and self-contained, has seemed so nearly successful. To an extent unparalleled in other fields, the development of an individual technical specialty can be understood without going beyond the literature of that specialty and a few of its near neighbors. Only occasionally need the historian take note of a particular concept, problem, or technique which entered the field from outside. Nevertheless, the apparent autonomy of the internal approach is misleading in essentials, and the passion sometimes expended in its defense has obscured important problems. The insulation of a mature scientific community suggested by Kuhn's analysis is an insulation primarily with respect to concepts and secondarily with respect to problem structure. There are, however, other aspects of scientific advance, such as its timing. These do depend critically on the factors emphasized by the external approach to scientific development. Particularly when the sciences are viewed as an interacting group rather than as a collection of specialties, the cumulative effects of external factors can be decisive.

Both the attraction of science as a career and the differential appeal of different fields are, for example, significantly conditioned by factors external to science. Furthermore, since progress in one field is sometimes dependent on the prior development of another, differential growth rates may affect an entire evolutionary pattern. Similar considerations, as noted above, play a major role in the inauguration and initial form of new sciences. In addition, a new technology or some other change in the conditions of society may selectively alter the felt importance of a specialty's problems or even create new ones for it. By doing so they may sometimes accelerate the discovery of areas in which an established theory ought to work but does not, thereby hastening its rejection and replacement by a new one. Occasionally, they may even shape the substance of that new theory by ensuring that the crisis to which it responds occurs in one problem area rather than another. Or again, through the crucial intermediary of institutional reform, external conditions may create new channels of communication between previously disparate specialties, thus fostering cross-fertilization which would otherwise have been absent or long delayed.

There are numerous other ways, including direct subsidy, in which the larger culture impinges on scientific development, but the preceding sketch should sufficiently display a direction in which the history of science must now develop. Though the internal and external approaches to the history of science have a sort of natural autonomy, they are, in fact, complementary concerns. Until they are practiced as such, each drawing from the other, important aspects of scientific development are unlikely to be understood. That mode of practice has hardly yet begun, as the response to the Merton thesis indicates, but perhaps the analytic categories it demands are becoming clear.

The Relevance of the History of Science

Turning in conclusion to the question about which judgments must be the most personal of all, one may ask about the potential harvest to be reaped from the work of this new profession. First and foremost will be more and better histories of science. Like any other scholarly discipline, the field's primary responsibility must be to itself. Increasing signs of its selective impact on other enterprises may, however, justify brief analysis.

Among the areas to which the history of science relates, the one least likely to be significantly affected is scientific research itself. Advocates of the history of science have occasionally described their field as a rich repository of forgotten ideas and methods, a few of which might well dissolve contemporary scientific dilemmas. When a new concept or theory is successfully deployed in a science, some previously ignored precedent is usually discovered in the earlier literature of the field. It is natural to wonder whether attention to history might not have accelerated the innovation. Almost certainly, however, the answer is no. The quantity of material to be searched, the absence of appropriate indexing categories, and the subtle but usually vast differences between the anticipation and the effective innovation, all combine to suggest that reinvention rather than rediscovery will remain the most efficient source of scientific novelty.

The more likely effects of the history of science on the fields it chronicles are indirect, providing increased understanding of the scientific enterprise itself. Though a clearer grasp of the nature of scientific development is unlikely to resolve particular puzzles of research, it may well stimulate reconsideration of such matters as science education, administration, and policy. Probably, however, the implicit insights which historical study can produce will first need to be made explicit by the intervention of other disciplines, of which three now seem particularly likely to be effective.

Though the intrusion still evokes more heat than light, the philosophy of science is today the field in which the impact of the history of science is most apparent. Feyerabend, Hanson, Hesse, and Kuhn have all recently insisted on the in appropriateness of the traditional philosopher's ideal image of science, and in search of an alternative they have all drawn heavily from history. Following directions pointed by the classic statements of Norman Campbell and Karl Popper (and sometimes also significantly influenced by Ludwig Wittgenstein), they have at least raised problems that the philosophy of science is no longer likely to ignore. The resolution of those problems is for the future, perhaps for the indefinitely distant future. There is as yet no developed and matured "new philosophy" of science. But already the questioning of older stereotypes, mostly positivistic, is proving a stimulus and release to some practitioners of those newer sciences which have most depended upon explicit canons of scientific method in their search for professional identity.

A second field in which the history of science is likely to have increasing effect is the sociology of science. Ultimately neither the concerns nor the techniques of that field need be historical. But in the present underdeveloped state of their specialty, sociologists of science can well learn from history something about the shape of the enterprise they investigate. The recent writings of Ben-David, Hagstrom, Merton, and others give evidence that they are doing so. Very likely it will be through sociology that the history of science has its primary impact on science policy and administration.

Closely related to the sociology of science (perhaps equivalent to it if the two are properly construed) is a field that, though it scarcely yet exists, is widely described as "the science of science." Its goal, in the words of its leading exponent, Derek Price, is nothing less than "the theoretic analysis of the structure and behavior of science itself," and its techniques are an eclectic combination of the historian's, the sociologist's, and the econometrician's. No one can yet guess to what extent that goal is attainable, but any progress toward it will inevitably and immediately enhance the significance both to social scientists and to society of continuing scholarship in the history of science.

Bibliography

Other relevant material may be found in the biographies of Koyré and Sarton.

Agassi, Joseph. 1963. *Towards an Historiography of Science*. History and Theory, vol. 2. The Hague: Mouton.

Ben-David, Joseph. 1960. "Scientific Productivity and Academic Organization in Nineteenth-century Medicine," *American Sociological Review* 25: 828–43.

Boas, Marie. 1958. *Robert Boyle and Seventeenth-Century Chemistry*. Cambridge: Cambridge University Press.

Boyer, Carl B. 1949. *The Concepts of the Calculus: A Critical and Historical Discussion of the Derivative and the Integral*. New York: Hafner. A paperback edition was published in 1959 by Dover as *The History of the Calculus and Its Conceptual Development*.

Butterfield, Herbert. 1957. *The Origins of Modern Science, 1300–1800*. 2d. ed., rev. New York: Macmillan. A paperback edition was published in 1962 by Collier.

Cardwell, Donald S. L. 1957. *The Organisation of Science in England: A Retrospect*. Melbourne and London: Heinemann.

Clagett, Marshall. 1959. *The Science of Mechanics in the Middle Ages*. Madison: University of Wisconsin Press.

Cohen, I. Bernard. 1956. *Franklin and Newton: An Inquiry into Speculative Newtonian Experimental Science and Franklin's Work in Electricity as an Example Thereof*. American Philosophical Society. Memoirs, vol. 43. Philadelphia: The Society.

Costabel, Pierre. 1960. *Leibniz et la dynamique: Les textes de 1692*. Paris: Hermann.

Crosland, Maurice. 1963. "The Development of Chemistry in the Eighteenth Century." *Studies on Voltaire and the Eighteenth Century* 24: 369–441.

Daumas, Maurice. 1955. *Lavoisier: Théoricien et expérimentateur*. Paris: Presses Universitaires de France.

Dijksterhuis, Edward J. 1961. *The Mechanization of the World Picture*. Oxford: Clarendon. First published in Dutch in 1950.

Dugas, René. 1955. *A History of Mechanics*. Neuchâtel: Editions du Griffon; New York: Central Book. First published in French in 1950.

Duhem, Pierre. 1906–13. *Études sur Léonard de Vinci*. 3 vols. Paris: Hermann.

Dupree, A. Hunter. 1957. *Science in the Federal Government: A History of Policies and Activities to 1940*. Cambridge, Mass.: Belknap.

——. 1959. *Asa Gray: 1810–1888*. Cambridge, Mass.: Harvard University Press.

Feyerabend, P. K. 1962. "Explanation, Reduction and Empiricism." In Herbert Feigl and Grover Maxwell, eds., *Scientific Explanation, Space, and Time*, pp. 28–97. Minnesota Studies in the Philosophy of Science, vol. 3. Minneapolis: University of Minnesota Press.

Gillispie, Charles C. 1960. *The Edge of Objectivity: An Essay in the History of Scientific Ideas*. Princeton, N.J.: Princeton University Press.

Guerlac, Henry. 1959. "Some French Antecedents of the Chemical Revolution." *Chymia* 5: 73–112.

——. 1961. *Lavoisier; the Crucial Year; The Background and Origin of His First Experiments on Combustion in 1772.* Ithaca, N.Y.: Cornell University Press.

Hagstrom, Warren O. 1965. *The Scientific Community.* New York: Basic Books.

Hanson, Norwood R. 1961. *Patterns of Discovery: An Inquiry into the Conceptual Foundations of Science.* Cambridge: Cambridge University Press.

Hesse, Mary B. 1963. *Models and Analogies in Science.* London: Sheed & Ward.

Hill, Christopher. 1965. "Debate: Puritanism, Capitalism and the Scientific Revolution." *Past and Present,* no. 29: 68–97. Articles relevant to the debate may also be found in numbers 28, 31, 32, and 33.

Hindle, Brooke. 1956. *The Pursuit of Science in Revolutionary America, 1735–1789.* Chapel Hill: University of North Carolina Press.

Institute for the History of Science, University of Wisconsin, 1957. 1959. *Critical Problems in the History of Science: Proceedings.* Edited by Marshall Clagett, Madison: University of Wisconsin Press.

Jammer, Max. 1961. *Concepts of Mass in Classical and Modern Physics.* Cambridge, Mass.: Harvard University Press.

Journal of the History of Ideas. 1957. *Roots of Scientific Thought: A Cultural Perspective.* Edited by Philip P. Wiener and Aaron Noland. New York; Basic Books. Selections from the first 18 volumes of the *Journal.*

Koyré, Alexandre. 1939. *Études galiléennes.* 3 vols. Actualités scientifiques et industrielles, nos. 852, 853, and 854. Paris: Hermann. Volume 1: *À l'aube de la science classique.* Volume 2: *La loi de la chute des corps: Descartes et Galilée.* Volume 3: *Galilée et la loi d'inertie.*

—— 1961. *La révolution astronomique: Copernic, Kepler, Borelli.* Paris: Hermann.

Kuhn, Thomas S. 1962. *The Structure of Scientific Revolutions.* Chicago: University of Chicago Press. A paperback edition was published in 1964.

Lilley, S. 1949. "Social Aspects of the History of Science." *Archives internationales d'histoire des sciences* 2: 376–443.

Maier, Anneliese. 1949–58. *Studien zur Naturphilosophie der Spätscholastik.* 5 vols. Rome: Edizioni de "Storia e Letteratura."

Merton, Robert K. 1967. *Science, Technology and Society in Seventeenth-Century England.* New York: Fertig.

——. 1957. "Priorities in Scientific Discovery: A Chapter in the Sociology of Science." *American Sociological Review* 22: 635–59.

Metzger, Hélène. 1930. *Newton, Stahl, Boerhaave et la doctrine chimique.* Paris: Alcan.

Meyerson, Émile. 1964. *Identity and Reality.* London: Allen & Unwin. First published in French in 1908.

Michel, Paul-Henri. 1950. *De Pythagore à Euclide.* Paris: Edition "Les Belles Lettres."

Needham, Joseph. 1954–1965. *Science and Civilisation in China.* 4 vols. Cambridge: Cambridge University Press.

Neugebauer, Otto. 1957. *The Exact Sciences in Antiquity.* 2d ed. Providence. R.I.: Brown University Press. A paperback edition was published in 1962 by Harper.

Nicolson, Marjorie H. 1960. *The Breaking of the Circle: Studies in the Effect of the "New Science" upon Seventeenth-Century Poetry.* Rev. ed. New York: Columbia University Press. A paperback edition was published in 1962.

O'Malley, Charles D. 1964. *Andreas Vesalius of Brussels, 1514–1564.* Berkeley and Los Angeles: University of California Press.

Panofsky, Erwin. 1954. *Galileo as a Critic of the Arts.* The Hague: Nijhoff.

Partington, James R. 1962–. *A History of Chemistry.* New York: St. Martins. Volumes 2–4 were published from 1962 to 1964; Volume 1 [was in preparation as of the publication of Kuhn's essay, part I was published in 1970, part II was never published].

Price, Derek J. de Solla. 1966. "The Science of Scientists." *Medical Opinion and Review* 1:81–97.

Roger, Jacques. 1963. *Les sciences de la vie dans la pensée française du XVIII siècle: La génération des animaux de Descartes à* l'Encyclopédie. Paris: Colin.

Sarton, George. 1927–48. *Introduction to the History of Science.* 3 vols. Baltimore: Williams & Wilkins.

Schofield, Robert E. 1963. *The Lunar Society of Birmingham: A Social History of Provincial Science and Industry in Eighteenth-Century England.* Oxford: Clarendon.

Shryock, Richard H. 1947. *The Development of Modern Medicine.* 2d ed. New York: Knopf.

Singer, Charles J. 1922. *The Discovery of the Circulation of the Blood.* London: Bell.

Stocking, George W. Jr. 1966. "Franz Boas and the Culture Concept in Historical Perspective." *American Anthropologist* New Series 68:867–82.

Taton, René, ed. 1964. *Enseignement et diffusion des sciences en France au XVIIIᵉ siècle.* Paris: Hermann.

Thorndike, Lynn. 1959–64. *A History of Magic and Experimental Science. 8 vols.* New York: Columbia University Press.

Truesdell, Clifford A. 1960. *The Rational Mechanics of Flexible or Elastic Bodies 1638–1788: Introduction to Leonhardi Euleri* Opera omnia *Vol. X et XI seriei secundae.* Leonhardi Euleri Opera omnia, Ser. 2, Vol. 11, part 2. Turin: Fussli.

Vucinich, Alexander S. 1963. *Science in Russian Culture.* Volume 1: *A History to 1860.* Stanford University Press.

Westfall, Richard S. 1958. *Science and Religion in Seventeenth Century England.* New Haven: Yale University Press.

Whittaker, Edmund, 1951–53. *A History of the Theories of Aether and Electricity.* 2 vols. London: Nelson. Volume 1: *The Classical Theories.* Volume 2: *The Modern Theories, 1900–1926.* Volume 1 is a revised edition of *A History of the Theories of Aether and Electricity from the Age of Descartes to the Close of the Nineteenth Century,* published in 1910. A paperback edition was published in 1960 by Harper.

Yates, Frances A. 1964. *Giordano Bruno and the Hermetic Tradition.* Chicago: University of Chicago Press.

CHAPTER 10

Scientific Research under a Historical Microscope

MARTIN RUDWICK

Pages 3–16 of *The Great Devonian Controversy.* Chicago: The University of Chicago Press, 1985.

1.1 Introducing the Devonian Controversy

In the early nineteenth century, geology was a new, exciting, and fashionable science.[1] It was experiencing its first and greatest boom in conceptual innovation, empirical expansion, and public approval and interest. It attracted some of the most talented in the scientific world, particularly those with a taste for travel and the outdoor life rather than for mathematics. But those with lesser talents or more limited opportunities could also hope to gain recognition and respect, for it was felt that any worthwhile grand conclusions in geology had to rest on the foundations of local details that required much time and patience to acquire. The empiricism esteemed by geologists bound leaders and locals into a symbiotic partnership that was seen as a model of the ideal community of science.

The most spectacular achievement of this fashionable new science was its disclosure of the vast history of the earth and of life on its surface. Human beings were discovered to be the merest newcomers at the end of a long saga of prehuman life. Bizarre extinct monsters were pieced together from scattered bones. But while that required all the skills of a trained anatomist, even the country lady or the schoolboy could hope to find traces of lesser denizens of "former worlds." Ammonites and trilobites might be less spectacular than mammoths and mosasaurs, but they were equally strange, and anyway more appropriate in a drawing-room cabinet. This new and exciting vista into the history of the earth was no fantasy devised by skeptical philosophies to confound the faithful; it was vouched for by leading geologists of impeccable piety as a disclosure that ought to evoke an enlarged sense of awe at the scale and diversity of the created world.

Geology was also popular and important from a quite different point of view. The growth of new forms of industrial production, though hardly yet perceived as the "industrial revolution" it was to become in retrospect, was transforming parts of Britain and spreading to parts of the Continent too. Heavy industry was increasingly dependent on expanding supplies of coal and metal ores; mines were meeting the demand by driving ever deeper below the surface and further from the shallow workings that had been sufficient in previous centuries. In doing so, the owners and managers of mines found their traditional empirical rules of operation increasingly inadequate. The extension of old workings and the discovery of new ones demanded new methods of prediction and exploration; geology was looked to as a potential source of methods that would be soundly based on scientific principles. Geology was therefore esteemed as much for its prospective contribution to the economy as for its enlarged view of the natural world.

The popular interest in geology, in both its theoretical and practical forms, was rooted securely in foundations laid by its leading practitioners. They claimed that the history of the earth and its inhabitants could be deciphered by learning to read "the record of the rocks," and that the succession of layers or strata in the earth's crust were Nature's own historical documents. Imperfect the record might be, but deceptive it was not. Correctly interpreted, much of the history seemed plain and straightforward. Each "formation" of strata was like a chapter in the volume of Nature's history; the sequence of chapters was clear, and each was characterized more or less by distinctive types of rock and by distinctive fossils. By hammering along a line of coastal cliffs or by going from quarry to quarry across country, the geologist could plot local sequences of strata and collect their fossils. By combining and coordinating such local sequences, the broad outlines of a global history could be discerned with ever increasing clarity, and predictions could be made about the likely extension of economically valuable deposits such as coal.

At the lower end of the whole pile of strata, however, the rocks that recorded the earlier chapters in the earth's history proved less easy to decipher. The geological record became ever more puzzling, the formations apparently chaotic, and their fossils rare and obscure. Yet it was here if anywhere that scientific knowledge might hope to reach back to some record of the creation of life itself; it was also here that many deposits of coal, and most of the valuable metal ores, were known to be situated.

By the 1830's this problem was attracting some of the best geologists of that generation. They began to unravel the sequence of these ancient strata, to discern what they believed to be the earliest history of life, and to infer the principles on which mineral exploration might be based. But in the midst of this work a controversy arose that threatened the stability and success of the whole enterprise. The Devonian controversy began as a dispute about the identification and correct sequence of strata in the county of Devon in southwest England. From the start, however, its implications were seen to be international, and indeed global. As it developed, the controversy drew in most of the leading geologists in Britain and increasingly those in other countries too. It was only resolved—to the satisfaction of most if not all of the participants—when it was exported from Britain to France, Belgium, and the German states, and finally to Russia, North America, and the rest of the world. The vast expansion of research in the century and a half since the controversy subsided has been regarded by geologists as confirming the validity of that solution in ways that would have delighted, but not surprised, those who first worked it out.

The Devonian controversy exposed the procedural roots of geological practice and subjected them to more probing scrutiny than ever before. The successful resolution of the controversy endowed geological practice with a new confidence in the reliability of its conclusions—a confidence it retains to the present day. This is the episode in scientific research practice that in this book will be put under a historical microscope. The remainder of this chapter explains programmatically what is meant by that metaphor. Readers who are more interested in the story itself than in the principles that the storytelling entails can turn to chapter two without further ado.

1.2 Watching the Natives at Work

Nearly twenty years ago Sir Peter Medawar urged those interested in understanding science to study in detail "what scientists *do*" in their research.[2] As a distinguished practicing scientist, he himself was well aware that what scientists *say* about their activity can never be taken at face value: their accounts are invaluable as source materials, but not necessarily reliable as interpretative conclusions. He might have added that what philosophers say should be handled even more gingerly, since at least until recently they seemed to be interested only in prescribing what scientists ought to be doing, and they showed indifference if not hostility to any truly empirical study of scientific activity. Medawar claimed in fact that "only unstudied evidence will do—and that means listening at a keyhole." The metaphor conveyed vividly the sense of intimate knowledge that he realized was needed, but it also implied a degree of observational and interpretative distance—not to mention improper intrusion—which was not really intrinsic to his suggestion.

Those analysts of science who have followed Medawar's suggestion, whether knowingly or not, have rightly chosen to open the laboratory door and to establish themselves as observers on the inside. They have done so in two distinct styles, each with its own limitations. Some have studied modern scientific research by being accepted as participant-observers within a laboratory. They have used the perspective and even the techniques of the anthropologist, treating the scientists as exotic natives with strange and puzzling customs. These ethnographers or microsociologists of science have given some detailed and illuminating accounts of routine procedures in scientific research. But with few exceptions they have described the research in a static manner, failing to show how the procedures are used in a *temporal process* to develop some new scientific conclusion. Furthermore, they often show an extreme skepticism—or at least an extreme agnosticism—about the status of the knowledge the scientists claim to be producing. In minimizing if not discounting its reference to any "real" external world of nature, their accounts of science open up a gulf in self-understanding between themselves and the scientists they observe—a gulf which surely no modern anthropologist would find tolerable in the interpretation of exotic cultures.[3]

Historians of science, on the other hand, have not in recent decades been neglectful of the "natives' point of view." Those who have studied past scientific knowledge in the making have not needed to listen at a keyhole or to peep through it: rather they have looked over the shoulder of their chosen subject and retraced the course of the research from laboratory notebooks, letters, and other texts.[4] But most historical studies of what scientists *did*, in some specific setting in the past, are focused on one person. The biographical genre rightly remains popular with historians of science, although it has become unfashionable among other historians. But even the best biographies are bound to distort to some extent the processes of scientific activity. By focusing on the work of one individual, they inevitably give less than adequate attention to the complex web of social and cognitive interactions that bind even the most distinguished or reclusive scientist into his or her immediate network of colleagues, in collaboration or rivalry or both.

What are needed, for a fuller understanding of the processes by which scientific knowledge is shaped, are empirical studies of science in the making—whether in the past or the present is of lesser consequence—which focus not on one individual scientist but on a specific scientific *problem* that brought together some *group* of individuals in an interacting network of exchange. Such studies need to pay attention to the role of all the participants, however minor their contributions may seem to be; to follow the dynamics of interaction with equal attention to exchanges both public and private, formal and informal, ritualized and spontaneous; and above all, by something akin to Gilbert Ryle's "thick description," to try to discern the *meaning*, for the participants themselves, of the social drama that is scientific research. For as Clifford Geertz commented of anthropological understanding, "the trick is not to get yourself into some inner correspondence of spirit with your informants . . . [but] to figure out what the devil they think they are up to."[5] This book is a contribution to that empirical study of scientific research, seen as the work of an interacting social group. It is an attempt in the first place to look over the shoulders of nineteenth-century geologists as they wrestled with the problems raised by the Devonian controversy, and to figure out what the devil they thought they were up to.

1.3 Applying a Historical Microscope

The historian of science who studies any period before the twentieth century is deprived of one of the sociologist's classic forms of evidence, the interview. But historians of science who have had to deal with its nearest equivalent, namely the written or taped reminiscences of scientists who are no longer alive, generally feel that such deprivation is less than disastrous. For the recollections of scientists—whether spoken or written; prompted or spontaneous—are notoriously and systematically unreliable, even when they are made only weeks or months after the events being recalled, let alone years or decades later. This is certainly not a matter of dishonesty or even primarily of imperfect memory. It is an inevitable consequence of the ever-changing *contexts* of meaning and use within which the events are retrospectively set, even by those with the most reliable memories. Charles Darwin's well-known statement that as a young man he had collected facts and worked "without any theory," referring to a period when he had actually compiled some of the most creatively theoretical notebooks known in the history of science, is but one outstanding example of a pervasive problem.[6] The historian of science is therefore on safer ground if able, by the nature of the record, to focus attention on documentary material that is strictly contemporary with the events being reconstructed and analyzed.

The character of that material determines what may be termed the "graininess" or degree of temporal "resolution" that can be reached in any given historical study. As the study of human anatomy can benefit from every degree of optical resolution, from naked-eye dissection to electronmicrography, so the empirical study of scientific activity needs research at every degree of temporal graininess. At one extreme is the historian's study of the *longue durée* of scientific practice over the centuries—a study, it should be said, that has hardly yet begun. At the other extreme are such disparate studies as the experimental psychologist's analysis of scientific thinking in a task measured in minutes, and the ethnomethodologist's analysis of scientists' conversations at the laboratory bench. Between those extremes lies what is perhaps the most promising level of description and analysis to which the historian can hope to contribute: namely, the level of reconstruction plotted neither in centuries and decades, nor in hours and minutes, but in years, months, and weeks, and—with a bit of luck—sometimes even in days.[7] This is, as it were, the high-quality "light microscope" of the analysis of scientific research practice. It shows somewhat less detail than the "electron microscope" of still more short-term or fine-grained studies; but what it does show is less confusing and much easier to relate to the larger-scale or long-term features that can be seen with the "dissecting microscope" and "naked-eye" studies of conventional historical analysis. At this degree of relatively fine-grained resolution, the historian can exploit to the full the rich and varied evidence that is sometimes available.

The Devonian controversy is an exceptionally favorable case for the application of this kind of historical microscope. For various contingent historical reasons, the period in which it took place is unusually well endowed with source materials. "Is it not true," Susan Cannon asked rhetorically some twenty years ago, "that for the middle of the nineteenth century we possess the most complete documentation, for selected individuals, not only that ever has existed *but that ever will exist?*"[8] The records of the Devonian controversy show that this perhaps surprising claim may be as valid for selected scientific *problems* as for some outstanding individuals. The contemporary evidence, as for much other scientific research in the past two centuries, ranges from published books and articles, through participants' accounts of meetings and public discussions, to the early drafts of scientific papers, exchanges of private letters, and finally the private notebooks of individuals. That spectrum runs broadly from public to private, but no category of material has any special privileged status above any other. All are needed for any adequate reconstruction of the process of scientific research.

The records of the Devonian controversy are not perfect; there are some important and tantalizing gaps, for example, in what are generally quite complete series of letters exchanged between certain pairs of participants. There is also a systematic bias in the documentation, in that the fullest records are those of the major participants, whereas the notebooks and correspondence of less important figures are generally underrepresented. Nevertheless, the historical traces of the Devonian controversy are so rich that the historian can be fairly confident of having reached the point of diminishing returns, where even the discovery of another bundle of letters or notebooks in someone's attic would be unlikely to cause any major alteration in the reconstruction of the whole controversy.[9]

The dramatis personae in any scientific episode are not, of course, known in advance to the historian. Nevertheless, the records themselves can be made to disclose who, in the opinion of contemporaries, were the most significant actors. The sociologist, investigating some analogous problem in modern science, can detect and map the network of those most intensely involved by using a technique of "snowball sampling," asking each inform ant to name the others most relevant to the debate, whether as supporters or as opponents.[10] In much the same way the historian can let the documents themselves, and especially the correspondence, lead to a snowballing reconstitution of a cast list that is likewise founded on the actors' own perceptions. That is what has been possible in the case of the Devonian controversy. Furthermore, the historian, unlike the overprudent sociologist, has no need to conceal historical personages behind the obstructive and often ineffectual veil of pseudonymity.[11] The historical natives can be themselves, in all their marvelous particularity, warts and all.

1.4 Research as Skilled Craftsmanship

It has been more than twenty years since Thomas Kuhn's *The Structure of Scientific Revolutions* first made widely known and readably explicit much that reflective practicing scientists had long known intuitively in their bones.[12] Leaving aside Kuhn's striking but more problematic claims about "revolutionary science" and the incommensurability of paradigms, much in his account of "normal science" seemed as descriptively correct and perceptive to practicing scientists as it seemed alarming and threatening to prescriptively inclined or moralizing philosophers. As Kuhn emphasized, the ordinary business of scientific research is carried on within a shared or collective framework of methodological assumptions, heuristic maxims, routine procedures, observational and experimental standards, criteria of interpretative judgment, and much else besides. Even before Kuhn's book was published, Michael Polanyi, drawing on his long, firsthand experience as a distinguished scientist, had likewise emphasized the communal framework of the pervasive *tacit* knowledge that underlies the widest range of human skills. He pointed out that in such skills, including those of scientific research, much cannot be fully specified, at least not by those who practice them. They are the skills of connoisseurship and other forms of personal judgment; like the skills of the craftsman, they are learned not from textbooks but by working alongside a more experienced practitioner within a living communal tradition.[13]

This picture of scientific work as skilled craftsmanship, practiced within a shared tradition that is maintained by a social collectivity, jarred not only against the fiercely held convictions of many philosophers but also against the routine assumptions of many historians of science. Even now, its validity would be more widely appreciated if those who analyze scientific work were not generally such narrowly bookish people, and if they had firsthand experience not only of scientific research itself but also of skilled *manual* crafts outside the intellectual or academic sphere altogether. In any case, it is clear that historians have not yet adjusted their modes of writing to take full account of the social dimension of scientific practice. It has become common, and in some circles fashionable, to explore the social impact of scientific work or, conversely, the penetration of wider social interests into scientific ideas and concepts.[14] But historical studies of the *detailed content* of scientific research are still generally individualistic in their approach. The work of individual scientists is often analyzed in admirable detail, and full weight may also be given to the ways in which that work was influenced by others. But few historians of science have set out deliberately to recover what such a network of individuals had in common, particularly what they held *tacitly* in common.[15]

This book is not primarily designed to remedy that defect, even by a single example. Nonetheless, any detailed account of a scientific episode such as the Devonian controversy must be preceded by at least a brief sketch of the social and intellectual framework within which the new scientific knowledge was shaped; making explicit, as far as possible, what for the actors remained largely tacit and taken for granted. To extend the dramaturgical metaphor, the scene must be set and the background of the plot summarized before the play itself can begin. Dramatists have used preludes, choruses, soliloquies, ignorant newcomers, and other devices for this purpose. I have chosen the form of a prelude, which is part one of this book. Here the social scenery of the Devonian controversy is sketched, and the tacit conventions of the actors' scientific practice are briefly made explicit. This leads into the first chapter of part two, in which the background history is summarized and the main actors introduced. The way is then clear for the rest of the narrative to unfold without repeated interruption for the explanations that would otherwise be necessary. But the description of the craft knowledge underlying the Devonian controversy cannot be confined to the introductory part of the book. It is within the main narrative that the collective tacit framework of early nineteenth-century geology must be shown in operation, if its role in the shaping of a new piece of scientific knowledge is to be adequately portrayed.

1.5 The Revival of Narrative

Narrative is out of fashion among historians generally. It survives in the traditional genres of political history and biography, but number-crunching cliometricians reject it as unscientific, and those who trace the *longue durée* of historical *structure* dismiss it as mere *histoire événementielle*. When Lawrence Stone discerned a recent movement toward a revival of narrative among the so-called new historians, the examples he was able to cite showed that the trend was defined more by the absence of quantification and analysis than by the positive presence of a strict temporal framework.[16] In the history of science, too, narrative is out of favor. Chronicles of disconnected events—usually a series of publications—still pass for the history of science in some quarters, but this only confirms the opinion of more sophisticated practitioners that narrative has no place in progressive historiography.

It is high time for a genuine revival of narrative to be set in train, but it must be narrative with a purpose, and no mere chronicle. In the fine-grained study of scientific research practice, narrative is not so much a literary convenience as a methodological necessity. If scientific knowledge is to be studied *in the making*, the closest attention must be paid to strict chronology, not only in description but also in analysis. If, to use Harry Collins's apt metaphor, the "ship in a bottle" of any established scientific conclusion is to be understood through its mode of construction, the *sequence* of manipulations by which the model is inserted and made to appear so permanent must be a primary object of attention.[17] A casual or anecdotal jumping back and forth over years or even decades, which is common in many accounts of the fortunes of both scientific concepts and scientific institutions, has no proper place in any fine-grained account of scientific knowledge in the making.

Sociologists of science such as Collins have been so concerned with avoiding retrospective judgments of scientific research—and rightly so—that they have deliberately chosen to study problems or controversies that are as yet unresolved, and where the science is therefore still in the making.[18] This is highly effective in preventing any possible use of hindsight, either by the sociologists themselves or by the scientists they interview. But it ensures methodological purity at the cost of comprehensive understanding, for only incomplete episodes can be analysed—at least until many years are past, when repeat interviews may reveal that some conclusion or consensus has at last been reached. This is where the historian of science can make a distinctive contribution; for despite some risk to purity, historical materials allow the study of *completed* episodes of scientific practice.

The risk is that the description and analysis may be irreparably distorted by the historian's or the readers' knowledge of the outcome of the episode or the "correct" solution of the controversy. That risk is not negligible. Many detailed historical studies—some of them otherwise admirable—analyze the earlier phases of specific scientific developments with repeated forward reference to problems that had not yet arisen, experiments not yet performed, theories not yet devised, and publications not yet composed. Even historians of science who are zealous in sniffing out the "presentist" or "whiggish" heresies of others are themselves often guilty of what may be termed the "second-order whiggism" of retrospective description. This may not be as blatant as the presentist interpretations of some scientists, with their repeated invocation of what "we now know" as an unproblematic standard for understanding the past history of their field. The forward reference may not be to present knowledge, but rather to the later and mature work of the same individual or to the later development of the same discipline—however unmodern that may still have been. But even this precludes any genuine understanding of the *processes* by which new knowledge is shaped.

Narrative in the service of understanding the shaping of knowledge must rigorously and self-consciously avoid hindsight. To paraphrase and extend an earlier quotation, the historian of science must try to figure out what the devil the scientists thought they were up to, not just in the general sense of empathizing with their worldview or comprehending the contemporary state of their discipline, but also in the far more specific sense of following and making sense of what they did and said and wrote and argued about, week by week and month by month. What one knows, as a historian, that certain scientists said or wrote in September must not be allowed to warp one's judgment of what the devil they thought they were up to back in June. As far as is humanly possible (which means imperfectly), the historian must shelve any knowledge of what was for the participants an unknown future, in order to reconstruct the processes by which they were later to reach it. It is not entirely frivolous to add that this feat is greatly facilitated if the historian has a poor memory.

A deliberate and consistent abstention from hindsight is only one way in which a narrative of the shaping of any piece of scientific knowledge is as contrived a literary form as any other kind of history. A proper ambition to tell the story "how it really was," at least in the sense of handling the extant records as conscientiously as possible, does not imply that the resultant narrative will be a plain, unvarnished chronicle. On the contrary, it must aim to be as carefully constructed for its purpose as any well-crafted, traditional novel. For example, if the narrative deals with an interacting group of scientists rather than a single individual, the separate activities of different members must be woven into the main plot in a way that respects the trajectories of their individual lives while breaching the strict chronology as little as possible. Likewise, there is no need for the undoubted tedium of long stretches of scientific work to be reproduced in stretches of equally tedious narrative, as Frederic Holmes too pessimistically concluded.[19] A narrative that does justice to the intricate twists and turns of research does indeed need to be long and detailed. But it can and should reflect the "subjective time" of the scientists themselves, expanding where the action seemed to them most intense and exciting, and contracting where they were bored, frustrated, or just diverted to other lines of work. The chronology that a narrative should punctiliously observe is primarily that of the sequence of events, rather than the lapse of "real time" as measured by the calendar on the laboratory wall. Like nineteenth-century geology, it is more concerned with a relative than with an absolute timescale.

No single narrative account of an episode in scientific research can claim to be definitive, and not only because new evidence may come to light subsequently. In principle the same story could be told from a number of different viewpoints. If pursued consistently this would generate a series of interlocking but separate narratives. In Alan Ayckbourn's dramatic trilogy *The Norman Conquests*, the same plot unfolds and the same actors perform their parts in each of three parallel plays, set in the dining room, sitting room, and garden of the same house over the same weekend.[20] By analogy, it would be possible in principle to write at least three parallel narratives of the Devonian controversy; for example, as it was experienced within the English-, French-, and German-speaking scientific worlds. I have chosen to tell it primarily from the first of those perspectives because that is the setting in which the dramatic action was most intense, and for which the documentation is most complete. It should be emphasized, however, that the possibility of telling the same story from several different perspectives does not reduce each indifferently to a mere "account," in the somewhat pejorative sense of that word favored by some sociologists. The materials for each narrative may be selected according to different criteria of significance, but all the alternative accounts are, or should be, under the same constraints in their proper use of documentary and other evidence.

Part two of this book comprises an attempt to write a narrative of the Devonian controversy in the manner just outlined. The narrative is as rigorously nonretrospective as I can make it, and it keeps as closely to the original sequence of events as is compatible with the simultaneous tracing of many individual trajectories. Generous quotations are included, not least because the tone of the participants' exchanges and the metaphors they chose to use are essential to an understanding of the ways in which the new scientific knowledge was shaped. Interpretative commentary is confined to the exposition of meanings and inferences that the participants might plausibly have recognized for themselves, at that particular moment in the development of the controversy, and to the recall of relevant earlier moments that the participants themselves might plausibly have remembered. The anticipation of later moments that they could not possibly have known in advance is rigorously excluded. The resultant narrative may at first sight seem unduly long. But any substantial contraction of its scale would have reduced this case study to the schematic level, draining it of any value as a portrayal of the complexities of real scientific research. [. . .]

A nonretrospective narrative is designed, by careful contrivance, to minimize the bias that a knowledge of the eventual outcome can have on the telling of the story. Herbert Butterfield's plea for a narrative political history applies with equal force to the history of science: "we must have the kind of story in which . . . we can never quite guess, at any given moment, what is going to happen next."[21] In the case of the Devonian controversy, an extra bonus comes fortuitously from the very obscurity into which the episode has fallen since it was successfully resolved. With luck, most readers of this book will not know who were the goodies and who the baddies—not that such labels have any rightful place in the history of science—nor will they know in advance how the plot was to end. I shall not tell them.

1.6 Beyond Earshot of the Natives

The goal of this study is to contribute toward making "small facts speak to large issues." Any worthwhile speculation about large issues must be empirically grounded in "long-term, mainly (though not exclusively) qualitative, highly participative, and almost obsessively fine-comb field study."[22] Geertz's prescription for anthropological research is directly applicable, *mutatis mutandis*, to the empirical study of scientific research practice and the shaping of scientific knowledge, past and present. It is an accurate characterization of the historical method that will be used in this account of the Devonian controversy. As Geertz recognized for his own discipline, however, the problem is to make the transition from microscopic description to large-scale interpretation. It is not enough to treat a fine-grained study as adequate in itself without further analysis, either as a microscopic "world in a teacup" or as a natural experiment or "sociological cloudchamber." Small facts can only speak to large issues if they are deliberately made to do so.

A nonretrospective narrative of any episode in the history of science should be couched in terms that the historical actors themselves could have recognized and appreciated with only minor cultural translation. But then the risk is that one is left "awash with immediacies" and trapped inside the conceptual world of the natives being studied. Most sociologists and philosophers of science have in practice tilted toward the opposite extreme, couching their conclusions in terms so distant from the natives' own perceptions and experience that one is left "stranded in abstractions."[23] The ideal for the investigator might be to turn the formal hermeneutical circle into a dynamic spiral of involvement and detachment, immediacy and abstraction. For purposes of presentation, however, it may be better to cut through the historian's knot of indecision about the proper balance between description and analysis, and simply to attempt both, separately and in succession. This is what I have chosen to do in this book: the narrative of the Devonian controversy in part two is followed in part three by an analysis of the case study and its implications.

In contrast with the formalistic hypothetical examples still favored by many philosophers, any real piece of scientific research must be described by the historian (or, for that matter, the sociologist) in a style that does justice to its real complexity, muddle, and messiness. As already noted, this means that any adequate narrative is bound to be long and detailed. Furthermore, if the narrative is to be fully nonretrospective, the historian must abstain not only from forward reference to later phases but also from imposing on the narrative the simplifying generalizations that may be apparent in retrospect. The reader may have to be left feeling awash in immediacies, with only such landmarks for orientation as were also available to the historical actors at the time. The proper moment for rescue from the rising tide of immediacies comes after the narrative and outside its framework. The terms of the discussion can then be those of late twentieth-century analysts of science—historical, sociological, and philosophical—rather than those of the early nineteenth-century scientists themselves. Retrospective reflection in the light of the known outcome of the controversy becomes at this point permissible, indeed indispensable. The scientific natives are out of earshot.

The analysis in part three of this book is structured by the conviction that the new scientific knowledge produced in most episodes of scientific research practice—including the Devonian controversy—should not be treated only or primarily as the creative achievement of one or a few outstanding individuals. It should be regarded rather as the outcome of processes of interaction within a group or cast list that included, in their diverse roles, not only star performers but also minor actors and walk-on parts. The interactions may include those of bitter rivalry and fierce antagonism as well as those of generous giving and amicable collaboration. But whatever their affective quality or moral status, they are the manifestations of what Bruno Latour and Steve Woolgar have aptly termed the "agonistic field" of intensive social negotiation in science.[24] In the course of research it is in the agonistic field that new interpretative schemes may slowly grow in plausibility and perceived solidity, forming the focus for a consensus, while other schemes gradually dissolve into implausibility and fade into oblivion, adhered to only by marginal individuals. This is what the historian of science should be able to trace and analyze in detail, perhaps even more reliably than the participant-observer in current research.

Notes

1. There is no adequate general account of the earth sciences in the early nineteenth century. Porter 1977 gives an important analysis of the preceding century and a half, with due attention to empirical, intellectual, and social aspects, but his account is confined to Britain. [Gillispie 1996/1951] gives a classic interpretation of the perceived implications of the new science of geology in early nineteenth-century Britain, but he does not set out to analyze the technical content of geological research in any detail. Rudwick 1985 describes the palaeontological branch of the science over a wider period, and Greene 1982 the structural or tectonic; both attempt to transcend the British bias of much historical writing in English, but neither deals with the central *stratigraphical* tradition within which the Devonian controversy developed.

2. Medawar 1967, p. 151.

3. [Latour and Woolgar 1986] and Knorr-Cetina 1981 are representative examples of the "ethnographic" approach: both are studies of research in large biological laboratories in California. For important collections of essays summarizing this kind of work more generally, see Knorr-Cetina, Krohn, and Whitley 1980 and Knorr-Cetina and Mulkay 1983. The importance of studying temporal processes in (even) conventional ethnography has been stressed particularly by Victor Turner (see, e.g., Turner 1974).

4. Of many fine examples, the following recent studies are representative: Gruber's 1981a and Kohn's 1980 reconstructions of Charles Darwin's early theoretical development; Holmes's 1974 analysis of some of Claude Bernard's laboratory research; and [Westfall's 1983] biographical study of Isaac Newton.

5. Geertz 1976, p. 224. On "thick description," see Geertz 1973, chap. 1, and Elkana 1981. The rejection by "discourse analysts" of any such interpretative goal (see, e.g., Mulkay and Gilbert 1982) seems to be based on the assumption that all other analysts of scientific work do, or must necessarily, take the "accounts" of participants at face value, and make an arbitrary selection from those accounts if they are inconsistent. No historian could possibly accept the validity of that assumption.

6. Barlow 1958, p. 119. For analyses of the theoretical thinking in the notebooks, see, e.g., Gruber 1981a and Kohn 1980. For a balanced appreciation of the oral testimony of living scientists, see Holmes 1984.

7. For the concept of "fine-grained" analyses, see Holmes 1981, 1984. The biographical studies mentioned earlier are good examples of this level of resolution, applied to individual scientists. The importance of working at different levels of resolution in the study of individuals is well argued by Gruber 1981b.

8. Cannon 1964, p. 30. Cannon may have had second thoughts about the validity of the claim, for it was omitted from the revised version of this seminal article: Cannon 1978, chap. 8. Although the prediction about the future was rash, the claim was in my opinion justified with respect to the present, provided the documentation is judged not just on bulk but also on the *quality* of the insights that the documents allow. The archives of distinguished twentieth-century scientists may be more voluminous, but also less revealing, than those of some of their nineteenth-century predecessors.

9. See introduction to list of manuscript sources (p. 463, Rudwick 1985).

10. See, e.g., Collins 1975, 1981.

11. The scientists in Latour and Woolgar 1979, as in many other sociological studies, are labeled with letters of the alphabet, although one of the participants in the research that is analyzed was a Nobel prize winner! One major actor ("O") in Collins's 1975 study of gravitational radiation was revealed by name in a follow-up study (1981). Likewise it is said that the pseudonymous characters in Goodfield (1981) are easily identifiable by specialists in the relevant scientific field.

12. Kuhn 1962. The comment is not intended to detract from the originality of Kuhn's work,

but on the contrary to point out that certain components of it were immediately recognized by many practicing scientists as an authentic characterization of their work.

13. Polanyi 1958, 1966. The term "craft knowledge" seems to have been first used explicitly in this context by Ravetz 1971, chap. 3, but it is certainly implicit in Polanyi's much earlier work. Fleck's prewar study (1935, trans. 1979), although important in retrospect, had little influence at the time.

14. For examples of the latter, based on thorough historical research rather than purely on political conviction, see the essays in Barnes and Shapin 1979 and Shapin's 1982 important review of this research.

15. An outstanding exception is Frank's 1980 fine study of the physiological research of William Harvey's followers in Oxford in the mid-seventeenth century. See also Geison's 1981 important review of the problems of studying research "schools" in science.

16. Stone 1979; see also White 1984.

17. Collins 1974; see also Brannigan 1981, chap. 3.

18. See, e.g., Collins 1975, 1981, Pinch 1981, and Pickering 1981.

19. Holmes 1974, pp. xvi, xvii.

20. Ayckbourn 1975.

21. Butterfield 1957, p. 106.

22. Geertz 1973, p. 23.

23. Geertz 1976.

24. Latour and Woolgar 1979, p. 237 and passim.

References

Ayckbourn, Alan. 1975. *The Norman Conquests*. London: Chatto and Windus.

Barlow, Nora (ed.). 1958. *The Autobiography of Charles Darwin 1809–1882*. London: Collins.

Barnes, Barry and Shapin, Steven. 1979. *Natural Order*. London: Sage Publications.

Brannigan, Augustine. 1981. *The Social Basis of Scientific Discoveries*. Cambridge: Cambridge University Press.

Butterfield, Herbert. 1957. *George III and the Historians*. London: Collins.

Cannon, Susan Faye. 1978. *Science in Culture*. New York: Science History Publications; Folkestone, England: Dawson.

——. 1964. "History in Depth," *History of Science* 3: 20–38.

Collins, Harry. 1981. "Son of Seven Sexes," *Social Studies of Science* 11 (1): 33–62.

——. 1975. "The Seven Sexes," *Sociology* 9 (2): 205–24.

Elkana, Yehuda. 1981. "A Programmatic Attempt at an Anthropology of Knowledge," pp. 1–68 in Mendelsohn, Everett and Elkana, Yehuda (eds.), *Sciences and Cultures: anthropological and historical studies of the sciences*, Dordrecht: Reidel.

Fleck, Ludwik. 1979. *Genesis and Development of a Scientific Fact*, trans. Fred Bradley and Thaddeus Trenn; T. Trenn and Robert Merton (eds.), "Foreword" by Thomas S. Kuhn, Chicago: Chicago University Press.

——. 1935. *Entstehung und Entwicklung einer wissenschaftlichen Tatsache. Einführung in die Lehre vom Denkstil und Denkkollektiv*. Basel: Benno Schwabe und Co.

Frank, Robert. 1980. *William Harvey and the Oxford Physiologists*. Berkeley: University of California Press.

Geertz, Clifford. 1976. "From the Native's Point of View," pp. 221–237 in *Meaning in Anthropology*, Keith Basso and Henry Selby (eds.). Albuquerque: University of New Mexico Press.

——. 1973. *The Interpretation of Cultures: Selected Essays*. New York: Basic Books.

Geison, Gerald. 1981. "Scientific Change, Emerging Specialties, and Research Schools," *History of Science* 19: 20–40.

Gillispie, Charles. 1996/1951. *Genesis and geology: a study in the relations of scientific thought, natural theology, and social opinion in Britain, 1790–1850.* Cambridge: Harvard University Press.

Greene, Mott. 1982. *Geology in the Nineteenth Century.* Ithaca: Cornell University Press.

Gruber, Howard. 1981a. *Darwin on Man.* Chicago: University of Chicago Press.

———. 1981b. "On the Relation between 'Aha' Experiences and the Construction of Ideas," *History of Science* 19: 41–59.

Holmes, Frederick. 1984. "Lavoisier and Krebs," *Isis* 75: 131–142.

———. "The fine structure of scientific creativity," *History of Science* 19: 60–70.

———. 1974. *Claude Bernard and Animal Chemistry.* Cambridge: Harvard University Press.

Knorr-Cetina, Karin. 1981. *The Manufacture of Knowledge.* Oxford: Pergamon Press.

Knorr-Cetina, Krohn, and Whitley (eds.). 1980. *The Social Process of Scientific Investigation.* Dordrecht: Reidel.

Knorr-Cetina and Mulkay (eds.). 1983. *Science Observed.* London: Sage Publications.

Kohn, David. 1980. "Theories to Work By," *Studies in the History of Biology* Volume 4: 67–170.

Kuhn, Thomas. 1962. *The Structure of Scientific Revolutions.* Chicago: University of Chicago Press.

Latour, Bruno and Woolgar, Steve. 1986/1979. *Laboratory Life.* First ed. Beverly Hills: Sage Publications; second rev. ed. Princeton: Princeton University Press.

Medawar, Peter. 1967. *The Art of the Soluble.* London: Methuen.

Mulkay, Michael and Gilbert, Nigel. 1982. "What is the Ultimate Question?" *Social Studies of Science* 12: 309–19.

Pickering, Andrew. 1981. "The hunting of the quark," *Isis* 72: 216 – 235.

Pinch, Trevor. 1981. "The sun-set," *Social Studies of Science* 11: 131–158.

Polanyi, Michael. 1966. *The Tacit Dimension.* Chicago: University of Chicago Press.

———. 1958. *Personal Knowledge: Towards a Post-Critical Philosophy.* Chicago: University of Chicago Press.

Porter, Roy. 1977. *The Making of Geology: Earth Science in Britain, 1660–1815.* Cambridge: Cambridge University Press.

Ravetz, Jerome. 1971. *Scientific Knowledge and Its Social Problems.* Oxford: Oxford University Press.

Rudwick, Martin. 1985. *The Meaning of Fossils: Episodes in the History of Palaeontology,* second edition. Chicago: University of Chicago Press.

Shapin, Steven. 1982. "History of science and its sociological reconstructions," *History of Science* 20: 157–211.

Stone, Lawrence. 1979. "The Revival of Narrative: Reflections on a New Old History," *Past and Present* 85: 3–24.

Turner, Victor. 1974. *Dramas, fields and metaphors.* Ithaca, NY: Cornell University Press.

Westfall, Richard. 1983/1980. *Never at Rest.* Cambridge: Cambridge University Press.

White, Hayden. 1984. "The question of narrative in contemporary historical theory," *History and Theory* 23: 1–33.

CHAPTER 11

The Polity of Science

STEVEN SHAPIN AND SIMON SCHAFFER

Pages 332–344 in *Leviathan and the Air Pump*. Princeton: Princeton University Press, 1985.

Lords and Commons of England, consider what Nation it is wherof ye are, and wherof ye are the Governours.

MILTON, Areopagitica

SOLUTIONS to the problem of knowledge are solutions to the problem of social order. That is why the materials in this book are contributions to political history as well as to the history of science and philosophy. Hobbes and Boyle proposed radically different solutions to the question of what was to count as knowledge: which propositions were to be accounted meaningful and which absurd, which problems were soluble and which not, how various grades of certainty were to be distributed among intellectual items, where the boundaries of authentic knowledge were to be drawn. In so doing, Hobbes and Boyle delineated the nature of the philosophical life, the ways in which it was permissible or obligatory for philosophers to deal with each other, what they were to question and what to take for granted, how their activities were to relate to proceedings in the wider society. In the course of offering solutions. to the question of what proper philosophical knowledge was and how it was to be achieved, Hobbes and Boyle specified the rules and conventions of differing philosophical forms of life. We conclude this book by developing some ideas about the relationships between knowledge and political organization.

There are three senses in which we want to say that the history of science occupies the same terrain as the history of politics. First, scientific practitioners have created, selected, and maintained a polity within which they operate and make their intellectual product; second, the intellectual product made within that polity has become an element in political activity in the state; third, there is a conditional relationship between the nature of the polity occupied by scientific intellectuals and the nature of the wider polity. We can elaborate each of these points by refining a notion we have used informally throughout this book: that of an intellectual *space*.[1]

Our previous usages of terminology such as "experimental space" or "philosophical space" have been twofold: we have referred to space in an abstract sense, as a cultural domain. This is the sense customarily intended when one speaks of the boundaries of disciplines or the overlap between areas of culture. The cartographic metaphor is a good one: it reminds us that there are, indeed, abstract cultural boundaries that exist in social space. Sanctions can be enforced by community members if the boundaries are transgressed. But we have also, at times, used the notion of space in a physically more concrete sense. The receiver of the air-pump circumscribed such a space, and we have shown the importance attached by Boyle to defending the integrity of that space. Yet we want to elaborate some notions concerning a rather larger-scale physical space. If someone were to be asked in 1660, "Where can I find a natural philosopher at work?", to what place would he be directed? For Hobbes there was to be no special space in which one did natural philosophy. Clearly, there were spaces that were deemed grossly inappropriate. Since philosophy was a noble activity, it was not to be done in the apothecary shop, in the garden, or in the tool room. He told his adversaries that philosophers were not "apothecaries," "gardeners," or any other sort of "workmen." Neither was philosophy to be withdrawn into the Inns of Court, the physicians' colleges, the clerics' convocations, or the universities. Philosophy was not the exclusive domain of the professional man, Any such withdrawal into special professional spaces threatened the public status of philosophy. Recall Hobbes's indictment of the Royal Society as yet another restricted professional space. He asked "Cannot anyone who wishes come?" and gave the answer. "The place where they meet is not public."[2] We have seen that the experimentalists also insisted upon the public nature of their activity, but Boyle's "public" and Hobbes's "public" were different usages. Hobbes's philosophy had to be public in the sense that it must not become the preserve of interested professionals. The special interests of professional groups had acted historically to corrupt knowledge. Geometry had escaped this appropriation only because, as a contingent historical matter, its theorems and findings had not been seen to have a bearing on such interests: "Because men care not, in that subject, what be truth, as a thing that crosses no man's ambition, profit or lust."[3] Hobbes's philosophy also had to be public because its purpose was the establishment of public peace and because it commenced with social acts of agreement: settling the meanings and proper uses of words. Its public was not a witnessing and believing, public, but an assenting and professing public: not a public of eyes and hands, but one of minds and tongues.

In Boyle's programme there was to be a special space in which experimental natural philosophy was done, in which experiments were performed and witnessed. This was the nascent *laboratory*. What kind of physical and social space was this laboratory? Consider the German experimental scene in figure 22. This picture comes from Caspar Schott's *Mechanica hydraulico-pneumatica* of 1657, and it shows experimental knowledge being constituted. This was the book that prompted Boyle's decision to begin the construction of an air-pump allegedly superior to Guericke's device shown here.[4] Guericke himself is shown in the left foreground. He holds a baton (possibly of his office as *Bürgermeister* in Magdeburg) in his right hand, and

Figure 22 Otto von Guericke's first pump demonstrated before witnesses. From Schott's Mechanica hydraulico-pneumatica (Würzburg, 1657), p. 445. (Courtesy of Cambridge University Library.)

with his left he points another stick at his machine; he is not shown actually touching the pump with his hand. He is not dressed in any special way, such as might be necessitated by actual manipulations with this rather messy machine; nor is he dressed differently from the witnesses to the experiment, assembled separately from Guericke in the right foreground. The architectural space in which the scene is set is a courtyard or forum. We do not know whether it is meant that these experiments were specially brought to this public place to be tried, or whether the artist or engraver was merely using artistic conventions familiar to him to situate the objects and actions he was told to depict. [. . .] This picture shows the natural philosopher as presiding officer, and it shows the experimental witnesses, but it does not show any human being actually doing an experiment. The machines are worked by *putti* (cherubs). This was a standard convention of baroque illustrations. Here and elsewhere, it was implied that the resulting knowledge was divine.

What little we do know about English experimental spaces in the middle part of the seventeenth century indicates that their status as private or public was intensely debated. We briefly noted [. . .] that the word "laboratory" arrived in English usage in the seventeenth century, carrying with it apparently hermetical overtones: the space so designated was private, inhabited by "secretists." During the 1650s and 1660s new open laboratories were developed, alongside Boyle's rhetorical efforts to lure the alchemists into public space and his assaults on the legitimacy of private practice. The public space insisted upon by experimental philosophers was a space for collective witnessing. We have shown the importance of witnessing for the constitution of the matter of fact. Witnessing was regarded as effective if two general conditions could be satisfied: first, the witnessing experience had to be made accessible; second, witnesses had to be reliable and their testimony had to be creditable. The first condition worked to open up experimental space, while the second acted to restrict entry. What in fact resulted was, so to speak, a public space with restricted access. (Arguably, this is an adequate characterization of the scientific laboratory of the late twentieth century: many laboratories have no legal sanction against public entry, but they are, as a practical matter, open only to "authorized personnel.") Restriction of access, we have indicated, was one of the positive recommendations of this new experimental space in Restoration culture. Either by decision or by tacit processes, the space was restricted to those who gave their assent to the legitimacy of the game being played within its confines.

[. . .] We described differences in the engagements Boyle conducted with two sorts of adversaries: those who disputed moves within the experimental game and those who disputed the game. The latter could be permitted entry to the experimental community only at the price of putting that community's life at risk. Public stipulations about the accessibility of the experimental laboratory were tempered by the practical necessity of disciplining the experimental collective. This tension meant that Hobbes's identification of the Royal Society as a restricted place was potentially damaging, just as it is damaging in modern liberal societies to remark upon the sequestration of science. Democratic ideals and the exigencies of professional expertise form an unstable compound.[5] Hobbes's identification of restrictions on the experimental public shows why virtual witnessing was so vitally important, and why troubles in the experimental programme of physical replication were so energetically dealt with. Virtual witnessing acted to ensure that witnesses to matters of fact could effectively be mobilized in abstract space, while securing adequate policing of the physical space occupied by local experimental communities.

For Hobbes, the activity of the philosopher was not bounded: there was no cultural space where knowledge could be had where the philosopher should not go.[6] The methods of the natural philosopher were, in crucial respects, identical to those of the civic philosopher, just as the purpose of each was the same: the achievement and protection of public peace. Hobbes's own career was a token of the philosophical enterprise so conceived. For Boyle and his colleagues, the topography of culture looked different. Their cultural terrain was vividly marked out with boundary-stones and warning notices. Most importantly, the experimental study of nature was to be visible withdrawn from "humane affairs." The experimentalists were not to "meddle with" affairs of "church and state." The study of nature occupied a quite different space from the study of men and their affairs: objects and subjects would not and could not be treated as part of the same philosophical enterprise. By erecting such boundaries, the experimentalists thought to create a quiet and a moral space for the natural philosopher: "civil war" within their ranks would be avoided by observing these boundaries and the conventions of discourse within them. They would not speak of that which could not be mobilized into a matter of fact by the conventionally agreed patterns of community activity—thus the importance of legislation against speech about entities that would not be made sensible: either those that indisputably *did* exist (e.g., God and immaterial spirits) or those that probably did not (e.g., the aether). As a practical matter, Hobbes could hardly deny that the experimentalists had established a community with some politically important characteristics: a community whose members endeavoured to avoid metaphysical talk and causal inquiry, and which displayed many of the attributes of internal peace. But this community was not a society of *philosophers.* In abandoning the philosophical quest, such a group was contributing to civil disorder. It was the philosopher's task to secure public peace; this he could only do by rejecting the boundaries the experimentalists proposed between the study of nature and the study of men and their affairs.

The politics that regulated transactions between the philosophical community and the state was important, for it acted to characterize and to protect the knowledge the philosopher produced. The politics that regulated transactions within the philosophical community was equally important, for it laid down the rules by which authentic knowledge was to be produced. We remarked [. . .] that Hobbes assumed philosophical places to have "masters": Father Mersenne had been such a master in Paris, and Hobbes spoke of Boyle and some few of his friends "as masters of the rest" in the Royal Society. It was fitting that philosophical places should have masters who determined right philosophy, just as it was right and necessary that the commonwealth should have such a master. Indeed, Leviathan could legitimately act as a philosophical master. Hobbes found it no argument against the King's right to determine religious principles that "priests were better instructed," and he also rejected the argument "that the authority of teaching *geometry* must not depend upon kings, except they themselves were geometricians."[7] Insofar as a philosophical master was not Leviathan, he was someone else who had found out fundamental matters: the correct principles upon which a unified philosophical enterprise could proceed. He was a master by virtue of his exercise of pure mind, not by his craft-skills or ingenuity. In the body politic of the Hobbesian philosophical place, the mind was the undisputed master of the eyes and the hands.

In the body politic of the experimental community, mastery was *constitutionally restricted*. We have seen how Hooke described the experimental body in terms of the relationships that ought to subsist between intellectual faculties: "The *Understanding* is to *order* all the inferior services of the lower Faculties; but yet it is to do this only as a *lawful Master*, and not as a *Tyrant*." The experimental polity was an organic community in which each element crucially depended upon all others, a community that rejected absolute hierarchical control by a master. Hooke continued:

> So many are the *links*, upon which the true Philosophy depends, of which, if any one be *loose*, or *weak*, the whole *chain* is in danger of being dissolv'd; it is to *begin* with the Hands and Eyes, and to *proceed* on through the Memory, to be *continued* by the Reason; nor is it to stop there, but to *come about* to the Hands and Eyes again, and so, by a *continual passage round* from one Faculty to another, it is to be maintained in life and strength, as much as the body of man is.[8]

The experimental polity was said to be composed of free men, freely acting, faithfully delivering what they witnessed and sincerely believed to be the case. It was a community whose freedom was responsibly used and which publicly displayed its capacity for self-discipline. Such freedom was safe. Even disputes within the community could be pointed to as models for innocuous and managed conflict. Moreover, such free action was said to be requisite for the production and protection of objective knowledge. Interfere with this form of life and you will interfere with the capacity of knowledge to mirror reality. Mastery, authority, and the exercise of arbitrary power all acted to distort legitimate philosophical knowledge. By contrast, Hobbes proposed that philosophers should have masters who enforced peace among them and who laid down the principies of their activity. Such mastery did not corrode philosophical authenticity. The Hobbesian form of life was not, after all, predicated upon a model of men as free-acting, witnessing, and believing individuals. Hobbesian man differed from Boylean man precisely in the latter's possession of free will and in the role of that will in constituting knowledge. Hobbesian philosophy did not seek the foundations of knowledge in witnessed and testified matters of fact: one did not ground philosophy in "dreams." We see that both games proposed for natural philosophers assumed a causal connection between the political structure of the philosophical community and the genuineness of the knowledge produced. Hobbes's philosophical truth was to be generated and sustained by absolutism. Boyle and his colleagues lacked a precise vocabulary for the polity they were attempting to erect. Almost all of the terms they used were highly contested in the early Restoration: "civil society," a "balance of powers," a "commonwealth." The experimental community was to be neither tyranny nor democracy. The "middle wayes" were to be taken.[9]

Scientific activity, the scientist's role, and the scientific community have always been dependent: they exist, are valued, and supported insofar as the state or its various agencies see point in them. What sustained the experimental space that was created in the mid-seventeenth century? The nascent laboratory of the Royal Society and other experimental spaces were producing things that were widely wanted in Restoration society. These wants did not simply preexist, waiting to be met; they were actively cultivated by the experimentalists. The experimentalists' task was to show others that their problems could be solved if they came to the experimental philosopher and to the space he occupied in Restoration culture.[10] If the experimentalists could effectively cultivate and satisfy these wants, the legitimacy of experimental activity and the integrity of laboratory and scientific role would be ensured. The wants addressed by the experimental community spread across Restoration economic, political, religious, and cultural activity. Did gunners want their artillery pieces to fire more accurately? Then they should bring their practical problems to the physicists of the Royal Society.

Did brewers want a more reliable ale? Then they should come to the chemists. Did physicians want a theoretical framework for the explanation and treatment of fever? Then they should inspect the wares of the mechanical philosopher. The experimental laboratory was advertised as a place where practically useful knowledge was produced.[11] But the laboratory could also supply solutions to less tangible problems. Did theologians desire facts and schemata that could be deployed to convince otherwise obdurate men of the existence and attributes of the Deity? They, too, should come to the laboratory where their wants would be satisfied. Through the eighteenth century one of the most important justifications for the natural philosopher's role was the spectacular display of God's power in nature.[12] Theologians could come to the place where the Leyden jar operated if they wanted to show cynics the reality of God's majesty; natural theologians could come to the astronomer's observatory if they wanted evidence of God's wise and regular arrangements for the order of nature; moralists could come to the natural historian if they wanted socially usable patterns of natural hierarchy, order, and the due submission of ranks. The scientific role could be institutionalized and the scientific community could be legitimized insofar as the experimental space became a place where this multiplicity of interests was addressed, acquitted, and drawn together. One of the more remarkable features of the early experimental programme was the intensity with which its proponents worked to publicize experimental spaces as useful: to identify problems in Restoration society to which the work of the experimental philosopher could provide the solutions.

There was another desideratum the experimental community sought to mobilize and satisfy in Restoration society. The experimental philosopher could be made to provide a model of the moral citizen, and the experimental community could be constituted as a model of the ideal polity. Publicists of the early Royal Society stressed that theirs was a community in which free discourse did not breed dispute, scandal, or civil war; a community that aimed at peace and had found out the methods for effectively generating and maintaining consensus; a community without arbitrary authority that had learnt to order itself. The experimental philosophers aimed to show those who looked at their community an idealized reflection of the Restoration settlement. Here was a functioning example of how to organize and sustain a peaceable society between the extremes of tyranny and radical individualism. Did civic philosophers and political actors wish to construct such a society? Then they should come to the laboratory to see how it worked.

This book has been concerned with the identification of alternative philosophical forms of life, with the display of their conventional bases, and with the analysis of what hinged upon the choice between them. We have not taken as one of our questions, "Why did Boyle win?" Obviously, many aspects of the programme he recommended continue to characterize modern scientific activity and philosophies of scientific method. Yet, an unbroken continuum between Boyle's interventions and twentieth-century science is highly unlikely. For example, the relationship between Boyle's experimental programme and Newton's "mathematical way" is yet to be fully explored. Nevertheless, modern historians who find in Boyle the "founder" of truly modern science can point to similar sentiments among late seventeenth-century and eighteenth-century commentators. Despite these qualifications the general form of an answer to the question of Boyle's "success" begins to emerge, and it takes a satisfyingly historical form. This experimental form of life achieved local success to the extent that the Restoration settlement was secured. Indeed, it was one of the important elements in that security.

Insofar as we have displayed the political status of solutions to problems of knowledge, we have not referred to politics as something that happens solely outside of science and which can, so to speak, press in upon it. The experimental community vigorously developed and deployed such boundary-speech, and we have sought to situate this speech historically and to explain why these conventionalized ways of talking developed. What we cannot do if we want to be serious about the historical nature of our inquiry is to use such actors' speech unthinkingly as an explanatory resource. The language that transports politics outside of science is precisely what we need to understand and explain. We find ourselves standing against much current sentiment in the history of science that holds that we should have less talk of the "insides" and "outsides" of science, that we have transcended such outmoded categories. Far from it; we have not yet begun to understand the issues involved. We still need to understand how such boundary-conventions developed: how, as a matter of historical record, scientific actors allocated items with respect to *their* boundaries (not ours), and how, as a matter of record, they behaved with respect to the items thus allocated. Nor should we take any one system of boundaries as belonging self-evidently to the thing that is called "science."

We have had three things to connect: (1) the polity of the intellectual community; (2) the solution to the practical problem of making and justifying knowledge; and (3) the polity of the wider society. We have made three connections: we have attempted to show (1) that the solution to the problem of knowledge is political; it is predicated upon laying down rules and conventions of relations between men in the intellectual polity; (2) that the knowledge thus produced and authenticated becomes an element in political action in the wider polity; it is impossible that we should come to understand the nature of political action in the state without referring to the products of the intellectual polity; (3) that the contest among alternative forms of life and their characteristic forms of intellectual product depends upon the political success of the various candidates in insinuating themselves into the activities of other institutions and other interest groups. He who has the most, and the most powerful, allies wins.

We have sought to establish that what the Restoration polity and experimental science had in common was a form of life. The practices involved in the generation and justification of proper knowledge were part of the settlement and protection of a certain kind of social order. Other intellectual practices were condemned and rejected because they were judged inappropriate or dangerous to the polity that emerged in the Restoration. It is, of course, far from original to notice an intimate and an important relationship between the form of life of experimental natural science and the political forms of liberal and pluralistic societies. During the Second World War, when liberal society in the West was undergoing its most virulent challenge, that perception was formed into part of the problematic of the academic study of science. What sort of society is able to sustain legitimate and authentic science? And what contribution does scientific knowledge make to the maintenance of liberal society?[13] The answer then given was unambiguous: an open and liberal society was the natural habitat of science, taken as the quest for objective knowledge. Such knowledge, in turn, constituted one of the sureties for the continuance of open and liberal society. Interfere with the one, and you will erode the other.

Now we live in a less certain age. We are no longer so sure that traditional characterizations of how science proceeds adequately describe its reality, just as we have come increasingly to doubt whether liberal rhetoric corresponds to the real nature of the society in which we now live. Our present-day problems of defining our knowledge, our society, and the relationships between them centre on the same dichotomies between the public and the private, between authority and expertise, that structured the disputes we have examined in this book. We regard our scientific knowledge as open and accessible in principle, but the public does not understand it. Scientific journals are in our public libraries, but they are written in a language alien to the citizenry. We say that our laboratories constitute some of our most open professional spaces, yet the public does not enter them. Our society is said to be democratic, but the public cannot call to account what they cannot comprehend. A form of knowledge that is the most open in principle has become the most closed in practice. To entertain these doubts about our science is to question the constitution of our society. It is no wonder that scientific knowledge is so difficult to hold up to scrutiny.

In this book we have examined the origins of a relationship between our knowledge and our polity that has, in its fundamentals, lasted for three centuries. The past offers resources for understanding the present, but not, we think, for foretelling the future. Nevertheless, we can venture one prediction as highly probable. The form of life in which we make our scientific knowledge will stand or fall with the way we order our affairs in the state.

We have written about a period in which the nature of knowledge, the nature of the polity, and the nature of the relationships between them were matters for wide-ranging and practical debate. A new social order emerged together with the rejection of an old intellectual order. In the late twentieth century that settlement is, in turn, being called into serious question. Neither our scientific knowledge, nor the constitution of our society, nor traditional statements about the connections between our society and our knowledge are taken for granted any longer. As we come to recognize the conventional and artifactual status of our forms of knowing, we put ourselves in a position to realize that it is ourselves and not reality that is responsible for what we know. Knowledge, as much as the state, is the product of human actions. Hobbes was right.

Notes

1. We are not aware of any specific debts for this usage. However, topographic sensibilities in the study of culture characterize a number of modern French sociologists and historians; see, for example Foucault, "Questions on Geography" [Pages 63–77 in C. Gordon, (ed.). *Power/Knowledge: Selected Interviews and Other Writings 1972–1977*. New York: Pantheon, 1980]; idem, "Médecins, juges et sorciers au 17e siècle" [*Médecine de France*, n°200, 1er trimestre 1969, pp. 121–128].
2. Hobbes, *Dialogus physicus*, translated in Shapin and Schaffer 1985, p. 240.
3. Hobbes, *Leviathan*, p. 91. Hobbes made no claim of the sort that geometry is essentially neutral.
4. Unfortunately, we have not been able to locate any picture of a seventeenth-century English experimental scene in pneumatics. Other diagrams, e.g., our figure 21, attempt to show the technical construction of Guericke's machine without depicting the experimental scene in which knowledge was constituted.
5. This has often been noted by historians dealing with widely differing settings; see, for example, Daniels, "The Pure-Science Ideal and Democratic Culture"; Ezrahi, "Science and the Problem of Authority in Democracy"; Fries, "The Ideology of Science during the Nixon Years"; Gillispie, "The *Encyclopédie* and the Jacobin Philosophy of Science."
6. According to Hobbes, men "cannot have any idea of [God] in their mind, answerable to his nature" ("Leviathan," p. 92), and, for that reason, theology was explicitly excluded from the philosophical enterprise ("Concerning Body," p. 10).
7. Hobbes, "Philosophical Rudiments," p. 247.
8. Hooke, *Micrographia*, "The Preface," sig b2^r.
9. The phrase is Hooke's: ibid., sig b1^v, similar locutions typify much Royal Society publicity.
10. For this section we are deeply indebted to recent work by Bruno Latour, especially his "Give Me a Laboratory" and *Les microbes: guerre et paix.*
11. From the best modern historical research it now appears that none of the utilitarian promissory notes could be, or were, cashed in the seventeenth century; see Westfall,

"Hooke, Mechanical Technology, and Scientific Investigation"; A. R. Hall, "Gunnery, Science, and the Royal Society." If science did not deliver technological utility, it becomes even more important to ask about its other perceived values, including social, political, and religious uses.

12. See particularly Schaffer, "Natural Philosophy"; idem, "Natural Philosophy and Public Spectacle."

13. Merton, *The Sociology of Science*, chaps. 12–13; Needham, *The Grand Titration*; Zilsel, *Die sozialen Ursprünge der neuzeitlichen Wissenschaft*.

Debates in History and Philosophy of Science

Themes

These sections deal with prominent debates, which are intended to raise larger questions. First, the debate between Mill and Whewell stimulates reflection on the role of **hypotheses in science** (Duhem), on **induction** (Mill, Whewell), and on the **logic of scientific discovery** (Mill, Whewell, Duhem, and Hanson). The discussion of of forces and causes in part B takes place within the tradition of **natural philosophy**, and raises questions of the **divisibility of matter** (Descartes, Leibniz, Newton, du Châtelet), the source of our knowledge of the ideas of **causality, force, and power** (Hume, Kant), **extension** as the essence or nature of body, and the possibility or impossibility of **atoms and the vacuum** (Descartes, Leibniz, Newton, du Châtelet). In part C, the debate between **catastrophism and uniformitarianism** (Whewell, de Buffon, Cuvier, Playfair) leads to larger questions about whether the **laws of nature** change over time (Cuvier, Playfair, Whewell), of what the archaeological and geological record tells us of the **succession** of animals, geological formations, and meteorological events on the earth (Whewell, Cuvier, Playfair, Darwin), and of how to assess the **true causes** of events (Whewell, Playfair, Darwin).

A
Hypotheses in Scientific Discovery

CHAPTER 12

Introduction to "Hypotheses in Scientific Discovery"

The debate between John Stuart Mill and William Whewell has been illuminated by much recent scholarship (see Achinstein 1992; Fisch 1991, 1985a, and 1985b; Fisch and Schaffer 1991; and Snyder 2006 and 2011, among others).

Francis Bacon influenced Whewell, Richard Jones, and John Herschel as undergraduates (Snyder 2006, chapter one). Whewell moves beyond Bacon's methods in the *Novum Organum* to develop his characteristic view of induction as involving both observation and reason, which Snyder, along with Fisch 1985a, calls his "antithetical epistemology," where the antitheses are observation and reason. This may appear to indicate a resemblance between his view and Kant's, but, unlike Kant, Whewell argues that the knowledge drawn from ideas or conceptions is knowledge of real phenomena as they are in themselves.

Whewell defends a version of "Discoverer's Induction" (see Snyder 2006 and 1997a), the fundamental operation of which is the "colligation" of facts. Colligation goes beyond just giving instances, to "superinduce" a conception on the facts. Note that, while this conception may be a creation of the mind, the connection we understand by means of the conception may be real.

In *A System of Logic*, first published in 1843, John Stuart Mill responded to Whewell on induction. Here, I excerpt the later, 1882 edition. "For the purposes of the present inquiry," Mill writes, "Induction may be defined, the operation of discovering and proving general propositions" (Mill 1882, 207).

In one of his critiques of Whewell in this text, Mill summarizes Whewell's account of colligation:

> Dr. Whewell maintains that the general proposition which binds together the particular facts, and makes them, as it were, one fact, is not the mere sum of those facts, but something more, since there is introduced a conception of the mind, which did not exist in the facts themselves.
>
> (Mill 1882, 215)

Mill goes on to cite Whewell from the *Novum Organum Renovatum*:

> The particular facts are not merely brought together, but there is a new element added to the combination by the very act of thought by which they are combined [. . .] When the Greeks, after long observing the motions of the planets, saw that these motions might be rightly considered as produced by the motion of one wheel revolving in the inside of another wheel, these wheels were creations of their minds, added to the facts which they perceived by sense. And even if the wheels were no longer supposed to be material, but were reduced to mere geometrical spheres or circles, they were not the less products of the mind alone—something additional to the facts observed. The same is the case in all other discoveries. The facts are known, but they are insulated and unconnected, till the discoverer supplies from his own store a principle of connection. The pearls are there, but they will not hang together till some one provides the string.
>
> (Whewell 1858b, 72–73)

Mill objects that colligation, the providing of a "principle of connection," the string for the pearls, is not true induction. Colligation is, rather, invention, and as Mill puts it, "Invention, though it can be cultivated, can not be reduced to rule" (Mill 1882, 208).

While Whewell argues that colligation can allow for induction by providing "something additional to the facts observed," a general conception or idea, Mill argues that induction is a kind of direct inference from the facts according to a rule, not an invention: "Induction [. . .] is that operation of the mind, by which we infer that what we know to be true in a particular case or cases, will be true in all cases which resemble the former in certain assignable respects." Mill's objection in this case is to Whewell's argument that a scientist's construction of a concept, a "supplying" of a principle of connection "from his own store," is an induction or the basis of one. Rather, the principles of induction must be "operations of the mind" that can be "reduced to rule." This rules out colligation as a principle of induction.

Mill cites Whewell's example of Johannes Kepler's use of Tycho Brahe's astronomical data to conceive that the planets follow elliptical orbits, and not circular orbits as in the Ptolemaic system. According to Mill, Whewell "expresses himself as if" Kepler's novel hypothesis of elliptical orbits "was something added to the facts," that is, a colligation of Tycho's data (Mill 1882, 215). For Whewell, the elliptical orbit hypothesis is the string that connects the pearls, where the pearls are the observations of the position of Mars.

According to Mill, however, the elliptical orbit hypothesis is not an induction, because it does not go beyond the facts: "Kepler did not *put* what he had conceived into the facts, but *saw* it in them" (Mill 1882, 216). According to Mill, Whewell's analysis of the Kepler case blends together a real induction and a false one. The "colligation" is a false induction, as it only summarizes the import of the facts themselves; just as a series of observations that one can sail all the way around a land mass in the ocean can be summarized using the concept "island." According to Mill, Kepler's use of the concept "ellipse" for the positions of Mars is the same as applying the concept "island" to that series of observations. The only real "induction" Kepler performed, according to Mill, was in predicting that Mars will *continue* in an elliptical orbit. That prediction goes beyond the observed facts, and uses a known rule of induction to do so.

In "Mr. Mill's Logic," Whewell responds:

> There is no definite and stable distinction between Facts and Theories; Facts and Laws; Facts and Inductions. Inductions, Laws, Theories, which are true, *are* Facts. Facts involve Inductions. It is a fact that the moon is attracted by the earth, just as much as it is a Fact that an apple falls from a tree.
>
> (Whewell 1860, §19)

Whewell can concede, then, that once we have sailed all around a land mass, it is "in the facts" that it is an island. But he argues that something can be an induction, or at least a result of induction, and a fact at the same time. It might take a particular conception to be able to *see* the facts as connected in particular way, and thus to recognize another true fact or state of affairs:

> To hit upon the right conception is a difficult step; and when this step is once made, the facts assume a different aspect from what they had before: that done, they are seen in a new point of view; and the catching this point of view, is a special mental operation, requiring special endowments and habits of thought. Before this, the facts are seen as detached, separate, lawless; afterwards, they are seen as connected,

> simple, regular; as parts of one general fact, and thereby possessing innumerable new relations before unseen. Kepler, then, I say, bound together the facts by superinducing upon them the conception of an ellipse; and this was an essential element in his Induction.
>
> (Whewell 1860, §24)

Finally, Whewell argues to an extent against Mill's claim that invention, unlike induction, cannot be "reduced to a rule." Whewell observes that there is a single "operation" "of the mind" involved in colligation. Although we may not be able to give the rule for that operation in any given case, it always follows the same method:

> And there is the same essential element in all Inductive discoveries. In all cases, facts, before detached and lawless, are bound together by a new thought. They are reduced to law, by being seen in a new point of view. To catch this new point of view, is an act of the mind, springing from its previous preparation and habits. The facts, in other discoveries, are brought together according to other relations, or, as I have called them, Ideas;—the Ideas of Time, of Force, of Number, of Resemblance, of Elementary Composition, of Polarity, and the like. But in all cases, the mind performs the operation by an apprehension of some such relations; by singling out the one true relation; by combining the apprehension of the true relation with the facts; by applying to them the Conception of such a relation.
>
> (Whewell 1860, §25)

One can give a general rule for the operation of the mind in the colligation of facts, even though one may not be able to give the rule for finding the conception necessary to colligate the facts.

One may not immediately think of Pierre Duhem in this context, but the well-known chapter of *The Aim and Structure of Physical Theory*, "Physical Theory and Experiment," in which Duhem presents the "Duhem problem" (see Part I Introduction) contains a discussion of more general problems in scientific discovery, induction, and hypotheses, and even of the Mill–Whewell debate and the Kepler example.

Duhemian "holism," the notion that an entire theory is confronted with evidence rather than particular statements, informs a third perspective on the Mill–Whewell debate. On the holist view of scientific theories, the role of the analysis of Whewellian conceptions is not restricted to the colligation of particular facts. Rather, the scientist's ingenuity is needed to compare entire theories in terms of how well they account for given facts.

Duhem uses the example of the move from Kepler's laws to Newton's laws. According to Duhem, *"The principle of universal gravity, very far from being derivable by generalization and induction from the observational laws of Kepler, formally contradicts these laws. If Newton's theory is correct, Kepler's laws are necessarily false"* (Duhem 1954 [1914], 193, emphasis in original). In establishing whether Newton's or Kepler's system is correct, "It is no longer a matter of taking, one by one, laws justified by observation, and raising each of them by induction and generalization to the rank of a principle; it is a matter of comparing the corollaries of a whole group of hypotheses to a whole group of facts" (*ibid.*, 194). With respect to the Whewell–Mill dispute over colligation, then, Duhem adds a wrinkle: the question is not restricted to whether Kepler "saw" the ellipse in the facts, or whether he "discovered the right conception" using his own invention. It is also a matter of whether the consequences of Kepler's entire system of laws accounts for the facts, and how it accounts for the facts; whether there are other systems that also account for them; and whether a rival such system may do so better.

Duhem's analysis also illuminates a further implication of the Mill–Whewell debate: the rival accounts of scientific discovery. Mill had argued that induction, but not invention, could be reduced to "rules." Whewell argued that colligation could be characterized broadly as an "operation of the mind," but was well aware of the diversity of scientific methods, as is made evident in his lengthy histories (e.g., Whewell 1847, 1971 [1860], 1857, and 1858a). Duhem denies that all scientific results can be derived "by generalization and induction," since some, including Newton's, involve "comparing the corollaries of a whole group of hypotheses to a whole group of facts."

Here, we might recall the discussion of Kuhn and the logical empiricists in Part I. The D-N (deductive-nomological) account has been used, in some but not all cases, to argue that scientific discovery proceeds by codifying observations with certain properties (such as regularity) as laws of nature by means of induction, and then justifying particular results by deduction. Kuhn argued, at least early on, that scientific discovery eludes rational reconstruction.

In 1960, one of the pioneers of the history and philosophy of science, Norwood Russell Hanson, asked "Is there a logic of scientific discovery?" Hanson is arguing for a fair hearing for those, including Aristotle and Charles Sanders Peirce, who claim that there can be a logic, not a psychology, a history, or a sociology, of scientific discovery (Hanson 1960, 92).

Hanson argues that we can give a logical account of why a certain *"kind*
of hypothesis" will be appropriate to solving a particular scientific problem (*ibid.*, 93ff.). He concedes that one might challenge him on the basis that the

reasons for considering certain kinds of hypotheses might indeed be socio-logical rather than logical: political or religious reasons, for instance (*ibid.*, 94). But all Hanson has to establish is that there are some cases in which this consideration is logical. He points out that analogies between types of phenomena (this planet ought to behave like other planets, for instance), or a "formal symmetry in sets of equations or arguments," have provided logi-cal reasons to consider certain kinds of hypotheses as appropriate to certain problems (*ibid.*, 97). These do not exhaust the possible logical reasons for considering, formulating, and testing hypotheses in a given context, but they are linked to concrete examples.

Hanson goes on to argue that accounts that neglect to investigate the reasons for formulating and considering hypotheses, that neglect how the logic of discovery unfolds in history, risk giving "misleading" and inac-curate accounts of scientific practice (Hanson 1960, 98–99). Hanson con-cludes with an analysis of why and how scientists *eliminate* whole classes of hypotheses, another relatively unexplored area of the history and philoso-phy of science in practice.

More recent work on the logic of discovery takes distinct approaches. One looks for repeatable steps in formulating theories and hypotheses. This approach sometimes is applied to causal modeling, for instance. While Spirtes, Glymour, and Scheines 2001 is not intended entirely as a text on the logic of discovery, it has obvious implications for the search for novel causal relationships. Other formal approaches include those of Simon 1973, Zytkow and Simon 1988 (here see, e.g., Downes 1990 and McLaughlin 1982), Kelly 1987 and 1996, and Leuridan 2009, among others. Another approach is more informally methodological, and searches for techniques that are used in scientific practice. Kelly 1987 and 1996 also contain some arguments along these lines, as does Hanson 1960 and 1958, along with the work cited here by Peirce, Lakatos, Whewell, and others. Feferman 1978, Lakatos 1976, and Yuxin 1990 explore the question of whether there is a logic of mathematical discovery.

There is also recent work critical of the notion of universal discoverable rules for induction or discovery. Norton 2010a, 2010b, and 2003 sets out the foundation of a theory of "material induction." Laudan 1971 and 1980 deal in part with the question of why the search for a logic of discovery stalled in the 1970s and 1980s. Zahar 1983, Turrisi 1990, Shah 2007, and others explore the question of whether this search should resume.

References and Further Reading

Achinstein, Peter. 1992. "Inference to the Best Explanation: Or, Who Won the Mill–Whewell debate?" *Studies in History and Philosophy of Science* 23 (2): 349–364.

Buchdahl, Gerd. 1991. "Deductivist versus Inductivist Approaches in the Philosophy of Science as Illustrated by Some Controversies Between Whewell and Mill," pp. 311–344 in Fisch and Schaffer (eds.).

Butts, Robert. 1973. "Whewell's Logic of Induction," pp. 53–85 in Ron Giere and Richard Westfall (eds.), *Foundations of Scientific Method*. Bloomington: Indiana University Press.

Downes, Stephen. 1990. "Herbert Simon's Computational Models of Scientific Discovery," *PSA: Proceedings of the Biennial Meeting of the Philosophy of Science Association* 1990: 97–108.

Feferman, Solomon. 1978. "The Logic of Mathematical Discovery vs. the Logical Structure of Mathematics," *PSA: Proceedings of the Biennial Meeting of the Philosophy of Science Association* 1978: 309–327.

Fisch, Menachem. 1991. *William Whewell, Philosopher of Science*, Oxford: Oxford University Press.

——. 1985a. "Necessary and Contingent Truth in William Whewell's Antithetical Theory of Knowledge," *Studies in History and Philosophy of Science* 16: 275–314.

——. 1985b. "Whewell's Consilience of Inductions: An Evaluation," *Philosophy of Science* 52: 239–255.

Fisch, Menachem and Simon Schaffer (eds.). 1991. *William Whewell: A Composite Portrait*. Oxford: Oxford University Press.

Hanson, Norwood Russell. 1958. "The Logic of Discovery," *Journal of Philosophy* 55 (25): 1073–1089.

——. 1960. "More on 'the Logic of Discovery,'" *Journal of Philosophy* 57 (6): 182–188.

Hesse, Mary. 1968. "Consilience of Inductions," pp. 232–247 in Imre Lakatos (ed.), *The Problem of Inductive Logic*. Amsterdam: North Holland Publication Co.

——. 1971. "Whewell's Consilience of Inductions and Predictions [Reply to Laudan]," *Monist* 55: 520–24.

Heidelberger, Michael and Schiemann, Gregor (eds.). 2009. *The Significance of the Hypothetical in the Natural Sciences*. Berlin: De Gruyter.

Hutton, R. H. 1850. "Mill and Whewell on the Logic of Induction," *The Prospective Review* 6: 77–111.

Kelly, Kevin T. 1996. *The Logic of Reliable Inquiry*. Oxford: Oxford University Press.

——. 1987. "The Logic of Discovery," *Philosophy of Science* 54 (3): 435–452.

Kleiner, Scott. 1988. "The Logic of Discovery and Darwin's Pre-Malthusian Researches," *Biology and Philosophy* 3 (3): 293–315.

Lakatos, Imre. 1976. *Proofs and Refutations: The Logic of Mathematical Discovery*. Cambridge: Cambridge University Press.

Laudan, Larry. 1971. "William Whewell on the Consilience of Inductions," *Monist* 55: 368–391.

——. 1980. "Why Was the Logic of Discovery Abandoned?" pp. 173–183, in T. Nickles (ed.), *Scientific Discovery, Logic, and Rationality*, Dordrecht: D. Reidel.

Leuridan, Bert. 2009. "Causal Discovery and the Problem of Ignorance: An Adaptive Logic Approach," *Journal of Applied Logic* 7 (2): 188–205.

Losee, John. 1983. "Whewell and Mill on the Relation between Science and Philosophy of Science," *Studies in History and Philosophy of Science* 14: 113–126.

McLaughlin, Robert. 1982. "Invention and Induction: Laudan, Simon, and the Logic of Discovery," *Philosophy of Science* 49 (2): 198–211.

Norton, John D. 2010a. "Deductively Definable Logics of Induction," *Journal of Philosophical Logic* 39 (6): 617–665.

——. 2010b. "There Are No Universal Rules for Induction," *Philosophy of Science* 77 (5): 765–777.

——. 2003. "A Material Theory of Induction," *Philosophy of Science* 70 (4): 647–670.

Peirce, Charles Sanders. 1931. *Collected Papers of Charles Sanders Peirce.* Cambridge, Mass.: Harvard University Press.

Rosenberg, Alex. 2013. *Philosophy of Science.* New York and London: Routledge.

Shah, Mehul. 2007. "Is It Justifiable to Abandon All Search for a Logic of Discovery?" *International Studies in the Philosophy of Science* 21 (3): 253–269.

Simon, Herbert A. 1973. "Does Scientific Discovery Have a Logic?" *Philosophy of Science* 40 (4): 471–480.

Snyder, Laura. 2011. *The Philosophical Breakfast Club: Four Remarkable Men Who Transformed Science and Changed the World.* New York: Broadway Books.

———. 2008. "The Whole Box of Tools: William Whewell and the Logic of Induction," pp. 165–230, in John Woods and Dov Gabbay (eds.), *The Handbook of the History of Logic* Vol. VIII. Dordrecht: Kluwer.

———. 2006. *Reforming Philosophy: A Victorian Debate on Science and Society.* Chicago: University of Chicago Press.

———. 2005. "Confirmation for a Modest Realism," *Philosophy of Science* 72: 839–849.

———. 1999. "Renovating the Novum Organum: Bacon, Whewell and Induction," *Studies in History and Philosophy of Science* 30: 531–557.

———. 1997a. "The Mill–Whewell Debate: Much Ado about Induction," *Perspectives on Science* 5: 159–198.

———. 1997b. "Discoverers' Induction," *Philosophy of Science* 64: 580–604.

———. 1994. "It's All Necessarily So: William Whewell on Scientific Truth," *Studies in History and Philosophy of Science* 25: 785–807.

Spirtes, Peter, Glymour, Clark, and Scheines, Richard. 2001. *Causation, Prediction, and Search,* second ed. Cambridge, Mass.: MIT Press.

Strong, E. W. 1955. "William Whewell and John Stuart Mill: Their Controversy over Scientific Knowledge," *Journal of the History of Ideas* 16: 209–231.

Turrisi, Patricia A. 1990. "Peirce's Logic of Discovery: Abduction and the Universal Categories," *Transactions of the Charles S. Peirce Society* 26 (4): 465–497.

Whewell, William. 1847. *The Philosophy of the Inductive Sciences, Founded Upon Their History.* 2nd edition, in two volumes, London: John W. Parker.

———. 1857. *History of the Inductive Sciences, from the Earliest to the Present Time,* 3rd edition, in two volumes, London: John W. Parker.

———. 1858a. *The History of Scientific Ideas,* in two volumes, London: John W. Parker.

———. 1858b. *Novum Organon Renovatum (being the second part of the philosophy of the inductive sciences),* third edition., London: John W. Parker.

———. 1971 [1860]. *On the Philosophy of Discovery: Chapters Historical and Critical.* New York: Lenox Hill. Originally published 1860, London: John W. Parker.

Yuxin, Zheng. 1990. "From the Logic of Mathematical Discovery to the Methodology of Scientific Research Programmes," *British Journal for the Philosophy of Science* 41 (3): 377–399.

Zahar, Elie. 1983. "Logic of Discovery or Psychology of Invention?" *British Journal for the Philosophy of Science* 34 (3): 243–261.

Zytkow, Jan and Simon, Herbert. 1988. "Normative Systems of Discovery and Logic of Search," *Synthese* 74 (1): 65–90.

CHAPTER 13

Of Inductions Improperly So Called

JOHN STUART MILL

Pages 207–228 of *A System of Logic, Ratiocinative and Inductive*. New York: Harper & Brothers, Publishers, 1882.

Preliminary Observations on Induction in General

§ 1. THE portion of the present inquiry upon which we are now about to enter, may be considered as the principal, both from its surpassing in intricacy all the other branches, and because it relates to a process which has been shown in the preceding Book to be that in which the investigation of nature essentially consists. We have found that all Inference, consequently all Proof, and all discovery of truths not self-evident, consists of inductions, and the interpretation of inductions: that all our knowledge, not intuitive, comes to us exclusively from that source. What Induction is, therefore, and what conditions render it legitimate, can not but be deemed the main question of the science of logic—the question which includes all others. It is, however, one which professed writers on logic have almost entirely passed over. The generalities of the subject have not been altogether neglected by metaphysicians; but, for want of sufficient acquaintance with the processes by which science has actually succeeded in establishing general truths, their analysis of the inductive operation, even when unexceptionable as to correctness, has not been specific enough to be made the foundation of practical rules, which might be for induction itself what the rules of the syllogism are for the interpretation of induction: while those by whom physical science has been carried to its present state of improvement—and who, to arrive at a complete theory of the process, needed only to generalize, and adapt to all varieties of problems, the methods which they themselves employed in their habitual pursuits—never until very lately made any serious attempt to philosophize on the subject, nor regarded the mode in which they arrived at their conclusions as deserving of study, independently of the conclusions themselves.

§ 2. For the purposes of the present inquiry, Induction may be defined, the operation of discovering and proving general propositions. It is true that (as already shown) the process of indirectly ascertaining individual facts, is as truly inductive as that by which we establish general truths. But it is not a different kind of induction; it is a form of the very same process: since, on the one hand, generals are but collections of particulars, definite in kind but indefinite in number; and on the other hand, whenever the evidence which we derive from observation of known eases justifies us in drawing an inference respecting even one unknown case, we should on the same evidence be justified in drawing a similar inference with respect to a wider class of cases. The inference either does not hold at all, or it holds in all cases of a certain description; in all cases which, in certain definable respects, resemble those we have observed.

If these remarks are just; if the principles and rules of inference are the same whether we infer general propositions or individual facts; it follows that a complete logic of the sciences would be also a complete logic of practical business and common life. Since there is no case of legitimate inference from experience, in which the conclusion may not legitimately be a general proposition; an analysis of the process by which general truths are arrived at, is virtually an analysis of all induction whatever. Whether we are inquiring into a scientific principle or into an individual fact, and whether we proceed by experiment or by ratiocination, every step in the train of inferences is essentially inductive, and the legitimacy of the induction depends in both cases on the same conditions.

True it is that in the case of the practical inquirer, who is endeavoring to ascertain facts not for the purposes of science but for those of business, such, for instance, as the advocate or the judge, the chief difficulty is one in which the principles of induction will afford him no assistance. It lies not in making his inductions, but in the selection of them; in choosing from among all general propositions ascertained to be true, those which furnish marks by which he may trace whether the given subject possesses or not the predicate in question. In arguing a doubtful question of fact before a jury, the general propositions or principles to which the advocate appeals are mostly, in themselves, sufficiently trite, and assented to as soon as stated: his skill lies in bringing his case under those propositions or principles; in calling to mind such of the known or received maxims of probability as admit of application to the case in hand, and selecting from among them those best adapted to his object. Success is here dependent on natural or acquired sagacity, aided by knowledge of the particular subject, and of subjects allied with it. Invention, though it can be cultivated, can not be reduced to rule; there is no science which will enable a man to bethink himself of that which will suit his purpose.

But when he *has* thought of something, science can tell him whether that which he has thought of will suit his purpose or not. The inquirer or arguer must be guided by his own knowledge and sagacity in the choice of the inductions out of which he will construct his argument. But the validity of the argument when constructed, depends on principles, and must be tried by tests which are the same for all descriptions of inquiries, whether the result be to give A an estate, or to enrich science with a new general truth. In the one case and in the other, the senses, or testimony, must decide on the individual facts; the rules of the syllogism will determine whether, those facts being supposed correct, the case really falls within the formulæ of the different inductions under which it has been successively brought; and finally, the legitimacy of the inductions themselves must be decided by other rules, and these it is now our purpose to investigate. If this third part of the operation be, in many of the questions of practical life, not the most, but the least arduous portion of it, we have seen that this is also the case in some great departments of the field of science; in all those which are principally deductive, and most of all in mathematics; where the inductions themselves are few in number, and so obvious and elementary that they seem to stand in no need of the evidence of experience, while to combine them so as to prove a given theorem or solve a problem, may call for the utmost powers of invention and contrivance with which our species is gifted.

If the identity of the logical processes which prove particular facts and those which establish general scientific truths, required any additional confirmation, it would be sufficient to consider that in many branches of science, single facts have to be proved, as well as principles; facts as completely individual as any that are debated in a court of justice; but which are proved in the same manner as the other truths of the science, and without disturbing in any degree the homogeneity of its method. A remarkable example of this is afforded by astronomy. The individual facts on which that science grounds its most important deductions, such facts as the magnitudes of the bodies of the solar system, their distances from one another, the figure of the earth, and its rotation, are scarcely any of them accessible to our means of direct observation: they are proved indirectly, by the aid of inductions founded on other facts which we can more easily reach. For example, the distance of the moon from the earth was determined by a very circuitous process. The share which direct observation had in the work consisted in ascertaining, at one and the same instant, the zenith distances of the moon, as seen from two points very remote from one another on the earth's surface. The ascertainment of these angular distances ascertained their supplements; and since the angle at the earth's centre subtended by the distance between the two places of observation was deducible by spherical trigonometry from the latitude and longitude of those places, the angle at the moon subtended by the same line became the fourth angle of a quadrilateral of which the other three angles were known. The four angles being thus ascertained, and two sides of the quadrilateral being radii of the earth; the two remaining sides and the diagonal, or, in other words, the moon's distance from the two places of observation and from the centre of the earth, could be ascertained, at least in terms of the earth's radius, from elementary theorems of geometry. At each step in this demonstration a new induction is taken in, represented in the aggregate of its results by a general proposition.

Not only is the process by which an individual astronomical fact was thus ascertained, exactly similar to those by which the same science establishes its general truths, but also (as we have shown to be the case in all legitimate reasoning) a general proposition might have been concluded instead of a single fact. In strictness, indeed, the result of the reasoning *is* a general proposition; a theorem respecting the distance, not of the moon in particular, but of any inaccessible object; showing in what relation that distance stands to certain other quantities. And although the moon is almost the only heavenly body the distance of which from the earth can really be thus ascertained, this is merely owing to the accidental circumstances of the other heavenly bodies, which render them incapable of affording such data as the application of the theorem requires; for the theorem itself is as true of them as it is of the moon.[1]

We shall fall into no error, then, if in treating of Induction, we limit our attention to the establishment of general propositions. The principles and rules of Induction as directed to this end, are the principles and rules of all Induction; and the logic of Science is the universal Logic, applicable to all inquiries in which man can engage.

Of Inductions Improperly So Called

§ 1. INDUCTION, then, is that operation of the mind, by which we infer that what we know to be true in a particular case or cases, will be true in all cases which resemble the former in certain assignable respects. In other words, Induction is the process by which we conclude that what is true of certain individuals of a class is true of the whole class, or that what is true at certain times will be true in similar circumstances at all times.

This definition excludes from the meaning of the term Induction, various logical operations, to which it is not unusual to apply that name.

Induction, as above defined, is a process of inference; it proceeds from the known to the unknown; and any operation involving no inference, any process in which what seems the conclusion is no wider than the premises from which it is drawn, does not fall within the meaning of the term. Yet in the common books of Logic we find this laid down as the most perfect, indeed the only quite perfect, form of induction. In those books, every process which sets out from a less general and terminates in a more general expression—which admits of being stated in the form, "This and that A are B, therefore every A is B"—is called an induction, whether any thing be really concluded or not: and the induction is asserted not to be perfect, unless every single individual of the class A is included in the antecedent, or premise: that is, unless what we affirm of the class has already been ascertained to be true of every individual in it, so that the nominal conclusion is not really a conclusion, but a mere re-assertion of the premises. If we were to say, All the planets shine by the sun's light, from observation of each separate planet, or All the Apostles were Jews, because this is true of Peter, Paul, John, and every other apostle—these, and such as these, would, in the phraseology in question, be called perfect, and the only perfect, Inductions. This, however, is a totally different kind of induction from ours; it is not an inference from facts known to facts unknown, but a mere short-hand registration of facts known. The two simulated arguments which we have quoted, are not generalizations; the propositions purporting to be conclusions from them, are not really general propositions. A general proposition is one in which the predicate is affirmed or denied of an unlimited number of individuals; namely, all, whether few or many, existing or capable of existing, which possess the properties connoted by the subject of the proposition. "All men are mortal" does not mean all now living, but all men past, present, and to come. When the signification of the term is limited so as to render it a name not for any and every individual falling under a certain general description, but only for each of a number of individuals, designated as such, and as it were counted off individually, the proposition, though it may be general in its language, is no general proposition, but merely that number of singular propositions, written in an abridged character. The operation may be very useful, as most forms of abridged notation are; but it is no part of the investigation of truth, though often bearing an important part in the preparation of the materials for that investigation.

As we may sum up a definite number of singular propositions in one proposition, which will be apparently, but not really, general, so we may sum up a definite number of general propositions in one proposition, which will be apparently, but not really, more general. If by a separate induction applied to every distinct species of animals, it has been established that each possesses a nervous system, and we affirm thereupon that all animals have a nervous system; this looks like a generalization, though as the conclusion merely affirms of all what has already been affirmed of each, it seems to tell us nothing but what we knew before. A distinction, however, must be made. If in concluding that all animals have a nervous system, we mean the same thing and no more as if we had said "all known animals," the proposition is not general, and the process by which it is arrived at is not induction. But if our meaning is that the observations made of the various species of animals have discovered to us a law of animal nature, and that we are in a condition to say that a nervous system will be found even in animals yet undiscovered, this indeed is an induction; but in this case the general proposition contains more than the sum of the special propositions from which it is inferred. The distinction is still more forcibly brought out when we consider, that if this real generalization be legitimate at all, its legitimacy probably does not require that we should bave examined without exception every known species. It is the number and nature of the instances, and not their being the whole of those which happen to be known, that makes them sufficient evidence to prove a general law: while the more limited assertion, which stops at all known animals, can not be made unless we have rigorously verified it in every species. In like manner (to return to a former example) we might have inferred, not that all *the* planets, but that all *planets*, shine by reflected light: the former is no induction; the latter is an induction, and a bad one, being disproved by the case of double stars—self-luminous bodies which are properly planets, since they revolve round a centre. [. . .]

§ 3. There remains a third improper use of the term Induction, which it is of real importance to clear up, because the theory of Induction has been, in no ordinary degree, confused by it, and because the confusion is exemplified in the most recent and elaborate treatise on the inductive philosophy which exists in our language. The error in question is that of confounding a mere description, by general terms, of a set of observed phenomena, with an induction from them.

Suppose that a phenomenon consists of parts, and that these parts are only capable of being observed separately, and as it were piecemeal. When the observations have been made, there is a convenience (amounting for many purposes to a necessity) in obtaining a representation of the phenomenon as a whole, by combining, or as we may say, piecing these detached fragments together. A navigator sailing in the midst of the ocean discovers land: he can not at first, or by any one observation, determine whether it is a continent or an island; but he coasts along it, and after a few days finds himself to have sailed completely round it: he then pronounces it an island. Now there was no particular time or place of observation at which he could perceive that this land was entirely surrounded by water: he ascertained the fact by a succession of partial observations, and then selected a general expression which summed up in two or three words the whole of what he so observed. But is there any thing of the nature of an induction in this process? Did he infer any thing that had not been observed, from something else which had? Certainly not. He had observed the whole of what the proposition asserts. That the land in question is an island, is not an inference from the partial facts which the navigator saw in the course of his circumnavigation; it is the facts themselves; it is a summary of those facts; the description of a complex fact, to which those simpler ones are as the parts of a whole.

Now there is, I conceive, no difference in kind between this simple operation, and that by which Kepler ascertained the nature of the planetary orbits: and Kepler's operation, all at least that was characteristic in it, was not more an inductive act than that of our supposed navigator.

The object of Kepler was to determine the real path described by each of the planets, or let us say by the planet Mars (since it was of that body that he first established the two of his three laws which did not require a comparison of planets). To do this there was no other mode than that of direct observation: and all which observation could do was to ascertain a great number of the successive places of the planet; or rather, of its apparent places. That the planet occupied successively all these positions, or at all events, positions which produced the same impressions on the eye, and that it passed from one of these to another insensibly, and without any apparent breach of continuity; thus much the senses, with the aid of the proper instruments, could ascertain. What Kepler did more than this, was to find what sort of a curve these different points would make, supposing them to be all joined together. He expressed the whole series of the observed places of Mars by what Dr. Whewell calls the general conception of an ellipse. This operation was far from being as easy as that of the navigator who expressed the series of his observations on successive points of the coast by the general conception of an island. But it is the very same sort of operation; and if the one is not an induction but a description, this must also be true of the other.

The only real induction concerned in the case, consisted in inferring that because the observed places of Mars were correctly represented by points in an imaginary ellipse, therefore Mars would continue to revolve in that same ellipse; and in concluding (before the gap had been filled up by further observations) that the positions of the planet during the time which intervened between two observations, must have coincided with the inter-mediate points of the curve. For these were facts which had not been directly observed. They were inferences from the observations; facts inferred, as distinguished from facts seen. But these inferences were so far from being a part of Kepler's philosophical operation, that they had been drawn long before he was born. Astronomers had long known that the planets periodically returned to the same places. When this had been ascertained, there was no induction left for Kepler to make, nor did he make any further induction. He merely applied his new conception to the facts inferred, as he did to the facts observed. Knowing already that the planets continued to move in the same paths; when he found that an ellipse correctly represented the past path, he knew that it would represent the future path. In finding a compendious expression for the one set of facts, he found one for the other: but he found the expression only, not the inference; nor did he (which is the true test of a general truth) add any thing to the power of prediction already possessed.

§ 4. The descriptive operation which enables a number of details to be summed up in a single proposition, Dr. Whewell, by an aptly chosen expression, has termed the Colligation of Facts. In most of his observations concerning that mental process I fully agree, and would gladly transfer all that portion of his book into my own pages. I only think him mistaken in setting up this kind of operation, which according to the old and received meaning of the term, is not induction at all, as the type of induction generally; and laying down, throughout his work, as principles of induction, the principles of mere colligation.

Dr. Whewell maintains that the general proposition which binds together the particular facts, and makes them, as it were, one fact, is not the mere sum of those facts, but something more, since there is introduced a conception of the mind, which did not exist in the facts themselves. "The particular facts," says he,[3] "are not merely brought together, but there is a new element added to the combination by the very act of thought by which they are combined. . . . When the Greeks, after long observing the motions of the planets, saw that these motions might be rightly considered as produced by the motion of one wheel revolving in the inside of another wheel, these wheels were creations of their minds, added to the facts which they perceived by sense. And even if the wheels were no longer supposed to be material, but were reduced to mere geometrical spheres or circles, they were not the less products of the mind alone—something additional to the facts observed. The same is the case in all other discoveries. The facts are known, but they are insulated and unconnected, till the discoverer supplies from his own store a principle of connection. The pearls are there, but they will not hang together till some one provides the string."

Let me first remark that Dr. Whewell, in this passage, blends together, indiscriminately, examples of both the processes which I am endeavoring to distinguish from one another. When the Greeks abandoned the supposition that the planetary motions were produced by the revolution of material wheels, and fell back upon the idea of "mere geometrical spheres or circles," there was more in this change of opinion than the mere substitution of an ideal curve for a physical one. There was the abandonment of a theory, and the replacement of it by a mere description. No one would think of calling the doctrine of material wheels a mere description. That doctrine was an attempt to point out the force by which the planets were acted upon, and compelled to move in their orbits. But when, by a great step in philosophy, the materiality of the wheels was discarded, and the geometrical forms alone retained, the attempt to account for the motions was given up, and what was left of the theory was a mere description of the orbits. The assertion that the planets were carried round by wheels revolving in the inside of other wheels, gave place to the proposition, that they moved in the same lines which would be traced by bodies so carried: which was a mere mode of representing the sum of the observed facts; as Kepler's was another and a better mode of representing the same observations.

It is true that for these simply descriptive operations, as well as for the erroneous inductive one, a conception of the mind was required. The conception of an ellipse must have presented itself to Kepler's mind, before he could identify the planetary orbits with it. According to Dr. Whewell, the conception was something added to the facts. He expresses himself as if Kepler had put something into the facts by his mode of conceiving them. But Kepler did no such thing. The ellipse was in the facts before Kepler recognized it; just as the island was an island before it had been sailed round. Kepler did not *put* what he had conceived into the facts, but *saw* it in them. A conception implies, and corresponds to, something conceived: and though the conception itself is not in the facts, but in our mind, yet if it is to convey any knowledge relating to them, it must be a conception *of* something which really is in the facts, some property which they actually possess, and which they would manifest to our senses, if our senses were able to take cognizance of it. If, for instance, the planet left behind it in space a visible track, and if the observer were in a fixed position at such a distance from the plane of the orbit as would enable him to see the whole of it at once, he would see it to be an ellipse; and if gifted with appropriate instruments and powers of locomotion, he could prove it to be such by measuring its different dimensions. Nay, further: if the track were visible, and he were so placed that he could see all parts of it in succession, but not all of them at once, he might be able, by piecing together his successive observations, to discover both that it was an ellipse and that the planet moved in it. The case would then exactly resemble that of the navigator who discovers the land to be an island by sailing round it. If the path was visible, no one I think would dispute that to identify it with an ellipse is to describe it: and I can not see why any difference should be made by its not being directly an object of sense, when every point in it is as exactly ascertained as if it were so.

Subject to the indispensable condition which has just been stated, I do not conceive that the part which conceptions have in the operation of studying facts, has ever been overlooked or undervalued. No one ever disputed that in order to reason about any thing we must have a conception of it; or that when we include a multitude of things under a general expression, there is implied in the expression a conception of something common to those things. But it by no means follows that the conception is necessarily pre-existent, or constructed by the mind out of its own materials. If the facts are rightly classed under the conception, it is because there is in the facts themselves something of which the conception is itself a copy; and which if we can not directly perceive, it is because of the limited power of our organs, and not because the thing itself is not there. The conception itself is often obtained by abstraction from the very facts which, in Dr. Whewell's language, it is afterward called in to connect. This he himself admits, when he observes (which he does on several occasions), how great a service would be rendered to the science of physiology by the philosopher "who should establish a precise, tenable, and consistent conception of life."[4] Such a conception can only be abstracted from the phenomena of life itself; from the very facts which it is put in requisition to connect. In other cases, no doubt, instead of collecting the conception from the very phenomena which we are attempting to colligate, we select it from among those which have been previously collected by abstraction from other facts. In the instance of Kepler's laws, the latter was the case. The facts being out of the reach of being observed, in any such manner as would have enabled the senses to identify directly the path of the planet, the conception requisite for framing a general description of that path could not be collected by abstraction from the observations themselves; the mind had to supply hypothetically, from among the conceptions it had obtained from other portions of its experience, some one which would correctly represent the series of the observed facts. It had to frame a supposition respecting the general course of the phenomenon, and ask itself, If this be the general description, what will the details be? and then compare these with the details actually observed. If they agreed, the hypothesis would serve for a description of the phenomenon: if not, it was necessarily abandoned, and another tried. It is such a case as this which gives rise to the doctrine that the mind, in framing the descriptions, adds something of its own which it does not find in the facts.

Yet it is a fact surely, that the planet does describe an ellipse; and a fact which we could see, if we had adequate visual organs and a suitable position. Not having these advantages, but possessing the conception of an ellipse, or (to express the meaning in less technical language) knowing what an ellipse was, Kepler tried whether the observed places of the planet were consistent with such a path. He found they were so; and he, consequently, asserted as a fact that the planet moved in an ellipse. But this fact, which Kepler did not add to, but found in, the motions of the planet, namely, that it occupied in succession the various points in the circumference of a given ellipse, was the very fact, the separate parts of which had been separately observed; it was the sum of the different observations.

Having stated this fundamental difference between my opinion and that of Dr. Whewell, I must add, that his account of the manner in which a conception is selected, suitable to express the facts, appears to me perfectly just. The experience of all thinkers will, I believe, testify that the process is tentative; that it consists of a succession of guesses; many being rejected, until one at last occurs fit to be chosen. We know from Kepler himself that before hitting upon the "conception" of an ellipse, he tried nineteen other imaginary paths, which, finding them inconsistent with the observations, he was obliged to reject. But as Dr. Whewell truly says, the successful hypothesis, though a guess, ought generally to be called, not a lucky, but a skillful guess. The guesses which serve to give mental unity and wholeness to a chaos of scattered particulars, are accidents which rarely occur to any minds but those abounding in knowledge and disciplined in intellectual combinations.

How far this tentative method, so indispensable as a means to the colligation of facts for purposes of description, admits of application to Induction itself, and what functions belong to it in that department, will be considered in the chapter of the present Book which relates to Hypotheses. On the present occasion we have chiefly to distinguish this process of Colligation from Induction properly so called; and that the distinction may be made clearer, it is well to advert to a curious and interesting remark, which is as strikingly true of the former operation, as it appears to me unequivocally false of the latter.

In different stages of the progress of knowledge, philosophers have employed, for the colligation of the same order of facts, different conceptions. The early rude observations of the heavenly bodies, in which minute precision was neither attained nor sought, presented nothing inconsistent with the representation of the path of a planet as an exact circle, having the earth for its centre. As observations increased in accuracy, facts were disclosed which were not reconcilable with this simple supposition: for the colligation of those additional facts, the supposition was varied; and varied again and again as facts became more numerous and precise. The earth was removed from the centre to some other point within the circle; the planet was supposed to revolve in a smaller circle called an epicycle, round an imaginary point which revolved in a circle round the earth: in proportion as observation elicited fresh facts contradictory to these representations, other epicycles and other eccentrics were added, producing additional complication; until at last Kepler swept all these circles away, and substituted the conception of an exact ellipse. Even this is found not to represent with complete correctness the accurate observations of the present day, which disclose many slight deviations from an orbit exactly elliptical. Now Dr. Whewell has remarked that these successive general expressions, though apparently so conflicting, were all correct: they all answered the purpose of colligation; they all enabled the mind to represent to itself with facility, and by a simultaneous glance, the whole body of facts at the time ascertained: each in its turn served as a correct description of the phenomena, so far as the senses had up to that time taken cognizance of them. If a necessity afterward arose for discarding one of these general descriptions of the planet's orbit, and framing a different imaginary line, by which to express the series of observed positions, it was because a number of new facts had now been added, which it was necessary to combine with the old facts into one general description. But this did not affect the correctness of the former expression, considered as a general statement of the only facts which it was intended to represent. And so true is this, that, as is well remarked by M. Comte, these ancient generalizations, even the rudest and most imperfect of them, that of uniform movement in a circle, are so far from being entirely false, that they are even now habitually employed by astronomers when only a rough approximation to correctness is required. "Modern astronomy, in irrevocably destroying primitive hypotheses that were seen as the real laws of the world, has maintained carefully their positive and permanent value: the property of conveniently representing phenomena when it is a matter of giving a first sketch. Our resources in this respect are still more extensive, precisely because we do not deceive ourselves as to the reality of hypotheses; this permits us to employ without scruple, case by case, the one that we judge the most advantageous."[5]

Dr. Whewell's remark, therefore, is philosophically correct. Successive expressions for the colligation of observed facts, or, in other words, successive descriptions of a phenomenon as a whole, which has been observed only in parts, may, though conflicting, be all correct as far as they go. But it would surely be absurd to assert this of conflicting inductions.

The scientific study of facts may be undertaken for three different purposes: the simple description of the facts; their explanation; or their prediction: meaning by prediction, the determination of the conditions under which similar facts may be expected again to occur. To the first of these three operations the name of Induction does not properly belong: to the other two it does. Now, Dr. Whewell's observation is true of the first alone. Considered as a mere description, the circular theory of the heavenly motions represents perfectly well their general features: and by adding epicycles without limit, those motions, even as now known to us, might be expressed with any degree of accuracy that might be required. The elliptical theory, as a mere description, would have a great advantage in point of simplicity, and in the consequent facility of conceiving it and reasoning about it; but it would not really be more true than the other. Different descriptions, therefore, may be all true: but not, surely, different explanations. The doctrine that the heavenly bodies moved by a virtue inherent in their celestial nature; the doctrine that they were moved by impact (which led to the hypothesis of vortices as the only impelling force capable of whirling bodies in circles), and the Newtonian doctrine, that they are moved by the composition of a centripetal with an original projectile force; all these are explanations, collected by real induction from supposed parallel cases; and they were all successively received by philosophers, as scientific truths on the subject of the heavenly bodies. Can it be said of these, as was said of the different descriptions, that they are all true as far as they go? Is it not clear that only one can be true in any degree, and the other two must be altogether false? So much for explanations: let us now compare different predictions: the first, that eclipses will occur when one planet or satellite is so situated as to cast its shadow upon another; the second, that they will occur when some great calamity is impending over mankind. Do these two doctrines only differ in the degree of their truth, as expressing real facts with unequal degrees of accuracy? Assuredly the one is true, and the other absolutely false.[6]

In every way, therefore, it is evident that to explain induction as the colligation of facts by means of appropriate conceptions, that is, conceptions which will really express them, is to confound mere description of the observed facts with inference from those facts, and ascribe to the latter what is a characteristic property of the former.

There is, however, between Colligation and Induction, a real correlation, which it is important to conceive correctly. Colligation is not always induction; but induction is always colligation. The assertion that the planets move in ellipses, was but a mode of representing observed facts; it was but a colligation; while the assertion that they are drawn, or tend, toward the sun, was the statement of a new fact, inferred by induction. But the induction, once made, accomplishes the purposes of colligation likewise. It brings the same facts, which Kepler had connected by his conception of an ellipse, under the additional conception of bodies acted upon by a central force, and serves, therefore, as a new bond of connection for those facts; a new principle for their classification.

Further, the descriptions which arc improperly confounded with induction, are nevertheless a necessary preparation for induction; no less necessary than correct observation of the facts themselves. Without the previous colligation of detached observations by means of one general conception, we could never have obtained any basis for an induction, except in the case of phenomena of very limited compass. We should not be able to affirm any predicates at all, of a subject incapable of being observed otherwise than piecemeal: much less could we extend those predicates by induction to other similar subjects. Induction, therefore, always presupposes, not only that the necessary observations are made with the necessary accuracy, but also that the results of these observations are, so far as practicable, connected together by general descriptions, enabling the mind to represent to itself as wholes whatever phenomena are capable of being so represented.

§ 5. Dr. Whewell has replied at some length to the preceding observations, restating his opinions, but without (as far as I can perceive) adding any thing material to his former arguments. Since, however, mine have not had the good fortune to make any impression upon him, I will subjoin a few remarks, tending to show more clearly in what our difference of opinion consists, as well as, in some measure, to account for it.

Nearly all the definitions of induction, by writers of authority, make it consist in drawing inferences from known cases to unknown; affirming of a class, a predicate which has been found true of some cases belonging to the class; concluding because some things have a certain property, that other things which resemble them have the same property—or because a thing has manifested a property at a certain time, that it has and will have that property at other times.

It will scarcely be contended that Kepler's operation was an Induction in this sense of the term. The statement, that Mars moves in an elliptical orbit, was no generalization from individual cases to a class of cases. Neither was it an extension to all time, of what had been found true at some particular time. The whole amount of generalization which the ease admitted of, was already completed, or might have been so. Long before the elliptic theory was thought of, it had been ascertained that the planets returned periodically to the same apparent places; the series of these places was, or might have been, completely determined, and the apparent course of each planet marked out on the celestial globe in an uninterrupted line. Kepler did not extend an observed truth to other cases than those in which it had been observed: he did not widen the *subject* of the proposition which expressed the observed facts. The alteration he made was in the predicate. Instead of saying, the successive places of Mars are so and so, he summed them up in the statement, that the successive places of Mars are points in an ellipse. It is true, this statement, as Dr. Whewell says, was not the sum of the observations *merely;* it was the sum of the observations *seen under a new point of view.*[7] But it was not the sum of *more* than the observations, as a real induction is. It took in no cases but those which had been actually observed, or which could have been inferred from the observations before the new point of view presented itself. There was not that transition from known cases to unknown, which constitutes Induction in the original and acknowledged meaning of the term.

Old definitions, it is true, can not prevail against new knowledge: and if the Keplerian operation, as a logical process, be really identical with what takes place in acknowledged induction, the definition of induction ought to be so widened as to take it in; since scientific language ought to adapt itself to the true relations which subsist between the things it is employed to designate. Here then it is that I am at issue with Dr. Whewell. He does think the operations identical. He allows of no logical process in any case of induction, other than what there was in Kepler's case, namely, guessing until a guess is found which tallies with the facts; and accordingly, as we shall see hereafter, he rejects all canons of induction, because it is not by means of them that we guess. Dr. Whewell's theory of the logic of science would be very perfect if it did not pass over altogether the question of Proof. But in my apprehension there is such a thing as proof, and inductions differ altogether from descriptions in their relation to that element. Induction is proof; it is inferring something unobserved from something observed: it requires, therefore, an appropriate test of proof; and to provide that test, is the special purpose of inductive logic. When, on the contrary, we merely collate known observations, and, in Dr. Whewell's phraseology, connect them by means of a new conception; if the conception does serve to connect the observations, we have all we want. As the proposition in which it is embodied pretends to no other truth than what it may share with many other modes of representing the same facts, to be consistent with the facts is all it requires: it neither needs nor admits of proof; though it may serve to prove other things, inasmuch as, by placing the facts in mental connection with other facts, not previously seen to resemble them, it assimilates the case to another class of phenomena, concerning which real Inductions have already been made. Thus Kepler's so-called law brought the orbit of Mars into the class ellipse, and by doing so, proved all the properties of an ellipse to be true of the orbit: but in this proof Kepler's law supplied the minor premise, and not (as is the case with real Inductions) the major.

Dr. Whewell calls nothing Induction where there is not a new mental conception introduced, and every thing induction where there is. But this is to confound two very different things, Invention and Proof. The introduction of a new conception belongs to Invention: and invention may be required in any operation, but is the essence of none. A new conception may be introduced for descriptive purposes, and so it may for inductive purposes. But it is so far from constituting induction, that induction does not necessarily stand in need of it. Most inductions require no conception but what was present in every one of the particular instances on which the induction is grounded. That all men are mortal is surely an inductive conclusion; yet no new conception is introduced by it. Whoever knows that any man has died, has all the conceptions involved in the inductive generalization. But Dr. Whewell considers the process of invention which consists in framing a new conception consistent with the facts, to be not merely a necessary part of all induction, but the whole of it.

The mental operation which extracts from a number of detached observations certain general characters in which the observed phenomena resemble one another, or resemble other known facts, is what Bacon, Locke, and most subsequent metaphysicians, have understood by the word Abstraction. A general expression obtained by abstraction, connecting known facts by means of common characters, but without concluding from them to unknown, may, I think, with strict logical correctness, be termed a Description; nor do I know in what other way things can ever be described. My position, however, does not depend on the employment of that particular word; I am quite content to use Dr. Whewell's term Colligation, or the more general phrases, "mode of representing, or of expressing, phenomena": provided it be clearly seen that the process is not Induction, but something radically different.

What more may usefully be said on the subject of Colligation, or of the correlative expression invented by Dr. Whewell, the Explication of Conceptions, and generally on the subject of ideas and mental representations as connected with the study of facts, will find a more appropriate place in the Fourth Book, on the Operations Subsidiary to Induction: to which I must refer the reader for the removal of any difficulty which the present discussion may have left.

Notes

1. Dr. Whewell thinks it improper to apply the term Induction to any operation not terminating in the establishment of a general truth. Induction, he says (*Philosophy of Discovery*, p. 243), "is not the same thing as experience and observation. Induction is experience or observation *consciously* looked at in a *general* form. This consciousness and generality are necessary parts of that knowledge which is science." And he objects (p. 241) to the mode in which the word Induction is employed in this work, as an undue extension of that term "not only to the cases in which the general induction is consciously applied to a particular instance, but to the cases in which the particular instance is dealt with by means of experience in that rude sense in which experience can be asserted of brutes, and in which of course we can in no way imagine that the law is possessed or understood as a general proposition." This use of the term he deems a "confusion of knowledge with practical tendencies."

I disclaim, as strongly as Dr. Whewell can do, the application of such terms as induction, inference, or reasoning, to operations performed by mere instinct, that is, from an animal impulse, without the exertion of any intelligence. But I perceive no ground for confining the use of those terms to cases in which the inference is drawn in the forms and with the precautions required by scientific propriety. To the idea of Science, an express recognition and distinct apprehension of general laws as such, is essential: but nine-tenths of the conclusions drawn from experience in the course of practical life, are drawn without any such recognition: they are direct inferences from known cases, to a case supposed to be similar. I have endeavored to show that this is not only as legitimate an operation, but substantially the same operation, as that of ascending from known cases to a general proposition; except that the latter process has one great security for correctness which the former does not possess. In science, the inference must necessarily pass through the intermediate stage of a general proposition, because Science wants its conclusions for record, and not for instantaneous use. But the inferences drawn for the guidance of practical affairs, by persons who would often be quite incapable of expressing in unexceptionable terms the corresponding generalizations, may and frequently do exhibit intellectual powers quite equal to any which have ever been displayed in science; and if these inferences are not inductive, what are they? The limitation imposed on the term by Dr. Whewell seems perfectly arbitrary: neither justified by any fundamental distinction between what he includes and what he desires to exclude, nor sanctioned by usage, at least from the time of Reid and Stewart, the principal legislators (as far as the English language is concerned) of modern metaphysical terminology.

3. *Novum Organum Renovatum*, pp. 72, 73.
4. *Norum Organum Renovatum*, p. 32.
5. *Cours de philosophie positive*, vol. ii., p. 202.
6. Dr. Whewell, in his reply, contests the distinction here drawn, and maintains, that not only different descriptions, but different explanations of a phenomenon, may all be true. Of the three theories respecting the motions of the heavenly bodies, he says (*Philosophy of Discovery*, p. 231): "Undoubtedly all these explanations may be true and consistent with each other, and would be so if each had been followed out so as to show in what manner it could be made consistent with the facts. And this was, in reality, in a great measure done. The doctrine that the heavenly bodies were moved by vortices was successfully modified, so that it came to coincide in its results with the doctrine of an inverse-quadratic centripetal force. . . . . . . When this point was reached, the vortex was merely a machinery, well or ill devised, for producing such a centripetal force, and therefore did not contradict the doctrine of a centripetal force. Newton himself does not appear to have been averse to explaining gravity by impulse. So little is it true that if one theory be true the other must be false. The attempt to explain gravity by the impulse of streams of particles flowing through the universe in all directions, which I have mentioned in the *Philosophy*, is so far from being inconsistent with the Newtonian theory, that it is founded entirely upon it. And even with regard to the doctrine, that the heavenly bodies move by an inherent virtue; if this doctrine had been maintained in any such way that it was brought to agree with the facts, the inherent virtue must have had its laws determined; and then it would have been found that the virtue had a reference to the central body; and so, the 'inherent virtue' must have coincided in its effect with the Newtonian force; and then, the two explanations would agree, except so far as the word 'inherent' was concerned. And if such a part of an earlier theory as this word *inherent* indicates, is found to be untenable, it is of course rejected in the transition to later and more exact theories, in Inductions of this kind, as well as in what Mr. Mill calls Descriptions. There is, therefore, still no validity discoverable in the distinction which Mr. Mill attempts to draw between descriptions like Kepler's law of elliptical orbits, and other examples of induction."

If the doctrine of vortices had meant, not that vortices existed, but only that the planets moved *in the same manner* as if they had been whirled by vortices; if the hypothesis had been merely a mode of representing the facts, not an attempt to account for them; if, in short, it had been only a Description; it would, no doubt, have been reconcilable with the Newtonian theory. The vortices, however, were not a mere aid to conceiving the motions of the planets, but a supposed physical agent, actively impelling them; a material fact, which might be true or not true, but could not be both true and not true. According to Descartes's theory it was true, according to Newton's it was not true. Dr. Whewell probably means that since the phrases, centripetal and projectile force, do not declare the nature but only the direction of the forces, the Newtonian theory does not absolutely contradict any hypothesis which may be framed respecting the mode of their production. The Newtonian theory, regarded as a mere *description* of the planetary motions, does not; but the Newtonian theory as an *explanation* of them does. For in what does the explanation consist? In ascribing those motions to a general law which obtains between all particles of matter, and in identifying this with the law by which bodies fall to the ground. If the planets are kept in their orbits by a force which draws the particles composing them toward every other particle of matter in the solar system, they are not kept in those orbits by the impulsive force of certain streams of matter which whirl them round. The one explanation absolutely excludes the other. Either the planets are not moved by vortices, or they do not move by a law common to all matter. It is impossible that both opinions can be true. As well might it be said that there is no contradiction between the assertions, that a man died because somebody killed him, and that he died a natural death.

So, again, the theory that the planets move by a virtue inherent in their celestial nature, is incompatible with either of the two others: either that of their being moved by vortices, or that which regards them as moving by a property which they have in common with the earth and all terrestrial bodies. Dr. Whewell says that the theory of an inherent virtue agrees with Newton's when the word inherent is left out, which of course it would be (he says) if "found to be untenable." But leave that out, and where is the theory? The word inherent *is* the theory. When that is omitted, there remains nothing except that the heavenly bodies move "by a virtue," *i.e.*, by a power of some sort; or by virtue of their celestial nature, which directly contradicts the doctrine that terrestrial bodies fall by the same law.

If Dr. Whewell is not yet satisfied, any other subject will serve equally well to test his doctrine. He will hardly say that there is no contradiction between the emission theory and the undulatory theory of light; or that there can be both one and two electricities; or that the hypothesis of the production of the higher organic forms by development from the lower, and the supposition of separate and successive acts of creation, are quite reconcilable; or that the theory that volcanoes are fed from a central fire, and the doctrines which ascribe them to chemical action at a comparatively small depth below the earth's surface, are consistent with one another, and all true as far as they go.

If different explanations of the same fact can not both be true, still less, surely, can different predictions. Dr. Whewell quarrels (on what ground it is not necessary here to consider) with the example I had chosen on this point, and thinks an objection to an illustration a sufficient answer to a theory. Examples not liable to his objection are easily found, if the proposition that conflicting predictions can not both be true, can be made clearer by any examples. Suppose the phenomenon to be a newly-discovered comet, and that one astronomer predicts its return once in every 300 years—another once in every 400: can they both be right? When Columbus predicted that by sailing constantly westward he should in time return to the point from which he set out, while others asserted that he could never do so except by turning back, were both he and his opponents true prophets? Were the predictions which foretold the wonders of railways and steamships, and those which averred that the Atlantic could never be crossed by steam navigation, nor a railway train propelled ten miles an hour, both (in Dr. Whewell's words) "true, and consistent with one another?"

Dr. Whewell sees no distinction between holding contradictory opinions on a question of fact, and merely employing different analogies to facilitate the conception of the same fact. The case of different Inductions belongs to the former class, that of different Descriptions to the latter.

7. *Phil. of Discov.*, p. 256.

References

Comte, Auguste. 1975/1830–1842. *Cours de philosophie positive*. Paris: Hermann, 2 vols. First edition was translated and abridged by Harriet Martineau as *The positive philosophy of Auguste Comte*, London: J. Chapman, 1853.

Whewell, William. 1847. *The Philosophy of the Inductive Sciences, Founded Upon Their History*. 2nd edition, in two volumes, London: John W. Parker.

——. 1857. *History of the Inductive Sciences, from the Earliest to the Present Time*, 3rd edition, in two volumes, London: John W. Parker.

——. 1858a. *The History of Scientific Ideas*, in two volumes, London: John W. Parker.

——. 1858b. *Novum Organon Renovatum*, London: John W. Parker.

——. 1971 [1860]. *On the Philosophy of Discovery: Chapters Historical and Critical.* New York: Lenox Hill. Originally published 1860, London: John W. Parker.

CHAPTER **14**

Mr. Mill's Logic

WILLIAM WHEWELL[1]

Pages 238–262 and 270–288 from *On the Philosophy of Discovery*. New York: Burt Franklin, 1860.

THE *History of the Inductive Sciences* was published in 1837, and the *Philosophy of the Inductive Sciences* in 1840. In 1843 Mr. Mill published his *System of Logic*, in which he states that without the aid derived from the facts and ideas in my volumes, the corresponding portion of his own would most probably not have been written, and quotes parts of what I have said with commendation. He also, however, dissents from me on several important and fundamental points, and argues against what I have said thereon. I conceive that it may tend to bring into a clearer light the doctrines which I have tried to establish, and the truth of them, if I discuss some of the differences between us, which I shall proceed to do.[2] [. . .]

1. *What is Induction?*—1. Confining myself, then, to the material sciences, I shall proceed to offer my remarks on Induction with especial reference to Mr. Mill's work. And in order that we may, as I have said, proceed as intelligibly as possible, let us begin by considering what we mean by *Induction*, as a mode of obtaining truth; and let us note whether there is any difference between Mr. Mill and me on this subject.

"For the purposes of the present inquiry," Mr. Mill says (i. 347[3]), "Induction may be defined the operation of discovering and forming general propositions"; meaning, as appears by the context, the discovery of them

182

from particular facts. He elsewhere (i. 370) terms it "generalization from experience": and again he speaks of it with greater precision as the inference of a more general proposition from less general ones.

2. Now to these definitions and descriptions I assent as far as they go; though, as I shall have to remark, they appear to me to leave unnoticed a feature which is very important, and which occurs in all cases of Induction, so far as we are concerned with it. Science, then, consists of general propositions, inferred from particular facts, or from less general propositions, by Induction; and it is our object to discern the nature and laws of *Induction* in this sense. That the propositions are general, or are more general than the facts from which they are inferred, is an indispensable part of the notion of Induction, and is essential to any discussion of the process, as the mode of arriving at Science, that is, at a body of general truths. [. . .] *Induction* must be confined to cases where we have in our minds general propositions, in order that the sciences, which are our most instructive examples of the process we have to consider, may be, in any definite and proper sense, *Inductive* Sciences.

10. Perhaps some persons may be inclined to say that this difference of opinion, as to the extent of meaning which is to be given to the term *Induction*, is a question merely of words; a matter of definition only. This is a mode in which men in our time often seem inclined to dispose of philosophical questions; thus evading the task of forming an opinion upon such questions, while they retain the air of looking at the subject from a more comprehensive point of view. But as I have elsewhere said, such questions of definition are never questions of definition merely. A proposition is always implied along with the definition; and the truth of the proposition depends upon the settlement of the definition. This is the case in the present instance. We are speaking of *Induction*, and we mean that kind of Induction by which the sciences now existing among men have been constructed. On this account it is, that we cannot include, in the meaning of the term, mere practical tendencies or practical habits; for science is not constructed of these. No accumulation of these would make up any of the acknowledged sciences. The elements of such sciences are something of a kind different from practical habits. The elements of such sciences are principles which we *know;* truths which can be contemplated as being *true.* [. . .]

II. *Induction or Description?*—15. In the cases hitherto noticed, Mr. Mill extends the term *Induction,* as I think, too widely, and applies it to cases to which it is not rightly applicable. I have now to notice a case of an opposite kind, in which he does not apply it where I do, and condemns me for using it in such a case. I had spoken of Kepler's discovery of the Law, that the planets move round the sun in ellipses, as an example of Induction. The separate facts of any planet (Mars, for instance,) being in certain places at certain times, are all included in the general proposition which Kepler discovered, that Mars describes an ellipse of a certain form and position. This appears to me a very simple but a very distinct example of the operation of discovering general propositions; general, that is, with reference to particular facts; which operation Mr. Mill, as well as myself, says is Induction. But Mr. Mill denies this operation in this case to be Induction at all (i. 357). I should not have been prepared for this denial by the previous parts of Mr. Mill's book, for he had said just before (i. 350), "such facts as the magnitudes of the bodies of the solar system, their distances from each other, the figure of the earth and its rotation . . . are proved indirectly, by the aid of inductions founded on other facts which we can more easily reach." If the figure of the earth and its rotation are proved by Induction, it seems very strange, and is to me quite incomprehensible, how the figure of the earth's orbit and its revolution (and of course, of the figure of Mars's orbit and his revolution in like manner,) are not also proved by Induction. No, says Mr. Mill, Kepler, in putting together a number of places of the planet into one figure, only performed an act of *description*. "This descriptive operation," he adds (i. 359), "Mr. Whewell, by an aptly chosen expression, has termed Colligation of Facts." He goes on to commend my observations concerning this process, but says that, according to the old and received meaning of the term, it is not Induction at all.

16. Now I have already shown that Mr. Mill himself, a few pages earlier, had applied the term *Induction* to cases undistinguishable from this in any essential circumstance. And even in this case, he allows that Kepler did really perform an act of Induction (i. 358), "namely, in concluding that, because the observed places of Mars were correctly represented by points in an imaginary ellipse, therefore Mars would continue to revolve in that same ellipse; and even in concluding that the position of the planet during the time which had intervened between the two observations must have coincided with the intermediate points of the curve." Of course, in Kepler's Induction, of which I speak, I include all this; all this is included in speaking of the *orbit* of Mars: a continuous line, a periodical motion, are implied in the term *orbit*. I am unable to see what would remain of Kepler's discovery, if we take from it these conditions. It would not only not be an induction, but it would not be a description, for it would not recognize that Mars moved in an orbit. Are particular positions to be conceived as points in a curve, without thinking of the intermediate positions as belonging to the same curve? If so, there is no law at all, and the facts are not bound together by any intelligible tie.

In another place (ii. 209) Mr. Mill returns to his distinction of Description and Induction; but without throwing any additional light upon it, so far as I can see.

17. The only meaning which I can discover in this attempted distinction of Description and Induction is, that when particular facts are bound together by their relation in *space*, Mr. Mill calls the discovery of the connexion *Description*, but when they are connected by other general relations, as time, cause and the like, Mr. Mill terms the discovery of the connexion *Induction*. And this way of making a distinction, would fall in with the doctrine of other parts of Mr. Mill's book, in which he ascribes very peculiar attributes to space and its relations, in comparison with other Ideas, (as I should call them). But I cannot see any ground for this distinction, of connexion according to space and other connexions of facts.

To stand upon such a distinction, appears to me to be the way to miss the general laws of the formation of science. For example: The ancients discovered that the planets revolved in recurring periods, and thus connected the observations of their motions according to the Idea of *Time*. Kepler discovered that they revolved in ellipses, and thus connected the observations according to the Idea of *Space*. Newton discovered that they revolved in virtue of the Sun's attraction, and thus connected the motions according to the Idea of *Force*. The first and third of these discoveries are recognized on all hands as processes of Induction. Why is the second to be called by a different name? or what but confusion and perplexity can arise from refusing to class it with the other two? It is, you say, Description. But such Description is a kind of Induction, and must be spoken of as Induction, if we are to speak of Induction as the process by which Science is formed: for the three steps are all, the second in the same sense as the first and third, in co-ordination with them, steps in the formation of astronomical science.

18. But, says Mr. Mill (i. 363), "it is a fact surely that the planet does describe an ellipse, and a fact which we could see if we had adequate visual organs and a suitable position." To this I should reply: "Let it be so; and it is a fact, surely, that the planet does move periodically: it is a fact, surely, that the planet is attracted by the sun. Still, therefore, the asserted distinction fails to find a ground." Perhaps Mr. Mill would remind us that the elliptical form of the orbit is a fact which we could see if we had adequate visual organs and a suitable position: but that force is a thing which we cannot see. But this distinction also will not bear handling. Can we not see a tree blown down by a storm, or a rock blown up by gunpowder? Do we not here see force:—see it, that is, by its effects, the only way in which we need to see it in the case of a planet, for the purposes of our argument? Are not such operations of force, Facts which may be the objects of sense? and is not the operation of the sun's Force a Fact of the same kind, just as much as the elliptical form of orbit which results from the action? If the latter be "surely a Fact," the former is a Fact no less surely.

19. In truth, as I have repeatedly had occasion to remark, all attempts to

frame an argument by the exclusive or emphatic appropriation of the term *Fact* to particular cases, are necessarily illusory and inconclusive. There is no definite and stable distinction between Facts and Theories; Facts and Laws; Facts and Inductions. Inductions, Laws, Theories, which are true, *are* Facts. Facts involve Inductions. It is a fact that the moon is attracted by the earth, just as much as it is a Fact that an apple falls from a tree. That the former fact is collected by a more distinct and conscious Induction, does not make it the less a Fact. That the orbit of Mars is a Fact—a true Description of the path—does not make it the less a case of Induction.

20. There is another argument which Mr. Mill employs in order to show that there is a difference between mere colligation which is description, and induction in the more proper sense of the term. He notices with commendation a remark which I had made (i. 364), that at different stages of the progress of science the facts had been successfully connected by means of very different conceptions, while yet the later conceptions have not contradicted, but included, so far as they were true, the earlier: thus the ancient Greek representation of the motions of the planets by means of epicycles and eccentrics, was to a certain degree of accuracy true, and is not negatived, though superseded, by the modern representation of the planets as describing ellipses round the sun. And he then reasons that this, which is thus true of Descriptions, cannot be true of Inductions. He says (i. 367), "Different descriptions therefore may be all true: but surely not different explanations." He then notices the various explanations of the motions of the planets—the ancient doctrine that they are moved by an inherent virtue; the Cartesian doctrine that they are moved by impulse and by vortices; the Newtonian doctrine that they are governed by a central force; and he adds, "Can it be said of these, as was said of the different descriptions, that they are all true as far as they go? Is it not true that one only can be true in any degree, and that the other two must be altogether false?"

21. And to this questioning, the history of science compels me to reply very distinctly and positively, in the way which Mr. Mill appears to think extravagant and absurd. I am obliged to say, Undoubtedly, all these explanations *may* be true and consistent with each other, and would be so if each had been followed out so as to show in what manner it could be made consistent with the facts. And this was, in reality, in a great measure done[4]. The doctrine that the heavenly bodies were moved by vortices was successively modified, so that it came to coincide in its results with the doctrine of an inverse-quadratic centripetal force, as I have remarked in the *History*.[5] When this point was reached, the vortex was merely a machinery, well or ill devised, for producing such a centripetal force, and therefore did not contradict the doctrine of a centripetal force. Newton himself does not appear to have been averse to explaining gravity by impulse. So little is it true that if

the one theory be true the other must be false. The attempt to explain gravity by the impulse of streams of particles flowing through the universe in all directions, which I have mentioned in the *Philosophy*,[6] is so far from being inconsistent with the Newtonian theory, that it is founded entirely upon it. And even with regard to the doctrine, that the heavenly bodies move by an inherent virtue; if this doctrine had been maintained in any such way that it was brought to agree with the facts, the inherent virtue must have had its laws determined; and then, it would have been found that the virtue had a reference to the central body; and so, the "inherent virtue" must have coincided in its effect with the Newtonian force; and then, the two explanations would agree, except so far as the word "inherent" was concerned. And if such a part of an earlier theory as this word *inherent* indicates, is found to be untenable, it is of course rejected in the transition to later and more exact theories, in Inductions of this kind, as well as in what Mr. Mill calls Descriptions. There is therefore still no validity discoverable in the distinction which Mr. Mill attempts to draw between "descriptions" like Kepler's law of elliptical orbits, and other examples of induction. [. . .]

III. *In Discovery a new Conception is introduced.*—24. There is a difference between Mr. Mill and me in our view of the essential elements of this Induction of Kepler, which affects all other cases of Induction, and which is, I think, the most extensive and important of the differences between us. I must therefore venture to dwell upon it a little in detail.

I conceive that Kepler, in discovering the law of Mars's motion, and in asserting that the planet moved in an ellipse, did this;—he bound together particular observations of separate places of Mars by the notion, or, as I have called it, the *conception*, of an *ellipse*, which was supplied by his own mind. Other persons, and he too, before he made this discovery, had present to their minds the facts of such separate successive positions of the planet; but could not bind them together rightly, because they did not apply to them this conception of an *ellipse*. To supply this conception, required a special preparation, and a special activity in the mind of the discoverer. He, and others before him, tried other ways of connecting the special facts, none of which fully succeeded. To discover such a connexion, the mind must be conversant with certain relations of space, and with certain kinds of figures. To discover the right figure was a matter requiring research, invention, resource. To hit upon the right conception is a difficult step; and when this step is once made, the facts assume a different aspect from what they had before: that done, they are seen in a new point of view; and the catching this point of view, is a special mental operation, requiring special endowments and habits of thought. Before this, the facts are seen as detached, separate, lawless; afterwards, they are seen as connected, simple, regular; as parts of one general fact, and thereby possessing innumerable new relations before unseen. Kepler, then, I say, bound together the facts by superinducing upon them the *conception* of an *ellipse;* and this was an essential element in his Induction.

25. And there is the same essential element in all Inductive discoveries. In all cases, facts, before detached and lawless, are bound together by a new thought. They are reduced to law, by being seen in a new point of view. To catch this new point of view, is an act of the mind, springing from its previous preparation and habits. The facts, in other discoveries, are brought together according to other relations, or, as I have called them, *Ideas;*—the Ideas of Time, of Force, of Number, of Resemblance, of Elementary Composition, of Polarity, and the like. But in all cases, the mind performs the operation by an apprehension of some such relations; by singling out the one true relation; by combining the apprehension of the true relation with the facts; by applying to them the Conception of such a relation.

26. In previous writings, I have not only stated this view generally, but I have followed it into detail, exemplifying it in the greater part of the History of the principal Inductive Sciences in succession. I have pointed out what are the Conceptions which have been introduced in every prominent discovery in those sciences; and have noted to which of the above Ideas, or of the like Ideas, each belongs. The performance of this task is the office of the greater part of my *Philosophy of the Inductive Sciences.* For that work is, in reality, no less historical than the *History* which preceded it. The *History of the Inductive Sciences* is the history of the discoveries, mainly so far as concerns the *Facts* which were brought together to form sciences. The *Philosophy* is, in the first ten Books, the history of the *Ideas* and *Conceptions*, by means of which the facts were connected, so as to give rise to scientific truths. It would be easy for me to give a long list of the Ideas and Conceptions thus brought into view, but I may refer any reader who wishes to see such a list, to the Tables of Contents of the *History*, and of the first ten Books of the *Philosophy.*

27. That these Ideas and Conceptions are really distinct elements of the scientific truths thus obtained, I conceive to be proved beyond doubt, not only by considering that the discoveries never were made, nor could be made, till the right Conception was obtained, and by seeing how difficult it often was to obtain this element; but also, by seeing that the Idea and the Conception itself, as distinct from the Facts, was, in almost every science, the subject of long and obstinate controversies;—controversies which turned upon the possible relations of Ideas, much more than upon the actual relations of Facts. The first ten Books of the *Philosophy* to which I have referred, contain the history of a great number of these controversies. These controversies make up a large portion of the history of each science; a portion quite as important as the study of the facts; and a portion, at every

stage of the science, quite as essential to the progress of truth. Men, in seek-ing and obtaining scientific knowledge, have always shown that they found the formation of right conceptions in their own minds to be an essential part of the process.

28. Moreover, the presence of a Conception of the mind as a special ele-ment of the inductive process, and as the tie by which the particular facts are bound together, is further indicated, by there being some special new *term* or *phrase* introduced in every induction; or at least some term or phrase thenceforth steadily applied to the facts, which had not been applied to them before; as when Kepler asserted that Mars moved round the sun in an *elliptical orbit,* or when Newton asserted that the planets *gravitate* towards the sun; these new terms, *elliptical orbit,* and *gravitate,* mark the new conceptions on which the inductions depend. I have in the *Philosophy*[7] further illustrated this application of "technical terms," that is, fixed and settled terms, in every inductive discovery; and have spoken of their use in enabling men to proceed from each such discovery to other discoveries more general. But I notice these terms here, for the purpose of showing the existence of a conception in the discoverer's mind, corresponding to the term thus introduced; which conception, the term is intended to convey to the minds of those to whom the discovery is communicated.

29. But this element of discovery,—right conceptions supplied by the mind in order to bind the facts together,—Mr. Mill denies to be an element at all. He says, of Kepler's discovery of the elliptical orbit (i. 363), "It super-added nothing to the particular facts which it served to bind together"; yet he adds, "except indeed the knowledge that a resemblance existed between the planetary orbit and other ellipses"; that is, except the knowledge that it *was* an ellipse;—precisely the circumstance in which the discovery con-sisted. Kepler, he says, "asserted as a fact that the planet moved in an ellipse. But this fact, which Kepler did not add to, but found in the motion of the planet . . . was the very fact, the separate parts of which had been separately observed; it was the sum of the different observations."

30. That the fact of the elliptical motion was not merely the *sum* of the different observations, is plain from this, that other persons, and Kepler himself before his discovery, did not find it by adding together the obser-vations. The fact of the elliptical orbit was not the sum of the observa-tions *merely;* it was the sum of the observations, *seen under a new point of view,* which point of view Kepler's mind supplied. Kepler found it in the facts, because it was there, no doubt, for one reason; but also, for another, because he had, in his mind, those relations of thought which enabled him to find it. We may illustrate this by a familiar analogy. We too find the law in Kepler's book; but if we did not understand Latin, we should not find it there. We must learn Latin in order to find the law in the book. In like

manner, a discoverer must know the language of science, as well as look at the book of nature, in order to find scientific truth. All the discussions and controversies respecting Ideas and Conceptions of which I have spoken, may be looked upon as discussions and controversies respecting the grammar of the language in which nature speaks to the scientific mind. Man is the *Interpreter* of Nature; not the Spectator merely, but the Interpreter. The study of the language, as well as the mere sight of the characters, is requisite in order that we may read the inscriptions which are written on the face of the world. And this study of the language of nature, that is, of the necessary coherencies and derivations of the relations of phenomena, is to be pursued by examining Ideas, as well as mere phenomena;—by tracing the formation of Conceptions, as well as the accumulation of Facts. And this is what I have tried to do in the books already referred to.

31. Mr. Mill has not noticed, in any considerable degree, what I have said of the formation of the Conceptions which enter into the various sciences; but he has, in general terms, denied that the Conception is anything different from the facts themselves. "If," he says (i. 301), "the facts are rightly classed under the conceptions, it is because there is in the facts themselves, something of which the conception is a copy." But it is a copy which cannot be made by a person without peculiar endowments; just as a person cannot copy an ill-written inscription, so as to make it convey sense, unless he understand the language. "Conceptions," Mr. Mill says (ii. 217), "do not develope themselves from within, but are impressed from without." But what comes from without is not enough: they must have both origins, or they cannot make knowledge. "The conception," he says again (ii. 221), "is not furnished *by* the mind till it has been furnished *to* the mind." But it is furnished to the mind by its own activity, operating according to its own laws. No doubt, the conception may be formed, and in cases of discovery, must be formed, by the suggestion and excitement which the facts themselves produce; and must be so moulded as to agree with the facts. But this does not make it superfluous to examine, out of what *materials* such conceptions are formed, and *how* they are capable of being moulded so as to express laws of nature; especially, when we see how large a share this part of discovery—the examination how our ideas can be modified so as to agree with nature,—holds, in the history of science.

32. I have already (Art. 28) given, as evidence that the conception enters as an element in every induction, the constant introduction in such cases, of a new fixed term or phrase. Mr. Mill (ii. 282) notices this introduction of a new phrase in such cases as important, though he does not appear willing to allow that it is necessary. Yet the necessity of the conception at least, appears to result from the considerations which he puts forward. "What darkness," he says, "would have been spread over geometrical demonstration, if wher-

ever the word *circle* is used, the definition of a circle was inserted instead of it." "If we want to make a particular combination of ideas permanent in the mind, there is nothing which clenches it like a name specially devoted to express it." In my view, the new conception is the *nail* which connects the previous notions, and the name, as Mr. Mill says, *clenches* the junction.

33. I have above (Art. 30) referred to the difficulty of getting hold of the right conception, as a proof that induction is not a mere juxtaposition of facts. Mr. Mill does not dispute that it is often difficult to hit upon the right conception. He says (i. 360), "that a conception of the mind is introduced, is indeed most certain, and Mr. Whewell has rightly stated elsewhere, that to hit upon the right conception is often a far more difficult, and more meritorious achievement, than to prove its applicability when obtained. But," he adds, "a conception implies and corresponds to something conceived; and although the conception itself is not in the facts, but in our mind, it must be a conception of something which really is in the facts." But to this I reply, that its being really in the facts, does not help us at all towards knowledge, if we cannot see it there. As the poet says,

It is the mind that sees: the outward eyes
Present the object, but the mind descries.

And this is true of the sight which produces knowledge, as well as of the sight which produces pleasure and pain, which is referred to in the Tale.

34. Mr. Mill puts his view, as opposed to mine, in various ways, but, as will easily be understood, the answers which I have to offer are in all cases nearly to the same effect. Thus, he says (ii. 216), "the tardy development of several of the physical sciences, for example, of Optics, Electricity, Magnetism, and the higher generalizations of Chemistry, Mr. Whewell ascribes to the fact that mankind had not yet possessed themselves of the idea of Polarity, that is, of opposite properties in opposite directions. But what was there to suggest such an idea, until by a separate examination of several of these different branches of knowledge it was shown that the facts of each of them did present, in some instances at least, the curious phenomena of opposite properties in opposite directions?" But on this I observe, that these facts did not, nor do yet, present this conception to ordinary minds. The opposition of properties, and even the opposition of directions, which are thus apprehended by profound cultivators of science, are of an abstruse and recondite kind; and to conceive any one kind of polarity in its proper generality, is a process which few persons hitherto appear to have mastered; still less, have men in general come to conceive of them all as modifications of a general notion of Polarity. The description which I have given of Polarity in general, "opposite properties in opposite directions," is of itself a very

imperfect account of the manner in which corresponding antitheses are involved in the portions of science into which Polar relations enter. In excuse of its imperfection, I may say, that I believe it is the first attempt to define Polarity in general; but yet, the conception of Polarity has certainly been strongly and effectively present in the minds of many of the sagacious men who have discovered and unravelled polar phenomena. They attempted to convey this conception, each in his own subject, sometimes by various and peculiar expressions, sometimes by imaginary mechanism by which the antithetical results were produced; their mode of expressing themselves being often defective or imperfect, often containing what was superfluous; and their meaning was commonly very imperfectly apprehended by most of their hearers and readers. But still, the conception was there, gradually working itself into clearness and distinctness, and in the mean time, directing their experiments, and forming an essential element of their discoveries. So far would it be from a sufficient statement of the case to say, that they conceived polarity because they saw it;—that they saw it as soon as it came into view;—and that they described it as they saw it.

35. The way in which such conceptions acquire clearness and distinctness is often by means of Discussions of Definitions. To define well a thought which already enters into trains of discovery, is often a difficult matter. The business of such definition is a part of the business of discovery. These, and other remarks connected with these, which I had made in the *Philosophy*, Mr. Mill has quoted and adopted (ii. 242). They appear to me to point very distinctly to the doctrine to which he refuses his assent,—that there is a special process in the mind, in addition to the mere observation of facts, which is necessary at every step in the progress of knowledge. The Conception must be *formed* before it can be *defined*. The Definition gives the last stamp of distinctness to the Conception; and enables us to express, in a compact and lucid form, the new scientific propositions into which the new Conception enters.

36. Since Mr. Mill assents to so much of what has been said in the *Philosophy*, with regard, to the process of scientific discovery, how, it may be asked, would he express these doctrines so as to exclude that which he thinks erroneous? If he objects to our saying that when we obtain a new inductive truth, we connect phenomena by applying to them a new Conception which fits them, in what terms would he describe the process? If he will not agree to say, that in order to discover the law of the facts, we must find an appropriate Conception, what language would he use instead of this? This is a natural question; and the answer cannot fail to throw light on the relation in which his views and mine stand to each other.

Mr. Mill would say, I believe, that when we obtain a new inductive law of facts, we find something in which the facts *resemble each other;* and that the business of making such discoveries is the business of discovering such resemblances. Thus, he says (of me,) (ii. 211), "his Colligation of Facts by means of appropriate Conceptions, is but the ordinary process of finding by a comparison of phenomena, in what consists their agreement or resemblance." And the Methods of experimental Inquiry which he gives (i. 450, &c.), proceed upon the supposition that the business of discovery may be thus more properly described.

37. There is no doubt that when we discover a law of nature by induction, we find some point in which all the particular facts agree. All the orbits of the planets agree in being ellipses, as Kepler discovered; all falling bodies agree in being acted on by a uniform force, as Galileo discovered; all refracted rays agree in having the sines of incidence and refraction in a constant ratio, as Snell discovered; all the bodies in the universe agree in attracting each other, as Newton discovered; all chemical compounds agree in being constituted of elements in definite proportions, as Dalton discovered. But it appears to me a most scanty, vague, and incomplete account of these steps in science, to say that the authors of them discovered something in which the facts in each case agreed. The point in which the cases agree, is of the most diverse kind in the different cases—in some, a relation of space, in others, the action of a force, in others, the mode of composition of a substance;—and the point of agreement, visible to the discoverer alone, does not come even into his sight, till after the facts have been connected by thoughts of his own, and regarded in points of view in which he, by his mental acts, places them. It would seem to me not much more inappropriate to say, that an officer, who disciplines his men till they move together at the word of command, does so by finding something in which they agree. If the power of consentaneous motion did not exist in the individuals, he could not create it: but that power being there, he finds it and uses it. Of course I am aware that the parallel of the two cases is not exact; but in the one case, as in the other, that in which the particular things are found to agree, is something formed in the mind of him who brings the agreement into view. [. . .]

Mr. Mill thinks that I have been too favourable to the employment of hypotheses, as means of discovering scientific truth; and that I have countenanced a laxness of method, in allowing hypotheses to be established, merely in virtue of the accordance of their results with the phenomena. I believe I should be as cautious as Mr. Mill, in accepting mere hypothetical explanations of phenomena, in any case in which we had the phenomena, and their relations, placed before both of us in an equally clear light. I have not accepted the Undulatory theory of Heat, though recommended by so many coincidences and analogies. But I see some grave reasons for not giving any great weight to Mr. Mill's admonitions;—reasons drawn from the language which he uses on the subject, and which appears to me inconsistent with the conditions of the cases to which he applies it. Thus, when he says (ii. 22) that the condition of a hypothesis accounting for all the known phenomena is "often fulfilled equally well by two conflicting hypotheses," I can only say that I know of no such case in the history of Science, where the phenomena are at all numerous and complicated; and that if such a case were to occur, one of the hypotheses might always be resolved into the other. When he says, that "this evidence (the agreement of the results of the hypothesis with the phenomena) cannot be of the smallest value, because we cannot have in the case of such an hypothesis the assurance that if the hypothesis be false it must lead to results at variance with the true facts," we must reply, with due submission, that we have, in the case spoken of, the most complete evidence of this; for any change in the hypothesis would make it incapable of accounting for the facts. When he says that "if we give ourselves the license of inventing the causes as well as their laws, a person of fertile imagination might devise a hundred modes of accounting for any given fact"; I reply, that the question is about accounting for a large and complex series of facts, of which the laws have been ascertained: and as a test of Mr. Mill's assertion, I would propose as a challenge to any person of fertile imagination to devise any *one* other hypothesis to account for the perturbations of the moon, or the coloured fringes of shadows, besides the hypothesis by which they have actually been explained with such curious completeness. This challenge has been repeatedly offered, but never in any degree accepted; and I entertain no apprehension that Mr. Mill's supposition will ever be verified by such a performance.

50. I see additional reason for mistrusting the precision of Mr. Mill's views of that accordance of phenomena with the results of a hypothesis, in several others of the expressions which he uses (ii. 23). He speaks of a hypothesis being a "*plausible* explanation of all or most of the phenomena"; but the case which we have to consider is where it gives an *exact* representation of all the phenomena in which its results can be traced. He speaks of its being certain that the laws of the phenomena are "*in some measure analogous*" to those given by the hypothesis; the case to be dealt with being, that they are in every way identical. He speaks of this analogy being certain, from the fact that the hypothesis can be "for a moment *tenable*"; as if any one had recommended a hypothesis which is tenable only while a small part of the facts are considered, when it is inconsistent with others which a fuller examination of the case discloses. I have nothing to say, and have said nothing, in favour of hypotheses which are *not* tenable. He says there are many such "*harmonies* running through the laws of phenomena in other respects radically distinct"; and he gives as an instance, the laws of light and heat. I have never alleged such harmonies as grounds of theory, unless they

should amount to identities; and if they should do this, I have no doubt that the most sober thinkers will suppose the causes to be of the same kind in the two harmonizing instances. If chlorine, iodine and brome, or sulphur and phosphorus, have, as Mr. Mill says, analogous properties, I should call these substances *analogous:* but I can see no temptation to frame an hypothesis that they are *identical* (which he seems to fear), so long as Chemistry proves them distinct. But any hypothesis of an analogy in the constitution of these elements (suppose, for instance, a resemblance in their atomic form or composition) would seem to me to have a fair claim to trial; and to be capable of being elevated from one degree of probability to another by the number, variety, and exactitude of the explanations of phenomena which it should furnish. [. . .]

IX. *Successive Generalizations.*—56. There is one feature in the construction of science which Mr. Mill notices, but to which he does not ascribe, as I conceive, its due importance: I mean, that process by which we not only ascend from particular facts to a general law, but when this is done, ascend from the first general law to others more general; and so on, proceeding to the highest point of generalization. This character of the scientific process was first clearly pointed out by Bacon, and is one of the most noticeable instances of his philosophical sagacity. "There are," he says, "two ways, and can be only two, of seeking and finding truth. The one from sense and particulars, takes a flight to the most general axioms, and from these principles and their truth, settled once for all, invents and judges of intermediate axioms. The other method collects axioms from sense and particulars, ascending *continuously and by degrees,* so that in the end it arrives at the most general axioms": meaning by *axioms,* laws or principles. The structure of the most complete sciences consists of several such steps, —*floors,* as Bacon calls them, of successive generalization; and thus this structure may be exhibited as a kind of scientific pyramid. I have constructed this pyramid in the case of the science of Astronomy[9]: and I am gratified to find that the illustrious Humboldt approves of the design, and speaks of it as executed with complete success.[10] The capability of being exhibited in this form of successive generalizations, arising from particulars upward to some very general law, is the condition of all tolerably perfect sciences; and the steps of the successive generalizations are commonly the most important events in the history of the science.

57. Mr. Mill does not reject this process of generalization; but he gives it no conspicuous place, making it only one of three modes of reducing a law of causation into other laws. "There is," he says (i. 555), "the *subsumption* of one law under another; . . . the gathering up of several laws into one more general law which includes them all. He adds afterwards, that the general law is the *sum* of the partial ones (i. 557), an expression which

appears to me inadequate, for reasons which I have already stated. The general law is not the mere sum of the particular laws. It is, as I have already said, their amount *in a new point of view*. A new conception is introduced; thus, Newton did not merely add together the laws of the motions of the moon and of the planets, and of the satellites, and of the earth; he looked at them altogether as the result of a universal force of mutual gravitation; and therein consisted his generalization. And the like might be pointed out in other cases. [. . .]

60. Also, the Science of Mechanics, although Mr. Mill more especially refers to it, as a case in which the highest generalizations (for example the Laws of Motion) were those earliest ascertained with any scientific exactness, will, I think, on a more careful examination of its history, be found remarkably to confirm Bacon's view. For, in that science, we have, in the first place, very conspicuous examples of the vice of the method pursued by the ancients in flying to the highest generalizations first; as when they made their false distinctions of the laws of *natural* and *violent* motions, and of *terrestrial* and *celestial* motions. Many erroneous laws of motion were asserted through neglect of facts or want of experiments. And when Galileo and his school had in some measure succeeded in discovering some of the true laws of the motions of terrestrial bodies, they did not at once assert them as general: for they did not at all apply those laws to the celestial motions. As I have remarked, all Kepler's speculations respecting the causes of the motions of the planets, went upon the supposition that the First Law of terrestrial Motion did not apply to celestial bodies; but that, on the contrary, some continual force was requisite to keep up, as well as to originate, the planetary motions. Nor did Descartes, though he enunciated the Laws of Motion with more generality than his predecessors, (but not with exactness,) venture to trust the planets to those laws; on the contrary, he invented his machinery of Vortices in order to keep up the motions of the heavenly bodies. Newton was the first who extended the laws of terrestrial motion to the celestial spaces; and in doing so, he used all the laws of the celestial motions which had previously been discovered by more limited inductions. To these instances, I may add the gradual generalization of the third Law of motion by Huyghens, the Bernoullis, and Herman, which I have described in the *History*[11] as preceding that Period of Deduction, to which the succeeding narrative[12] is appropriated. In Mechanics, then, we have a cardinal example of the historically gradual and successive ascent of science from particulars to the most general laws.

61. The Science of Hydrostatics may appear to offer a more favourable example of the ascent to the most general laws, without going through the intermediate particular laws; and it is true, with reference to this science, as I have observed,[13] that it does exhibit the *peculiarity* of our possessing

the most general principles on which the phenomena depend, and from which many cases of special facts are explained by deduction; while other cases cannot be so explained, from the want of principles intermediate between the highest and the lowest. And I have assigned, as the reason of this peculiarity, that the general principles of the Mechanics of Fluids were not obtained with reference to the science itself, but by extension from the sister science of the Mechanics of Solids. The two sciences are parts of the same Inductive Pyramid; and having reached the summit of this Pyramid on one side, we are tempted to descend on the other from the highest generality to more narrow laws. Yet even in this science, the best part of our knowledge is mainly composed of inductive laws, obtained by inductive examination of particular classes of facts. The mere mathematical investigations of the laws of waves, for instance, have not led to any results so valuable as the experimental researches of Bremontier, Emy, the Webers, and Mr. Scott Russell. And in like manner in Acoustics, the Mechanics of Elastic Fluids,[14] the deductions of mathematicians made on general principles have not done so much for our knowledge, as the cases of vibrations of plates and pipes examined experimentally by Chladni, Savart, Mr. Wheatstone and Mr. Willis. We see therefore, even in these sciences, no reason to slight the wisdom which exhorts us to ascend from particulars to intermediate laws, rather than to hope to deduce these latter better from the more general laws obtained once for all.

62. Mr. Mill himself indeed, notwithstanding that he slights Bacon's injunction to seek knowledge by proceeding from less general to more general laws, has given a very good reason why this is commonly necessary and wise. He says (ii. 526), "Before we attempt to explain deductively, from more general laws, any new class of phenomena, it is desirable to have gone as far as is practicable in ascertaining the empirical laws of these phenomena; so as to compare the results of deduction, not with one individual instance after another, but with general propositions expressive of the points of agreement which have been found among many instances. For," he adds with great justice, "if Newton had been obliged to verify the theory of gravitation, not by deducing from it Kepler's laws, but by deducing all the observed planetary positions which had served Kepler to establish those laws, the Newtonian theory would probably never have emerged from the state of an hypothesis." To which we may add, that it is certain, from the history of the subject, that in that case the hypothesis would never have been framed at all.

X. *Mr. Mill's Hope from Deduction.*—63. Mr. Mill expresses a hope of the efficacy of Deduction, rather than Induction, in promoting the future progress of Science; which hope, so far as the physical sciences are concerned, appears to me at variance with all the lessons of the history of those sciences. He says (i. 579), "that the advances henceforth to be expected even in physical, and still more in mental and social science, will be chiefly the result of deduction, is evident from the general considerations already adduced": these considerations being, that the phenomena to be considered are very complex, and are the result of many known causes, of which we have to disentangle the results.

64. I cannot but take a very different view from this. I think that any one, looking at the state of physical science, will see that there are still a vast mass of cases, in which we do not at all know the causes, at least, in their full generality; and that the knowledge of new causes, and the generalization of the laws of those already known, can only be obtained by new *inductive* discoveries. Except by new Inductions, equal, in their efficacy for grouping together phenomena in new points of view, to any which have yet been performed in the history of science, how are we to solve such questions as those which, in the survey of what we already know, force themselves upon our minds? Such as, to take only a few of the most obvious examples—What is the nature of the connexion of heat and light? How does heat produce the expansion, liquefaction and vaporization of bodies? What is the nature of the connexion between the optical and the chemical properties of light? What is the relation between optical, crystalline and chemical polarity? What is the connexion between the atomic constitution and the physical qualities of bodies? What is the tenable definition of a mineral species? What is the true relation of the apparently different types of vegetable life (monocotyledons, dicotyledons, and cryptogamous plants)? What is the relation of the various types of animal life (vertebrates, articulates, radiates, &c.)? What is the number, and what are the distinctions of the Vital Powers? What is the internal constitution of the earth? These, and many other questions of equal interest, no one, I suppose, expects to see solved by deduction from principles already known. But we can, in many of them, see good hope of progress by a large use of induction; including, of course, copious and careful experiments and observations.

65. With such questions before us, as have now been suggested, I can see nothing but a most mischievous narrowing of the field and enfeebling of the spirit of scientific exertion, in the doctrine that "Deduction is the great scientific work of the present and of future ages"; and that "A revolution is peaceably and progressively effecting itself in philosophy the reverse of that to which Bacon has attached his name." I trust, on the contrary, that we have many new laws of nature still to discover; and that our race is destined to obtain a sight of wider truths than any we yet discern, including, as cases, the general laws we now know, and obtained from these known laws as they must be, by Induction.

66. I can see, however, reasons for the comparatively greater favour with which Mr. Mill looks upon Deduction, in the views to which he has mainly directed his attention. The explanation of remarkable phenomena by known laws of Nature, has, as I have already said, a greater charm for many minds than the discovery of the laws themselves. In the case of such explanations, the problem proposed is more definite, and the solution more obviously complete. For the process of induction includes a mysterious step, by which we pass from particulars to generals, of which step the reason always seems to be inadequately rendered by any words which we can use; and this step to most minds is not demonstrative, as to few is it given to perform it on a great scale. But the process of explanation of facts by known laws is deductive, and has at every step a force like that of demonstration, producing a feeling peculiarly gratifying to the clear intellects which are most capable of following the process. We may often see instances in which this admiration for deductive skill appears in an extravagant measure; as when men compare Laplace with Newton. Nor should I think it my business to argue against such a preference, unless it were likely to leave us too well satisfied with what we know already, to chill our hope of scientific progress, and to prevent our making any further strenuous efforts to ascend, higher than we have yet done, the mountain-chain which limits human knowledge.

Notes

1. [*A System of Logic, Ratiocinative and Inductive, being a connected view of the Principles of Evidence, and of the Methods of Scientific Investigation.* By John Stuart Mill.]
2. These Remarks were published in 1849, under the title *Of Induction, with especial reference to Mr. J. Mill's System of Logic.*
3. My references are throughout (except when otherwise expressed) to the volume and the page of Mr. Mill's first edition of his *Logic.*
4. On this subject see an Essay *On the Transformation of Hypotheses,* given in the Appendix.
5. B. vii. c. iii. sect. 3.
6. B. iii. c. ix. art. 7.
7. B. i. c. iii.
12. I have also, in the same place, given the Inductive Pyramid for the science of Optics. These Pyramids are necessarily inverted in their form, in order that, in reading in the ordinary way, we may proceed *to* the vertex. *Phil. Ind. Sc.* b. xi. c. vi.
13. *Cosmos,* vol. ii. note 35.
15. B. vi. c. v.
16. c. vi.
17. *Hist.* b. vi c. vi. sect 13.
18. *Hist. Ind. Sc. b.* viii.

CHAPTER 15

Physical Theory and Experiment

PIERRE DUHEM

Pages 180–205 and 208–211 in *The Aim and Structure of Physical Theory*, trans. Phillip Wiener. Princeton University Press, 1954 1914.

1 The Experimental Testing of a Theory Does Not Have the Same Logical Simplicity in Physics as in Physiology

The sole purpose of physical theory is to provide a representation and classification of experimental laws; the only test permitting us to judge a physical theory and pronounce it good or bad is the comparison between the consequences of this theory and the experimental laws it has to represent and classify. Now that we have minutely analyzed the characteristics of a physical experiment and of a physical law, we can establish the principles that should govern the comparison between experiment and theory; we can tell how we shall recognize whether a theory is confirmed or weakened by facts.

When many philosophers talk about experimental sciences, they think only of sciences still close to their origins, e.g., physiology or certain branches of chemistry where the experimenter reasons directly on the facts by a method which is only common sense brought to greater attentiveness but where mathematical theory has not yet introduced its symbolic representations. In such sciences the comparison between the deductions of a theory and the facts of experiment is subject to very simple rules. These rules were formulated in a particularly forceful manner by Claude Bernard, who would condense them into a single principle, as follows:

"The experimenter should suspect and stay away from fixed ideas, and always preserve his freedom of mind.

"The first condition that has to be fulfilled by a scientist who is devoted to the investigation of natural phenomena is to preserve a complete freedom of mind based on philosophical doubt."[1]

If a theory suggests experiments to be done, so much the better: ". . . we can follow our judgment and our thought, give free rein to our imagination provided that all our ideas are only pretexts for instituting new experiments that may furnish us probative facts or unexpected and fruitful ones."[2] Once the experiment is done and the results clearly established, if a theory takes them over in order to generalize them, coordinate them, and draw from them new subjects for experiment, still so much the better: ". . . if one is imbued with the principles of experimental method, there is nothing to fear; for so long as the idea is a right one, it will go on being developed; when it is an erroneous idea, experiment is there to correct it."[3] But so long as the experiment lasts, the theory should remain waiting, under strict orders to stay outside the door of the laboratory; it should keep silent and leave the scientist without disturbing him while he faces the facts directly; the facts must be observed without a preconceived idea and gathered with the same scrupulous impartiality, whether they confirm or contradict the predictions of the theory. The report that the observer will give us of his experiment should be a faithful and scrupulously exact reproduction of the phenomena, and should not let us even guess what system the scientist places his confidence in or distrusts.

"Men who have an excessive faith in their theories or in their ideas are not only poorly disposed to make discoveries but they also make very poor observations. They necessarily observe with a preconceived idea and, when they have begun an experiment, they want to see in its results only a confirmation of their theory. Thus they distort observation and often neglect very important facts because they go counter to their goal. That is what made us say elsewhere that we must never do experiments in order to confirm our ideas but merely to check them. . . . But it quite naturally happens that those who believe too much in their own theories do not sufficiently believe in the theories of others. Then the dominant idea of these condemners of others is to find fault with the theories of the latter and to seek to contradict them. The setback for science remains the same. They are doing experiments only in order to destroy a theory instead of doing them in order to look for the truth. They also make poor observations because they take into the results of their experiments only what fits their purpose, by neglecting what is unrelated to it, and by very carefully avoiding, whatever might go in the direction of the idea they wish to combat. Thus one is led by two parallel paths to the same result, that is to say, to falsifying science and the facts.

"The conclusion of all this is that it is necessary to obliterate one's opinion as well as that of others when faced with the decisions of the experiment; . . . we must accept the results of experiment just as they present themselves with all that is unforeseen and accidental in them."[4]

Here, for example, is a physiologist who admits that the anterior roots of the spinal nerve contain the motor nerve-fibers and the posterior roots the sensory fibers. The theory he accepts leads him to imagine an experiment: if he cuts a certain anterior root, he ought to be suppressing the mobility of a certain part of the body without destroying its sensibility; after making the section of this root, when he observes the consequences of his operation and when he makes a report of it, he must put aside all his ideas concerning the physiology of the spinal nerve; his report must be a raw description of the facts; he is not permitted to overlook or fail to mention any movement or quiver contrary to his predictions or to attribute it to some secondary cause unless some special experiment has given evidence of this cause; he must, if he does not wish to be accused of scientific bad faith, establish an absolute separation or watertight compartment between the consequences of his theoretical deductions and the establishing of the facts shown by his experiments.

Such a rule is not by any means easily followed; it requires of the scientist an absolute detachment from his own thought and a complete absence of animosity when confronted with the opinion of another person; neither vanity nor envy ought to be countenanced by him. As Bacon put it, he should never show eyes lustrous with human passions. Freedom of mind, which constitutes the sole principle of experimental method, according to Claude Bernard, does not depend merely on intellectual conditions, but also on moral conditions, making its practice rarer and more meritorious.

But if experimental method as just described is difficult to practice, the logical analysis of it is very simple. This is no longer the case when the theory to be subjected to test by the facts is not a theory of physiology but a theory of physics. In the latter case, in fact, it is impossible to leave outside the laboratory door the theory that we wish to test, for without theory it is impossible to regulate a single instrument or to interpret a single reading. We have seen that in the mind of the physicist there are constantly present two sorts of apparatus: one is the concrete apparatus in glass and metal, manipulated by him, the other is the schematic and abstract apparatus which theory substitutes for the concrete apparatus and on which the physicist does his reasoning. For these two ideas are indissolubly connected in his intelligence, and each necessarily calls on the other; the physicist can no sooner conceive the concrete apparatus without associating with it the idea of the schematic apparatus than a Frenchman can conceive an idea without associating it with the French word expressing it. This radical impossibility, preventing one from dissociating physical theories from the experimental procedures appropriate for testing these theories, complicates this test in a singular way, and obliges us to examine the logical meaning of it carefully.

Of course, the physicist is not the only one who appeals to theories at the very time he is experimenting or reporting the results of his experiments. The chemist and the physiologist when they make use of physical instruments, e.g., the thermometer, the manometer, the calorimeter, the galvanometer, and the saccharimeter, implicitly admit the accuracy of the theories justifying the use of these pieces of apparatus as well as of the theories giving meaning to the abstract ideas of temperature, pressure, quantity of heat, intensity of current, and polarized light, by means of which the concrete indications of these instruments are translated. But the theories used, as well as the instruments employed, belong to the domain of physics; by accepting with these instruments the theories without which their readings would be devoid of meaning, the chemist and the physiologist show their confidence in the physicist, whom they suppose to be infallible. The physicist, on the other hand, is obliged to trust his own theoretical ideas or those of his fellow-physicists. From the standpoint of logic, the difference is of little importance; for the physiologist and chemist as well as for the physicist, the statement of the result of an experiment implies, in general, an act of faith in a whole group of theories.

2 An Experiment in Physics Can Never Condemn an Isolated Hypothesis but Only a Whole Theoretical Group

The physicist who carries out an experiment, or gives a report of one, implicitly recognizes the accuracy of a whole group of theories. Let us accept this principle and see what consequences we may deduce from it when we seek to estimate the role and logical import of a physical experiment.

In order to avoid any confusion we shall distinguish two sorts of experiments: experiments of *application*, which we shall first just mention, and experiments of *testing*, which will be our chief concern.

You are confronted with a problem in physics to be solved practically; in order to produce a certain effect you wish to make use of knowledge acquired by physicists; you wish to light an incandescent bulb; accepted theories indicate to you the means for solving the problem; but to make use of these means you have to secure certain information; you ought, I suppose, to determine the electromotive force of the battery of generators at your disposal; you measure this electromotive force: that is what I call an experiment of application. This experiment does not aim at discovering whether accepted theories are accurate or not; it merely intends to draw on these theories. In order to carry it out, you make use of instruments that these same theories legitimize; there is nothing to shock logic in this procedure.

But experiments of application are not the only ones the physicist has to perform; only with their aid can science aid practice, but it is not through them that science creates and develops itself; besides experiments of application, we have experiments of testing.

A physicist disputes a certain law; he calls into doubt a certain theoretical point. How will he justify these doubts? How will he demonstrate the inaccuracy of the law? From the proposition under indictment he will derive the prediction of an experimental fact; he will bring into existence the conditions under which this fact should be produced; if the predicted fact is not produced, the proposition which served as the basis of the prediction will be irremediably condemned.

F. E. Neumann assumed that in a ray of polarized light the vibration is parallel to the plane of polarization, and many physicists have doubted this proposition. How did O. Wiener undertake to transform this doubt into a certainty in order to condemn Neumann's proposition? He deduced from this proposition the following consequence: If we cause a light beam reflected at 45° from a plate of glass to interfere with the incident beam polarized perpendicularly to the plane of incidence, there ought to appear alternately dark and light interference bands parallel to the reflecting surface; he brought about the conditions under which these bands should have been produced and showed that the predicted phenomenon did not appear, from which he concluded that Neumann's proposition is false, viz., that in a polarized ray of light the vibration is not parallel to the plane of polarization.

Such a mode of demonstration seems as convincing and as irrefutable as the proof by reduction to absurdity customary among mathematicians; moreover, this demonstration is copied from the reduction to absurdity, experimental contradiction playing the same role in one as logical contradiction plays in the other.

Indeed, the demonstrative value of experimental method is far from being so rigorous or absolute: the conditions under which it functions are much more complicated than is supposed in what we have just said; the evaluation of results is much more delicate and subject to caution.

A physicist decides to demonstrate the inaccuracy of a proposition; in order to deduce from this proposition the prediction of a phenomenon and institute the experiment which is to show whether this phenomenon is or is not produced, in order to interpret the results of this experiment and establish that the predicted phenomenon is not produced, he does not confine himself to making use of the proposition in question; he makes use also of a whole group of theories accepted by him as beyond dispute. The prediction of the phenomenon, whose nonproduction is to cut off debate, does not derive from the proposition challenged if taken by itself, but from the proposition at issue joined to that whole group of theories; if the predicted phenomenon is not produced, not only is the proposition questioned at fault, but so is the whole theoretical scaffolding used by the physicist. The only thing the experiment teaches us is that among the propositions used to predict the phenomenon and to establish whether it would be produced, there is at least one error; but where this error lies is just what it does not tell us. The physicist may declare that this error is contained in exactly the proposition he wishes to refute, but is he sure it is not in another proposition? If he is, he accepts implicitly the accuracy of all the other propositions he has used, and the validity of his conclusion is as great as the validity of his confidence.

Let us take as an example the experiment imagined by Zenker and carried out by O. Wiener. In order to predict the formation of bands in certain circumstances and to show that these did not appear, Wiener did not make use merely of the famous proposition of F. E. Neumann, the proposition which he wished to refute; he did not merely admit that in a polarized ray vibrations are parallel to the plane of polarization; but he used, besides this, propositions, laws, and hypotheses constituting the optics commonly accepted: he admitted that light consists in simple periodic vibrations, that these vibrations are normal to the light ray, that at each point the mean kinetic energy of the vibratory motion is a measure of the intensity of light, that the more or less complete attack of the gelatine coating on a photographic plate indicates the various degrees of this intensity. By joining these propositions, and many others that would take too long to enumerate, to Neumann's proposition, Wiener was able to formulate a forecast and establish that the experiment belied it. If he attributed this solely to Neumann's proposition, if it alone bears the responsibility for the error this negative result has put in evidence, then Wiener was taking all the other propositions he invoked as beyond doubt. But this assurance is not imposed as a matter of logical necessity; nothing stops us from taking Neumann's proposition as accurate and shifting the weight of the experimental contradiction to some other proposition of the commonly accepted optics; as H. Poincaré has shown, we can very easily rescue Neumann's hypothesis from the grip of Wiener's experiment on the condition that we abandon in exchange the hypothesis which takes the mean kinetic energy as the measure of the light intensity; we may, without being contradicted by the experiment, let the vibration be parallel to the plane of polarization, provided that we measure the light intensity by the mean potential energy of the medium deforming the vibratory motion.

These principles are so important that it will be useful to apply them to another example; again we choose an experiment regarded as one of the most decisive ones in optics.

We know that Newton conceived the emission theory for optical phenomena. The emission theory supposes light to be formed of extremely thin projectiles, thrown out with very great speed by the sun and other sources of light; these projectiles penetrate all transparent bodies; on account of the various parts of the media through which they move, they undergo attractions and repulsions; when the distance separating the acting particles is very small these actions are very powerful, and they vanish when the masses between which they act are appreciably far from each other. These essential hypotheses joined to several others, which we pass over without mention, lead to the formulation of a complete theory of reflection and refraction of light; in particular, they imply the following proposition: The index of refraction of light passing from one medium into another is equal to the velocity of the light projectile within the medium it penetrates, divided by the velocity of the same projectile in the medium it leaves behind.

This is the proposition that Arago chose in order to show that the theory of emission is in contradiction with the facts. From this proposition a second follows: Light travels faster in water than in air. Now Arago had indicated an appropriate procedure for comparing the velocity of light in air with the velocity of light in water; the procedure, it is true, was inapplicable, but Foucault modified the experiment in such a way that it could be carried out; he found that the light was propagated less rapidly in water than in air. We may conclude from this, with Foucault, that the system of emission is incompatible with the facts.

I say the *system* of emission and not the *hypothesis* of emission; in fact, what the experiment declares stained with error is the whole group of propositions accepted by Newton, and after him by Laplace and Biot, that is, the whole theory from which we deduce the relation between the index of refraction and the velocity of light in various media. But in condemning this system as a whole by declaring it stained with error, the experiment does not tell us where the error lies. Is it in the fundamental hypothesis that light consists in projectiles thrown out with great speed by luminous bodies? Is it in some other assumption concerning the actions experienced by light corpuscles due to the media through which they move? We know nothing about that. It would be rash to believe, as Arago seems to have thought, that Foucault's experiment condemns once and for all the very hypothesis of emission, i.e., the assimilation of a ray of light to a swarm of projectiles. If physicists had attached some value to this task, they would undoubtedly have succeeded in founding on this assumption a system of optics that would agree with Foucault's experiment.

In sum, the physicist can never subject an isolated hypothesis to experimental test, but only a whole group of hypotheses; when the experiment is in disagreement with his predictions, what he learns is that at least one of the hypotheses constituting this group is unacceptable and ought to be modified; but the experiment does not designate which one should be changed.

We have gone a long way from the conception of the experimental method arbitrarily held by persons unfamiliar with its actual functioning. People generally think that each one of the hypotheses employed in physics can be taken in isolation, checked by experiment, and then, when many varied tests have established its validity, given a definitive place in the system of physics. In reality, this is not the case. Physics is not a machine which lets itself be taken apart; we cannot try each piece in isolation and, in order to adjust it, wait until its solidity has been carefully checked. Physical science is a system that must be taken as a whole; it is an organism in which one part cannot be made to function except when the parts that are most remote from it are called into play, some more so than others, but all to some degree. If something goes wrong, if some discomfort is felt in the functioning of the organism, the physicist will have to ferret out through its effect on the entire system which organ needs to be remedied or modified without the possibility of isolating this organ and examining it apart. The watchmaker to whom you give a watch that has stopped separates all the wheelworks and examines them one by one until he finds the part that is defective or broken. The doctor to whom a patient appears cannot dissect him in order to establish his diagnosis; he has to guess the seat and cause of the ailment solely by inspecting disorders affecting the whole body. Now, the physicist concerned with remedying a limping theory resembles the doctor and not the watchmaker.

3 A "Crucial Experiment" Is Impossible in Physics

Let us press this point further, for we are touching on one of the essential features of experimental method, as it is employed in physics.

Reduction to absurdity seems to be merely a means of refutation, but it may become a method of demonstration: in order to demonstrate the truth of a proposition it suffices to corner anyone who would admit the contradictory of the given proposition into admitting an absurd consequence. We know to what extent the Greek geometers drew heavily on this mode of demonstration.

Those who assimilate experimental contradiction to reduction to absurdity imagine that in physics we may use a line of argument similar to the one Euclid employed so frequently in geometry. Do you wish to obtain from a group of phenomena a theoretically certain and indisputable explanation?

Enumerate all the hypotheses that can be made to account for this group of phenomena; then, by experimental contradiction eliminate all except one; the latter will no longer be a hypothesis, but will become a certainty.

Suppose, for instance, we are confronted with only two hypotheses. Seek experimental conditions such that one of the hypotheses forecasts the production of one phenomenon and the other the production of quite a different effect; bring these conditions into existence and observe what happens; depending on whether you observe the first or the second of the predicted phenomena, you will condemn the second or the first hypothesis; the hypothesis not condemned will be henceforth indisputable; debate will be cut off, and a new truth will be acquired by science. Such is the experimental test that the author of the *Novum Organum* [Francis Bacon] called the *"fact of the cross,"* borrowing this expression from the crosses which at an intersection indicate the various roads.

We are confronted with two hypotheses concerning the nature of light; for Newton, Laplace, or Biot light consisted of projectiles hurled with extreme speed, but for Huygens, Young, or Fresnel light consisted of vibrations whose waves are propagated within an ether. These are the only two possible hypotheses as far as one can see: either the motion is carried away by the body it excites and remains attached to it, or else it passes from one body to another. Let us pursue the first hypothesis; it declares that light travels more quickly in water than in air; but if we follow the second, it declares that light travels more quickly in air than in water. Let us set up Foucault's apparatus; we set into motion the turning mirror; we see two luminous spots formed before us, one colorless, the other greenish. If the greenish band is to the left of the colorless one, it means that light travels faster in water than in air, and that the hypothesis of vibrating waves is false. If, on the contrary, the greenish band is to the right of the colorless one, that means that light travels faster in air than in water, and that the hypothesis of emissions is condemned. We look through the magnifying glass used to examine the two luminous spots, and we notice that the greenish spot is to the right of the colorless one; the debate is over; light is not a body, but a vibratory wave motion propagated by the ether; the emission hypothesis has had its day; the wave hypothesis has been put beyond doubt, and the crucial experiment has made it a new article of the scientific credo.

What we have said in the foregoing paragraph shows how mistaken we should be to attribute to Foucault's experiment so simple a meaning and so decisive an importance; for it is not between two hypotheses, the emission and wave hypotheses, that Foucault's experiment judges trenchantly; it decides rather between two sets of theories each of which has to be taken as a whole, i.e., between two entire systems, Newton's optics and Huygens' optics.

But let us admit for a moment that in each of these systems everything is compelled to be necessary by strict logic, except a single hypothesis; consequently, let us admit that the facts, in condemning one of the two systems, condemn once and for all the single doubtful assumption it contains. Does it follow that we can find in the "crucial experiment" an irrefutable procedure for transforming one of the two hypotheses before us into a demonstrated truth? Between two contradictory theorems of geometry there is no room for a third judgment; if one is false, the other is necessarily true. Do two hypotheses in physics ever constitute such a strict dilemma? Shall we ever dare to assert that no other hypothesis is imaginable? Light may be a swarm of projectiles, or it may be a vibratory motion whose waves are propagated in a medium; is it forbidden to be anything else at all? Arago undoubtedly thought so when he formulated this incisive alternative: Does light move more quickly in water than in air? "Light is a body. If the contrary is the case, then light is a wave." But it would be difficult for us to take such a decisive stand; Maxwell, in fact, showed that we might just as well attribute light to a periodical electrical disturbance that is propagated within a dielectric medium.

Unlike the reduction to absurdity employed by geometers, experimental contradiction does not have the power to transform a physical hypothesis into an indisputable truth; in order to confer this power on it, it would be necessary to enumerate completely the various hypotheses which may cover a determinate group of phenomena; but the physicist is never sure he has exhausted all the imaginable assumptions. The truth of a physical theory is not decided by heads or tails.

4 Criticism of the Newtonian Method. First Example: Celestial Mechanics

It is illusory to seek to construct by means of experimental contradiction a line of argument in imitation of the reduction to absurdity; but the geometer is acquainted with other methods for attaining certainty than the method of reducing to an absurdity; the direct demonstration in which the truth of a proposition is established by itself and not by the refutation of the contradictory proposition seems to him the most perfect of arguments. Perhaps physical theory would be more fortunate in its attempts if it sought to imitate direct demonstration. The hypotheses from which it starts and develops its conclusions would then be tested one by one; none would have to be accepted until it presented all the certainty that experimental method can confer on an abstract and general proposition; that is to say, each would necessarily be either a law drawn from observation by the sole use of those two intellectual operations called induction and generalization, or else a corollary mathematically deduced from such laws. A theory based on such hypotheses would then not present anything arbitrary or doubtful; it would deserve all the confidence merited by the faculties which serve us in formulating natural laws.

It was this sort of physical theory that Newton had in mind when, in the "General Scholium" which crowns his *Principia*, he rejected so vigorously as outside of natural philosophy any hypothesis that induction did not extract from experiment; when he asserted that in a sound physics every proposition should be drawn from phenomena and generalized by induction.

The ideal method we have just described therefore deserves to be named the Newtonian method. Besides, did not Newton follow this method when he established the system of universal attraction, thus adding to his precepts the most magnificent of examples? Is not his theory of gravitation derived entirely from the laws which were revealed to Kepler by observation, laws which problematic reasoning transforms and whose consequences induction generalizes?

This first law of Kepler's, "The radial vector from the sun to a planet sweeps out an area proportional to the time during which the planet's motion is observed," did, in fact, teach Newton that each planet is constantly subjected to a force directed toward the sun.

The second law of Kepler's, "The orbit of each planet is an ellipse having the sun at one focus," taught him that the force attracting a given planet varies with the distance of this planet from the sun, and that it is in an inverse ratio to the square of this distance.

The third law of Kepler's, "The squares of the periods of revolution of the various planets are proportional to the cubes of the major axes of their orbits," showed him that different planets would, if they were brought to the same distance from the sun, undergo in relation to it attractions proportional to their respective masses.

The experimental laws established by Kepler and transformed by geometric reasoning yield all the characteristics present in the action exerted by the sun on a planet; by induction Newton generalized the result obtained; he allowed this result to express the law according to which any portion of matter acts on any other portion whatsoever, and he formulated this great principle: "Any two bodies whatsoever attract each other with a force which is proportional to the product of their masses and in inverse ratio to the square of the distance between them." The principle of universal gravitation was found, and it was obtained, without any use having been made of any fictive hypothesis, by the inductive method the plan of which Newton outlined.

Let us again examine this application of the Newtonian method, this time more closely; let us see if a somewhat strict logical analysis will leave intact the appearance of rigor and simplicity that this very summary exposition attributes to it.

In order to assure this discussion of all the clarity it needs, let us begin by recalling the following principle, familiar to all those who deal with mechanics: We cannot speak of the force which attracts a body in given circumstances before we have designated the supposedly fixed term of reference to which we relate the motion of all bodies; when we change this point of reference or term of comparison, the force representing the effect produced on the observed body by the other bodies surrounding it changes in direction and magnitude according to the rules stated by mechanics with precision.

That posited, let us follow Newton's reasoning.

Newton first took the sun as the fixed point of reference; he considered the motions affecting the different planets by reference to the sun; he admitted Kepler's laws as governing these motions, and derived the following proposition: If the sun is the point of reference in relation to which all forces are compared, each planet is subjected to a force directed toward the sun, a force proportional to the mass of the planet and to the inverse square of its distance from the sun. Since the latter is taken as the reference point, it is not subject to any force.

In an analogous manner Newton studied the motion of the satellites and for each of these he chose as a fixed reference point the planet which the satellite accompanies, the earth in the case of the moon, Jupiter in the case of the masses moving around Jupiter. Laws just like Kepler's were taken as governing these motions, from which it follows that we can formulate the following proposition: If we take as a fixed reference point the planet accompanied by a satellite, this satellite is subject to a force directed toward the planet varying inversely with the square of the distance. If, as happens with Jupiter, the same planet possesses several satellites, these satellites, were they at the same distance from the planet, would be acted on by the latter with forces proportional to their respective masses. The planet is itself not acted on by the satellite.

Such, in very precise form, are the propositions which Kepler's laws of planetary motion and the extension of these laws to the motions of satellites authorize us to formulate. For these propositions Newton substituted another which may be stated as follows: Any two celestial bodies whatsoever exert on each other a force of attraction in the direction of the straight line joining them, a force proportional to the product of their masses and to the inverse square of the distance between them. This statement presupposes all motions and forces to be related to the same reference point; the latter is an ideal standard of reference which may well be conceived by the geometer but which does not characterize in an exact and concrete manner the position in the sky of any body.

Is this principle of universal gravitation merely a generalization of the two statements provided by Kepler's laws and their extension to the motion of satellites? Can induction derive it from these two statements? Not at all. In fact, not only is it more general than these two statements and unlike them, but it contradicts them. The student of mechanics who accepts the principle of universal attraction can calculate the magnitude and direction of the forces between the various planets and the sun when the latter is taken as the reference point, and if he does he finds that these forces are not what our first statement would require. He can determine the magnitude and direction of each of the forces between Jupiter and its satellites when we refer all the motions to the planet, assumed to be fixed, and if he does he notices that these forces are not what our second statement would require.

The principle of universal gravity, very far from being derivable by generalization and induction from the observational laws of Kepler, formally contradicts these laws. If Newton's theory is correct, Kepler's laws are necessarily false.

Kepler's laws based on the observation of celestial motions do not transfer their immediate experimental certainty to the principle of universal weight, since if, on the contrary, we admit the absolute exactness of Kepler's laws, we are compelled to reject the proposition on which Newton based his celestial mechanics. Far from adhering to Kepler's laws, the physicist who claims to justify the theory of universal gravitation finds that he has, first of all, to resolve a difficulty in these laws: he has to prove that his theory, incompatible with the exactness of Kepler's laws, subjects the motions of the planets and satellites to other laws scarcely different enough from the first laws for Tycho Brahé, Kepler, and their contemporaries to have been able to discern the deviations between the Keplerian and Newtonian orbits. This proof derives from the circumstances that the sun's mass is very large in relation to the masses of the various planets and the mass of a planet is very large in relation to the masses of its satellites.

Therefore, if the certainty of Newton's theory does not emanate from the certainty of Kepler's laws, how will this theory prove its validity? It will calculate, with all the high degree of approximation that the constantly perfected methods of algebra involve, the perturbations which at each instant remove every heavenly body from the orbit assigned to it by Kepler's laws; then it will compare the calculated perturbations with the perturbations observed by means of the most precise instruments and the most scrupulous methods. Such a comparison will not only bear on this or that part of the Newtonian principle, but will involve all its parts at the same time; with those it will also involve all the principles of dynamics; besides, it will call in the aid of all the propositions of optics, the statics of gases, and the theory of heat, which are necessary to justify the properties of telescopes in their construction, regulation, and correction, and in the elimination of the errors caused by diurnal or annual aberration and by atmospheric refraction. It is no longer a matter of taking, one by one, laws justified by observation, and raising each of them by induction and generalization to the rank of a principle; it is a matter of comparing the corollaries of a whole group of hypotheses to a whole group of facts.

Now, if we seek out the causes which have made the Newtonian method fail in this case for which it was imagined and which seemed to be the most perfect application for it, we shall find them in that double character of any law made use of by theoretical physics: This law is symbolic and approximate.

Undoubtedly, Kepler's laws bear quite directly on the very objects of astronomical observation; they are as little symbolic as possible. But in this purely experimental form they remain inappropriate for suggesting the principle of universal gravitation; in order to acquire this fecundity they must be transformed and must yield the characters of the forces by which the sun attracts the various planets.

Now this new form of Kepler's laws is a symbolic form; only dynamics gives meanings to the words "force" and "mass," which serve to state it, and only dynamics permits us to substitute the new symbolic formulas for the old realistic formulas, to substitute statements relative to "forces" and "masses" for laws relative to orbits. The legitimacy of such a substitution implies full confidence in the laws of dynamics.

And in order to justify this confidence let us not proceed to claim that the laws of dynamics were beyond doubt at the time Newton made use of them in symbolically translating Kepler's laws; that they had received enough empirical confirmation to warrant the support of reason. In fact, the laws of dynamics had been subjected up to that time to only very limited and very crude tests. Even their enunciations had remained very vague and involved; only in Newton's *Principia* had they been for the first time formulated in a precise manner. It was in the agreement of the facts with the celestial mechanics which Newton's labors gave birth to that they received their first convincing verification.

Thus the translation of Kepler's laws into symbolic laws, the only kind useful for a theory, presupposed the prior adherence of the physicist to a whole group of hypotheses. But, in addition, Kepler's laws being only approximate laws, dynamics permitted giving them an infinity of different symbolic translations. Among these various forms, infinite in number, there is one and only one which agrees with Newton's principle. The observations of Tycho Brahé, so felicitously reduced to laws by Kepler, permit the theorist to choose this form, but they do not constrain him to do so, for there is an infinity of others they permit him to choose. [. . .]

8 Are Certain Postulates of Physical Theory Incapable of Being Refuted by Experiment?

We recognize a correct principle by the facility with which it straightens out the complicated difficulties into which the use of erroneous principles brought us.

If, therefore, the idea we have put forth is correct, namely, that comparison is established necessarily between the *whole* of theory and the *whole* of experimental facts, we ought in the light of this principle to see the disappearance of the obscurities in which we should be lost by thinking that we are subjecting each isolated theoretical hypothesis to the test of facts.

Foremost among the assertions in which we shall aim at eliminating the appearance of paradox, we shall place one that has recently been often formulated and discussed. Stated first by G. Milhaud in connection with the "*pure bodies*" of chemistry,[5] it has been developed at length and forcefully by H. Poincaré with regard to principles of mechanics;[6] Edouard Le Roy has also formulated it with great clarity.[7]

That assertion is as follows: Certain fundamental hypotheses of physical theory cannot be contradicted by any experiment, because they constitute in reality *definitions*, and because certain expressions in the physicist's usage take their meaning only through them.

Let us take one of the examples cited by Le Roy:

When a heavy body falls freely, the acceleration of its fall is constant. Can such a law be contradicted by experiment? No, for it constitutes the very definition of what is meant by "falling freely." If while studying the fall of a heavy body we found that this body does not fall with uniform acceleration, we should conclude not that the stated law is false, but that the body does not fall freely, that some cause obstructs its motion, and that the deviations of the observed facts from the law as stated would serve to discover this cause and to analyze its effects.

Thus, M. Le Roy concludes, "laws are verifiable, taking things strictly . . ., because they constitute the very criterion by which we judge appearances as well as the methods that it would be necessary to utilize in order to submit them to an inquiry whose precision is capable of exceeding any assignable limit."

Let us study again in greater detail, in the light of the principles previously set down, what this comparison is between the law of falling bodies and experiment.

Our daily observations have made us acquainted with a whole category of motions which we have brought together under the name of motions of heavy bodies; among these motions is the falling of a heavy body when it is not hindered by any obstacle. The result of this is that the words "free fall of a heavy body" have a meaning for the man who appeals only to the knowledge of common sense and who has no notion of physical theories.

On the other hand, in order to classify the laws of motion in question the physicist has created a theory, the theory of weight, an important application of rational mechanics. In that theory, intended to furnish a symbolic representation of reality, there is also the question of "free fall of a heavy body," and as a consequence of the hypotheses supporting this whole scheme free fall must necessarily be a uniformly accelerated motion.

The words "free fall of a heavy body" now have two distinct meanings. For the man ignorant of physical theories, they have their *real* meaning, and they mean what common sense means in pronouncing them; for the physicist they have a *symbolic* meaning, and mean "uniformly accelerated motion." Theory would not have realized its aim if the second meaning were not the sign of the first, if a fall regarded as free by common sense were not also regarded as uniformly accelerated, or *nearly* uniformly accelerated, since common-sense observations are essentially devoid of precision, according to what we have already said.

This agreement, without which the theory would have been rejected without further examination, is finally arrived at: a fall declared by common sense to be nearly free is also a fall whose acceleration is nearly constant. But noticing this crudely approximate agreement does not satisfy us; we wish to push on and surpass the degree of precision which common sense can claim. With the aid of the theory that we have imagined, we put together apparatus enabling us to recognize with sensitive accuracy whether the fall of a body is or is not uniformly accelerated; this apparatus shows us that a certain fall regarded by common sense as a free fall has a slightly variable acceleration. The proposition which in our theory gives its symbolic meaning to the words "free fall" does not represent with sufficient accuracy the properties of the real and concrete fall that we have observed.

Two alternatives are then open to us.

In the first place, we can declare that we were right in regarding the fall studied as a free fall and in requiring that the theoretical definition of these words agree with our observations. In this case, since our theoretical definition does not satisfy this requirement, it must be rejected; we must construct another mechanics on new hypotheses, a mechanics in which the words "free fall" no longer signify "uniformly accelerated motion," but "fall whose acceleration varies according to a certain law."

In the second alternative, we may declare that we were wrong in establishing a connection between the concrete fall we have observed and the symbolic free fall defined by our theory, that the latter was too simplified a scheme of the former, that in order to represent suitably the fall as our experiments have reported it the theorist should give up imagining a weight falling freely and think in terms of a weight hindered by certain obstacles like the resistance of the air, that in picturing the action of these obstacles by means of appropriate hypotheses he will compose a more complicated scheme than a free weight but one more apt to reproduce the details of the experiment; in short, . . . we may seek to eliminate by means of suitable "corrections" the "causes of error," such as air resistance, which influenced our experiment.

M. Le Roy asserts that we shall prefer the second to the first alternative, and he is surely right in this. The reasons dictating this choice are easy to perceive. By taking the first alternative we should be obliged to destroy from top to bottom a very vast theoretical system which represents in a most satisfactory manner a very extensive and complex set of experimental laws. The second alternative, on the other hand, does not make us lose anything of the terrain already conquered by physical theory; in addition, it has succeeded in so large a number of cases that we can bank with interest on a new success. But in this confidence accorded the law of fall of weights, we see nothing analogous to the certainty that a mathematical definition draws from its very essence, that is, to the kind of certainty we have when it would be foolish to doubt that the various points on a circumference are all equidistant from the center.

We have here nothing more than a particular application of the principle set down in Section 2 of this chapter. A disagreement between the concrete facts constituting an experiment and the symbolic representation which theory substitutes for this experiment proves that some part of this symbol is to be rejected. But which part? This the experiment does not tell us; it leaves to our sagacity the burden of guessing. Now among the theoretical elements entering into the composition of this symbol there is always a certain number which the physicists of a certain epoch agree in accepting without test and which they regard as beyond dispute. Hence, the physicist who wishes to modify this symbol will surely bring his modification to bear on elements other than those just mentioned.

But what impels the physicist to act thus is *not* logical necessity. It would be awkward and ill inspired for him to do otherwise, but it would not be doing something logically absurd; he would not for all that be walking in the footsteps of the mathematician mad enough to contradict his own definitions. More than this, perhaps some day by acting differently, by refusing to invoke causes of error and take recourse to corrections in order to reestablish agreement between the theoretical scheme and the fact, and by resolutely carrying out a reform among the propositions declared untouchable by common consent, he will accomplish the work of a genius who opens a new career for a theory.

Indeed, we must really guard ourselves against believing forever warranted those hypotheses which have become universally adopted conventions, and whose certainty seems to break through experimental contradiction by throwing the latter back on more doubtful assumptions. The history of physics shows us that very often the human mind has been led to overthrow such principles completely, though they have been regarded by common consent for centuries as inviolable axioms, and to rebuild its physical theories on new hypotheses. [...]

10 Good Sense Is the Judge of Hypotheses Which Ought to Be Abandoned

When certain consequences of a theory are struck by experimental contradiction, we learn that this theory should be modified but we are not told by the experiment what must be changed. It leaves to the physicist the task of finding out the weak spot that impairs the whole system. No absolute principle directs this inquiry, which different physicists may conduct in very different ways without having the right to accuse one another of illogicality. For instance, one may be obliged to safeguard certain fundamental hypotheses while he tries to reestablish harmony between the consequences of the theory and the facts by complicating the schematism in which these hypotheses are applied, by invoking various causes of error, and by multiplying corrections. The next physicist, disdainful of these complicated artificial procedures, may decide to change some one of the essential assumptions supporting the entire system. The first physicist does not have the right to condemn in advance the boldness of the second one, nor does the latter have the right to treat the timidity of the first physicist as absurd. The methods they follow are justifiable only by experiment, and if they both succeed in satisfying the requirements of experiment each is logically permitted to declare himself content with the work that he has accomplished.

That does not mean that we cannot very properly prefer the work of one of the two to that of the other. Pure logic is not the only rule for our judgments; certain opinions which do not fall under the hammer of the principle of contradiction are in any case perfectly unreasonable. These motives which do not proceed from logic and yet direct our choices, these "reasons which reason does not know" and which speak to the ample "mind of finesse" but not to the "geometric mind," constitute what is appropriately called good sense.

Now, it may be good sense that permits us to decide between two physicists. It may be that we do not approve of the haste with which the second one upsets the principles of a vast and harmoniously constructed theory whereas a modification of detail, a slight correction, would have sufficed to put these theories in accord with the facts. On the other hand, it may be that we may find it childish and unreasonable for the first physicist to maintain obstinately at any cost, at the price of continual repairs and many tangled-up stays, the worm-eaten columns of a building tottering in every part, when by razing these columns it would be possible to construct a simple, elegant, and solid system.

But these reasons of good sense do not impose themselves with the same implacable rigor that the prescriptions of logic do. There is something vague and uncertain about them; they do not reveal themselves at the same time with the same degree of clarity to all minds. Hence, the possibility of lengthy quarrels between the adherents of an old system and the partisans of a new doctrine, each camp claiming to have good sense on its side, each party finding the reasons of the adversary inadequate. The history of physics would furnish us with innumerable illustrations of these quarrels at all times and in all domains. Let us confine ourselves to the tenacity and ingenuity with which Biot by a continual bestowal of corrections and accessory hypotheses maintained the emissionist doctrine in optics, while Fresnel opposed this doctrine constantly with new experiments favoring the wave theory.

In any event this state of indecision does not last forever. The day arrives when good sense comes out so clearly in favor of one of the two sides that the other side gives up the struggle even though pure logic would not forbid its continuation. After Foucault's experiment had shown that light traveled faster in air than in water, Biot gave up supporting the emission hypothesis; strictly, pure logic would not have compelled him to give it up, for Foucault's experiment was *not* the crucial experiment that Arago thought he saw in it, but by resisting wave optics for a longer time Biot would have been lacking in good sense.

Since logic does not determine with strict precision the time when an inadequate hypothesis should give way to a more fruitful assumption, and since recognizing this moment belongs to good sense, physicists may hasten this judgment and increase the rapidity of scientific progress by trying consciously to make good sense within themselves more lucid and more vigilant. Now nothing contributes more to entangle good sense and to disturb its insight than passions and interests. Therefore, nothing will delay the decision which should determine a fortunate reform in a physical theory more than the vanity which makes a physicist too indulgent towards his own system and too severe towards the system of another. We are thus led to the conclusion so clearly expressed by Claude Bernard: The sound experimental criticism of a hypothesis is subordinated to certain moral conditions; in order to estimate correctly the agreement of a physical theory with the facts, it is not enough to be a good mathematician and skillful experimenter; one must also be an impartial and faithful judge.

Notes

1. Claude Bernard, *Introduction à [l'étude de] la Médecine expérimentale* (Paris, 1865), p. 63. (Translator's note: Translated into English by H. C. Greene, *An Introduction to the Study of Experimental Medicine* [New York: Henry Schuman, 1949].)
2. *ibid.*, p. 64.
3. *ibid.*, p. 70.
4. *ibid.*, p. 67.
5. G. Milhaud, "La Science rationnelle," *Revue de Métaphysique et de Morale*, IV (1896), 280. Reprinted in *Le Rationnel* (Paris, 1898), p. 45.
6. H. Poincaré, "Sur les Principes de la Mécanique," *Bibliothèque du Congrès international de Philosophie*, III: *Logique et Histoire des Sciences* (Paris, 1901), p. 457; "Sur la valeur objective des théories physiques," *Revue de Métaphysique et de Morale*, X (1902), 263; *La Science et l'Hypothèse*, p. 110. The edition of *La Science et l'Hypothèse* cited is the 1902 edition, Paris: Flammarion. This edition is translated in Poincaré 1913, *The Foundations of Science*, University Press of America.
7. E. Le Roy, "Un positivisme nouveau," *Revue de Métaphysique et de Morale*, IX (1901), 143–144.

CHAPTER 16

Is There a Logic Of Scientific Discovery?

NORWOOD RUSSELL HANSON

Pages 91–106, *Australasian Journal of Philosophy*, Volume 38, Issue 2, 1960.

The approved answer to this is "No". Thus Popper argues (*The Logic of Scientific Discovery*) "The initial stage, the act of conceiving or inventing a theory, seems to me neither to call for logical analysis nor to be susceptible of it." (p. 31.) Again, ". . . there is no such thing as a logical method of having new ideas, or a logical reconstruction of this process." (p. 32.) Reichenbach writes that philosophy of science ". . . cannot be concerned with [reasons for suggesting hypotheses], but only with [reasons for accepting hypotheses]." (*Experience and Prediction*, p. 382.) Braithwaite elaborates: "The solution of these historical problems involves the individual psychology of thinking and the sociology of thought. None of these questions are our business here." (*Scientific Explanation*, pp. 20, 21.)

Against this negative chorus, the 'Ayes' have not had it. Aristotle (*Prior Analytics*, II, 25), and Peirce (*Collected Papers*, I, Sec. 188) hinted that in science there may be more problems for the logician than just analyzing the arguments supporting already-invented hypotheses. But contemporary philosophers are unreceptive to this. Let us try once again to discuss the distinction F. C. S. Schiller made between the 'Logic of Proof and the 'Logic of Discovery'. (Cf. *Studies in the History and the Methods of the Sciences*, ed. Charles Singer.) We may be forced, with the majority, to conclude 'Nay.' But only after giving Aristotle and Peirce a sympathetic hearing. Is there *anything* in the idea of a 'logic of discovery' which merits the attention of a tough-minded, analytic logician?

It is unclear what a logic of discovery is a logic of. Schiller intended nothing more than "a logic of inductive inference". Doubtless his colleagues were so busy sectioning syllogisms, that they ignored inferences which mattered in science. All the attention philosophers now give to inductive reasoning, probability, and the principles of theory-construction, would have pleased Schiller. But, for Peirce, the work of Popper, Reichenbach and Braithwaite would read less like a *Logic of Discovery* than like a *Logic of the Finished Research Report.* Contemporary logicians of science have described how one sets out reasons in support of an hypothesis once proposed. They have said nothing about the conceptual context within which such an hypothesis is initially proposed. Both Aristotle and Peirce insisted that the proposal of an hypothesis can be a reasonable affair. One can have good reasons, or bad, for suggesting one kind of hypothesis initially, rather than some other kind. These reasons may differ in type from those which lead one to accept an hypothesis once suggested. (This is not to deny that one's reasons for proposing an hypothesis initially may be identical with his reasons for later accepting it.)

One thing must be stressed. When Popper, Reichenbach, and Braithwaite urge that there is no logical analysis appropriate to the psychological complex which attends the conceiving of a new idea, they are saying nothing which Aristotle or Peirce would reject. The latter did not think themselves to be writing manuals to help scientists make discoveries. There could be no such manual. ("There is no science which will enable a man to bethink himself of that which will suit his purpose", J. S. Mill, *A System of Logic*, III, Chapter I.) Apparently they felt that there is a *conceptual* inquiry, one properly called "a logic of discovery", which is *not* to be confounded with the psychology and sociology appropriate to understanding how some investigator stumbled on to an improbable idea in unusual circumstances. There are factual discussions such as these latter. Historians like Sarton and Clagett have undertaken such circumstantial inquiries. Others, e.g., Hadamard and Poincaré, have dealt with the psychology of discovery. But these are not logical discussions. They do not even turn on conceptual distinctions. Aristotle and Peirce thought they were doing something other than psychology, sociology, or history of discovery; they purported to be concerned with a *logic* of discovery.

This suggests caution for those who reject wholesale any notion of a logic of discovery on the grounds that such an inquiry can *only* be psychology, sociology, or history. That Aristotle and Peirce deny just this has made no impression. Perhaps Aristotle and Peirce were wrong. Perhaps there is no room for logic between the psychological dawning of a discovery and the justification of that discovery *via* successful predictions. But this should come as the conclusion of a discussion, not as its preamble. If Peirce is correct, nothing written by Popper, Reichenbach or Braithwaite cuts against him. Indeed, these authors do not discuss what Peirce wishes to discuss.

Let us begin this uphill argument by distinguishing

(1) reasons for accepting an hypothesis H, from
(2) reasons for suggesting H in the first place.

This distinction is in the spirit of Peirce's thesis. Despite his arguments, most philosophers deny any *logical* difference between these two. This must be faced. But let us shape the distinction before denting it with criticism.

What would be our reasons for accepting H? These will be those we might have for thinking H true. But the reasons for suggesting H originally, or for formulating H in one way rather than another, may not be those one requires before thinking H true. They are, rather, those reasons which make H a *plausible type of conjecture*. Now, no one will deny *some* differences between what is required to show H true, and what is required for deciding that H constitutes a plausible kind of conjecture. The question is: are these logical in nature, or more properly called "psychological" or "sociological"?

Or, one might urge (as does Professor Feigl) that the difference is just one of refinement, degree, and intensity. Feigl argues that considerations which settle whether H constitutes a plausible conjecture are of the *same type* as those which settle whether H is true. But since the initial proposal of an hypothesis is a groping affair, involving guesswork amongst sparse data, there *is* a distinction to be drawn; but this, Feigl urges, concerns two ends of a spectrum ranging all the way from inadequate and badly selected data, to that which is abundant, well-diversified, and buttressed by a battery of established theories. The issue therefore remains: is the difference between reasons for accepting H and reasons for suggesting it originally, one of logical type, or one of degree, or of psychology, or of sociology?

Already a refinement is necessary if our original distinction is to survive. The distinction just drawn must be re-set in the following, more guarded, language. Distinguish now

(1') reasons for accepting a particular, minutely-specified hypothesis H, from
(2') reasons for suggesting that, whatever specific claim the successful H will make, it will nonetheless be an hypothesis of one *kind* rather than another.

Neither Aristotle, nor Peirce, nor (if you will excuse the conjunction) myself in earlier papers, sought this distinction on these grounds. The earlier notion was that it was some particular, minutely-specified H which was being looked at in two ways: (1) what would count for the acceptance of that H, and (2) what would count in favour of suggesting that same H initially.

This way of putting it is objectionable. The issue is, whether (*before* having hit on an hypothesis which succeeds in its predictions) one can have good reasons for anticipating that the hypothesis will be one of some particular *kind*. Could Kepler, e.g., have had good reasons (*before* his elliptical orbit hypothesis was established) for supposing that the successful hypothesis concerning Mars' orbit would be of the non-circular kind?[1] He *could* have argued that, whatever path the planet *did* describe, it would be a closed, smoothly-curving, plane geometrical figure. Only this *kind* of hypothesis could entail such observation statements as that Mars' apparent velocities at 90° and at 270° (of excentric anomaly) were greater than any circular-type H could explain. Other *kinds* of hypotheses were available to Kepler: e.g., that Mars' *colour* is responsible for its high velocities, or that the dispositions of Jupiter's moons are responsible. But these would not have struck Kepler as capable of explaining such surprising phenomena. Indeed, he would have thought it unreasonable to develop such hypotheses at all, and would have argued thus. (Braithwaite counters: "But exactly which hypothesis was to be rejected was a matter for the 'hunch' of the physicists" (*Scientific Explanation*, p. 20). However, which *type* of hypothesis Kepler chose to reject was not just a matter of 'hunch'.)

I may still be challenged. Some will continue to berate my distinction between reasons for suggesting which type of hypothesis H will be, and reasons for accepting H ultimately.[2] There may indeed be "psychological" factors, the opposition concedes, which make certain types of hypothesis 'look' as if they might explain phenomena. Ptolemy knew, as well as did Aristarchus before him and Copernicus after him, that a kind of astronomy which displaced the earth would be theoretically simpler, and easier to manage, than the hypothesis of a geocentric, geostatic universe. *But,* philosophers challenge, for psychological, sociological, or historical reasons, alternatives to geocentricism did not 'look' as if they could explain the absence of stellar parallax. This cannot be a matter of logic, since for Copernicus one such alternative *did* 'look' as if it could explain this. In so far as scientists have *reasons* for formulating types of hypotheses (as opposed to hunches, and intuitions), these are just the kinds of reasons which later show a particular H to be true. Thus, if the absence of stellar parallax constitutes more than a psychological reason for Ptolemy's resistance to alternatives to geocentricism, then, in so far, it *is* his reason for rejecting such alternatives as *false*. Conversely, his reason for developing a geostatic type of hypothesis (again, absence of parallax) was his reason for taking some such hypothesis as *true*. Again, Kepler's reasons for rejecting Mars' colour or Jupiter's moons as indicating the kinds of hypotheses responsible for Mars' accelerations, were reasons which also served later in establishing some hypothesis of the non-circularity type.

So the objection to my distinction is: the only *logical* reason for proposing that H will be of a certain type is that *data* incline us to think some *particular* H true. What Hanson advocates is psychological, sociological, or historical in nature; it has no logical import for the differences between proposing and establishing hypotheses.

Kepler again illustrates the objection. Every historian of science knows how the idea of uniform circular motion affected astronomers before 1600. Indeed, in 1591 Kepler abandons an hypothesis because it entails other-than-uniform circular orbits—simply inconceivable for him. So psychological pressure against forming alternative types of hypothesis was great. But *logically* Kepler's reasons for entertaining a type of Martian motion other than uniformly circular *were* his reasons for accepting that as astronomical truth. He first encountered this type of hypothesis on perceiving that no simple adjustment of epicycle, deferent, and excentric could square Mars' observed distances, velocities, and apsidal positions. These were also reasons which led him to assert that the planet's orbit is not the effect of circular motions, but of an elliptical path. Even after other inductive reasons confirmed the truth of the latter hypothesis, these early reasons were *still* reasons for accepting H as true. So they cannot have been reasons merely for proposing which type of hypothesis H would be, and nothing more.

This objection has been made strong. If the following cannot weaken it, then we shall have to accept it; we shall have to grant that there is *no* aspect of discovery which has to do with logical or conceptual considerations.

When Kepler published *De Motibus Stellae Martis* he had established that Mars' orbit was an ellipse, inclined to the ecliptic, the sun in one of the foci. Later (in the *Harmonices Mundi*) he generalized this for other planets. Consider the hypothesis H': *Jupiter's* orbit is of the non-circular type.

The reasons which led Kepler to formulate H' were many. But they included this: that H (the hypothesis that *Mars'* orbit is elliptical) is true. Since Eudoxos, Mars had been the typical planet. (*We* know why. Mars' retrogradations and its movement around the empty focus—all this Earth observes with clarity because of our spatial relations with Mars.) Now, Mars' dynamical properties are usually found in the other planets. If its orbit is ellipsoidal, then it is reasonable to expect that, whatever the exact shape of the other orbits (e.g., Jupiter's) they will all be of the non-circular type.

But such reasons would not *establish* H'. Because what makes it reasonable to anticipate that H' will be of a certain type is *analogical* in character. (Mars does x; Mars is a typical planet; so, perhaps, all planets do the same kind of thing as x.) Analogies cannot establish hypotheses, not even *kinds* of hypotheses. Only observations can do that; in this the Hypothetico-Deductive account (of Popper, Reichenbach and Braithwaite) is correct. To establish H' requires plotting its successive positions on a smooth curve whose equations can be determined. It may then be possible to assert that Jupiter's orbit is, e.g., an ellipse, an oviform, an epicycloid, or whatever. But it would not be reasonable to expect this when discussing only what type of hypothesis is likely to describe Jupiter's orbit. Nor is it right to character-ize this difference between H-as-illustrative-of-a-type-of-hypothesis, and H-as-empirically-established, as a difference of psychology only. *Logically*, Kepler's analogical reasons for proposing that H' would be of a certain type were good reasons. But, logically, they would not then have been good rea-sons for asserting the truth of a specific value for H'—something which could be done only years later.

What are and are not good reasons for reaching a certain conclusion is a logical matter. No further observations are required to settle such issues, any more than we require experiments to decide, on the basis of one's bank statements, whether one is bankrupt. Similarly, whether or not Kepler's reasons for anticipating that H' will be of a certain kind are good reasons, this is a logical inquiry.

Thus, the differences between reasons for expecting that some as-yet-undiscovered H will be of a certain type, and those which establish that H, are greater than is conveyed by calling them "psychological", "sociological", or "historical".

Kepler reasoned initially by analogy. Another kind of reason which makes it plausible to propose that an H, once discovered, will be of a certain type, could be the detection of a formal symmetry in sets of equations or argu-ments. At important junctures Clerk Maxwell and Einstein detected such structural symmetries. This allowed them to argue, before getting their final answers, that those answers would be of a clearly describable type.

In the late 1920's, before anyone had explained the "negative-energy" solutions in Dirac's electron theory, good analogical reasons could have been advanced for the claim that, whatever specific assertion the ultimate H assumed, it would be of the Lorentz-invariant type. It could have been conjectured that the as-yet-undiscovered H would be compatible with the Dirac explanation of Compton scattering and doublet atoms, and would fail to confirm Schrödinger's hunch that the phase waves within configu-ration space actually described observable physical phenomena. All this could have been said before Weyl, Oppenheimer, and Dirac formulated the "Hole-theory of the positive electron". Good analogical reasons for sup-posing that the *type* of H which would succeed would be along these lines could have been, and were, advanced. Indeed, Schrödinger's attempt to rewrite the Dirac theory so that the negative-energy solutions disappeared was *rejected* for failing to preserve Lorentz-invariance.

Thus, reasoning from observations of As as Bs to the proposal "all As are Bs" is different in type from reasoning analogically from the fact that Cs are Ds to the proposal "the hypothesis relating As and Bs will be the same type as that relating Cs and Ds". (Here it is the *way* Cs are Ds which seems analogous to the way As are Bs.) And both of these are typically different from reasoning involving the detection of symmetries in equations describing As and Bs.

Indeed, put in this way, what could an objection *to* the foregoing consist in? Establishing an hypothesis, and proposing by analogy that an hypothesis is likely to be of a particular type: surely these follow reasoning which is different in type. Moreover, both procedures have a fundamentally logical or conceptual interest.

An objection: "Analogical arguments, and those based on the recognition of formal symmetries, are used because of inductively established beliefs in the reliability of arguments of that type. So the cash value of such appeals ultimately collapses into just those accounts given by H-D theorists."

Agreed. But we are not discussing the *genesis* of our faith in these types of arguments, only the *logic* of the arguments themselves. *Given* an analogical premise, or one based on symmetry considerations—or even on enumeration of particulars—one argues *from* these in logically different ways. Consider what further moves are necessary to convince one who doubted such arguments. A challenge to "All As are Bs", when this is based on induction by enumeration, could only be a challenge to justify induction, or at least to show that the particulars are being correctly described. This is inappropriate when the arguments rest on analogies, or on the recognition of formal symmetries.

Another objection: "Analogical reasons, and those based on symmetry—these are *still* reasons for H even after it is (inductively) established. They are reasons *both* for proposing that H will be of a certain type, and for accepting H."

Agreed, again. But, analogical and symmetry arguments could never *by themselves* establish particular Hs. They can only make it plausible to suggest that H (when discovered) will be of a certain type. However, inductive arguments can, by themselves, establish particular hypotheses. So they must differ from arguments of the analogical or symmetrical sort.

H-D philosophers have been most articulate on these matters. So, let us draw out a related issue on which Popper, Reichenbach and Braithwaite seem to me not to have said the last word.

J. S. Mill was wrong about Kepler (*A System of Logic*, III, 2–3). It is impossible to reconcile the delicate adjustment between theory, hypothesis, and observation recorded in *De Motibus Stellae Martis* with Mill's statement that Kepler's First Law is but "a compendious expression for the one set of directly observed facts". Mill did not understand Kepler. (As Peirce notes, *Collected Papers*, I, p. 31.) It is equally questionable whether Reichenbach understood him: "Kepler's laws of the elliptic motion of celestial bodies were inductive generalizations of observed fact . . . [he] observed a series of . . . positions of the planets Mars and found that they may be connected by a mathematical relation" (*Experience and Prediction*, p. 371). Mill's *Logic* is as misleading about scientific discovery as any account proceeding *via* what Bacon calls *"inductio per enumerationem simplicem ubi non reperitur instantia contradictoria"*. (Indeed Reichenbach observes: "It is the great merit of John Stuart Mill to have pointed out that all empirical inferences are reducible to the *inductio per enumerationem simplicem . . ."* (*op. cit.*, p. 389).) The accounts of H-D theorists are equally misleading.

An H-D account of Kepler's First Law would treat it as a high-level hypothesis in an H-D system. (This is Braithwaite's language.) It is regarded as a quasi-axiom, from whose assumption observation-statements follow. If these are true—if, e.g., they imply that Uranus' orbit is an ellipse and that its apparent velocity at 90° is greater than at aphelion—then in so far is the First Law confirmed. (Thus Braithwaite writes: "A scientific system consists of a set of hypotheses which form a deductive system . . . arranged in such a way that from some of the hypotheses as premises all the other hypotheses logically follow . . . the establishment of a system as a set of true propositions depends upon the establishment of its lowest level hypotheses . . ." (*Scientific Explanation*, pp. 12–13).)

This describes physical theory more adequately than did pre-Baconian accounts in terms of simple enumeration, or even post-Millian accounts in terms of ostensibly not-so-simple enumerations. It tells us about the logic of laws, and what they do in finished arguments and explanations. H-D accounts do not, however, tell us anything about the context in which laws are proposed in the first place; nor, perhaps, were they even intended to.

The induction-by-enumeration story *did* intend to do this. *It* sought to describe good reasons for initially proposing H. The H-D account must be silent on this point. Indeed, the two accounts are not strict alternatives. (As Braithwaite suggests they are when he remarks of a certain higher-level hypothesis that it "will not have been established by induction by simple enumeration; it will have been obtained by the hypothetico-deductive method . . ." (*op. cit.*, p. 303).) They are thoroughly compatible. Acceptance of the second is no reason for rejecting the first. A law *might* have been inferred from just an enumeration of particulars (e.g., Boyle's law in the 17th century, Bode's in the 18th, the laws of Ampère and Faraday in the 19th, and much of Meson theory now). It could *then* be built into an H-D system as a higher order proposition. If there is anything wrong with the older view, H-D accounts do not reveal this.

There *is* something wrong. It is false. Scientists do not always discover every feature of a law by enumerating and summarizing observables. (Thus even Braithwaite says: "Sophisticated generalizations (such as that about the proton-electron constitution of the hydrogen atom) . . . [were] certainly not derived by simple enumeration of instances . . ." (*op. cit.*, p. 11).) But *this* does not strengthen the H-D account as against the inductive view. There is *no* H-D account of how "sophisticated generalizations" are *derived*. On his own principles, the H-D theorist's lips are sealed on this matter. But there are conceptual considerations which help us understand the *reasoning* which is sometimes successful in determining the type of an as-yet-undiscovered hypothesis.

Were the H-D account construed as a description of scientific practice, it would be misleading. (Braithwaite's use of "derived" is thus misleading. So is his announcement (p. 11) that he is going to explain "how we *come to make* use of sophisticated generalizations".) Natural scientists do not "start from" hypotheses. They start from data. And even then not from commonplace data—but from surprising anomalies. Thus Aristotle remarks (*Metaphysics* 982 b 11ff.) that knowledge begins in astonishment. Peirce makes perplexity the trigger of scientific inquiry. (*Collected Papers*, II, Book III, ch. 2, Part III.) And James and Dewey treat intelligence as the result of mastering problem situations. (Dewey, *How We Think*, pp. 12ff.)

By the time a law gets fixed into an H-D system, the *original* scientific thinking is over. The pedestrian process of deducing observation-statements begins only after the physicist is convinced that the proposed hypothesis is at least of the right type to explain the initially perplexing data. Kepler's assistant could work out the consequences of H′, and check its validity by seeing whether Jupiter behaved as H′ predicts. This was possible because of Kepler's argument that what H had done for Mars, H′ might do for Jupiter. The H-D account is helpful here; it analyzes *the argument of a completed research report*. It helps us see how experimentalists elaborate a theoretician's hypotheses. And the H-D account illuminates yet another aspect of science, but its proponents have not stressed it. Scientists often dismiss explanations alternative to that which has won their provisional assent along lines that typify the H-D method. Examples are in Ptolemy's *Almagest*, when (on observational grounds) he rules out a moving earth; in Copernicus' *De Revolutionibus . . .*, when he rejects Ptolemy's lunar theory; in Kepler's *De Motibus Stellae Martis*, when he denies that the planes of the planetary orbits intersect in the centre of the ecliptic; and in Newton's *Principia*, when he discounts the idea that the gravitational force law might be of an inverse cube nature. These mirror formal parts of Mill's *System of Logic* or Braithwaite's *Scientific Explanation*.

Still, the H-D analysis remains silent on reasoning which often conditions the discovery of laws—reasoning which determines which type of hypothesis it is likely to be most fruitful to propose.

The induction-by-enumeration story views scientific inference as being from observations to the law, from particulars to the general. There is something true about this which the H-D account must ignore. Thus Newton wrote: "the main business of natural philosophy is to argue from phenomena. . . ." (*Principia*, Preface.)

This inductive view ignores what Newton never did: the inference is also from *explicanda* to an *explicans*. Why a bevelled mirror shows spectra in sunlight is not explained by saying that all bevelled mirrors do this. Why Mars moves more rapidly at 270° and 90° than could be expected of circular-uniform motions is not explained by saying that Mars (or even all planets) always move thus. On the induction view, these latter might count as laws. But only when it is explained why bevelled mirrors show spectra and why planets apparently accelerate at 90° will we have laws of the type suggested: Newton's Laws of Refraction and Kepler's First Law. And even before such discoveries were made, arguments in favour of those *types* of law were possible.

So the inductive view rightly suggests that laws are somehow related to inferences *from* data. It wrongly suggests that the resultant law is but a summary of these data, instead of being an explanation of these data. A logic of discovery, then, might consider the structure of arguments in favour of one *type* of possible explanation in a given context as opposed to other *types*.

H-D accounts all agree that laws explain data. (Thus Braithwaite says: "A hypothesis to be regarded as a natural law must be a general proposition which can be thought to explain its instances; if the reason for believing the general proposition is solely direct knowledge of the truth of its instances, it will be felt to be a poor sort of explanation of these instances . . ." (*op. cit.*, p. 302).) However, they obscure the initial connection between thinking about data and thinking about what kind of hypothesis will be most likely to lead to a law. They suggest that the fundamental inference in science is from higher-order hypotheses to observation-statements. This may characterize the setting out of one's reasons for making a prediction after H is formulated and provisionally established. It need not be a way of setting out reasons in favour of proposing originally what type H is likely to be.

Yet the original suggestion of an hypothesis-type is often a reasonable affair. It is not as dependent on intuition, hunches, and other imponderables as historians and philosophers suppose when they make it the province of genius but not of logic. If the establishment of H through its predictions has a logic, so has the initial suggestion that H is likely to be of one kind rather than another. To form the first specific idea of an elliptical planetary orbit, or of constant acceleration, or of universal gravitational attraction does indeed require genius: nothing less than a Kepler, a Galileo, or a Newton. But this does not entail that reflections leading to these ideas are non-rational. Perhaps only Kepler, Galileo, and Newton had intellects mighty enough to fashion these notions initially; to concede this is not to concede that their reasons for first entertaining concepts of such a type surpass rational inquiry.

H-D accounts begin with the hypothesis as given, as cooking recipes begin with the trout. However, recipes sometimes suggest "First catch your trout". The H-D account is a recipe physicists often use after catching hypotheses. However, the conceptual boldness which marks the history of physics shows more in the ways in which scientists *caught* their hypotheses than in the ways in which they elaborated these once caught.

To study only the verification of hypotheses leaves a vital part of the story untold—namely, what were the reasons Kepler, Galileo, and Newton had for thinking their hypotheses would be of one kind rather than another. In a letter to Fabricius, Kepler underlines this.

> Prague, July 4, 1603 Dear Fabricius,
> . . . You believe that I start with imagining some pleasant hypothesis and please myself in embellishing it, examining it only later by observations. In this you are very much mistaken. The truth is that after having built up an hypothesis on the ground of observations and given it proper foundations, I feel a peculiar desire to investigate whether I might discover some natural, satisfying combination between the two. . . .

If any H-D theorist has ever sought to give an account of the way in which hypotheses in science *are discovered*, Kepler's words are for him. Doubtless H-D philosophers have tried to give just such an account. Thus, Braithwaite writes: "Every *science* proceeds . . . by thinking of general hypotheses . . . from which particular consequences are deduced which can be tested by observation . . .", and again, "Galileo's deductive system was . . . presented as deducible from . . . Newton's laws of motion and . . . his law of universal gravitation . . ." (*Op. cit.*, pp. xv, xi, 18.)

How would an H-D theorist analyze the law of gravitation?

1. First, the hypothesis H: that between any two particles in the universe exists an attracting force varying inversely as the square of the distance between them ($F = \gamma\, Mm/r^2$).
2. Deduce from this (in accordance with the *Principia*)
   a. *Kepler's* Laws, and
   b. *Galileo's* Laws.

3. But particular instances of a. and b. square with what is observed.
4. Therefore H is, in so far, confirmed.

The H-D account says nothing about how H was first puzzled out. But consider why here the H-D account is *prima facie* plausible.

Historians remark that Newton's reflections on this problem began in 1680 when Halley asked: If between a planet and the sun there exists an attraction varying inversely as the square of their distance, what then would be the path of the planet? Halley was astonished by the immediate answer: "An ellipse". The astonishment arose not because Newton *knew* the path of a planet, but because he had apparently deduced this from the hypothesis of universal gravitation. Halley begged for the proof; but it was lost in the chaos of Newton's room. Sir Isaac's promise to work it out anew terminated in the writing of the *Principia* itself. Thus the story unfolds as an H-D plot: (1) from the suggestion of an hypothesis (whose genesis is a matter of logical indifference, i.e., psychology, sociology or history) to (2) the deduction of observation statements (the laws of Kepler and Galileo)—which turn out true, thus (3) establishing the hypothesis.

Indeed, the entire *Principia* unfolds as the plot requires—from propositions of high generality through those of restricted generality, terminating in observation-statements. Thus Braithwaite observes: "Newton's *Principia* [was] modelled on the Euclidian analogy and professed to prove [its] later propositions—those which were confirmed by confrontation with experience—by deducing them from original first principles . . .". (*Scientific Explanation*, p. 352.)

Despite this, the orthodox account is suspicious. The answer Newton gave Halley is not unique. He could have said "a circle" or "a parabola", and have been equally correct. The general answer is: "A conic section". The greatest mathematician of his time is not likely to have dealt with so mathematical a question concerning whether a formal demonstration is possible with an answer which is but a single value of the correct answer.

Yet the reverse inference, the *retro*duction, *is* unique. Given that the planetary orbits are ellipses, and allowing Huygens's law of centripetal force and Kepler's rule (that the square of a planet's period of revolution is proportional to the cube of its distance from the sun)—from this the *type* of the law of gravitation can be inferred. Thus the question "If the planetary orbits are ellipses what form will the force law take?" invites the unique answer "an inverse square type of law".

Given the datum that Mars moves in an ellipse, one can (by way of Huygens's Law and Kepler's Third Law) explain this uniquely by suggesting how it might follow from a law of the inverse square type, such as the law of universal gravitation was later discovered to be.

The rough idea behind all this is: Given an ellipsoidal eggshell, imagine a pearl moving inside it along the maximum elliptical orbit. What *kind* of force must the egg-shell exert on the pearl to keep the latter in this path? Huygens's weights, when whirled on strings, required a force in the string, and in Huygens's arm, of $F_{(k)} \propto r/t^2$ (where r signifies distance, T time, and k is a constant of proportionality). This restraining force kept the weights from flying away like stones from David's sling. And something like this force would be expected in the egg-shell. Kepler's Third Law gives $T^2 \propto r^3$. Hence $F_{(k)} \propto r/r^3 \propto 1/r^2$. The force the shell exerts on the pearl will be of a kind which varies inversely as the square of the distance of the pearl from that focus of the ellipsoidal egg-shell where the force may be supposed to be centred. This is not yet the law of gravitation. But it certainly is an argument which suggests that the Law is likely to be of an inverse square type. This follows by what Peirce called 'retroductive reasoning'. But what *is* this retroductive reasoning whose superiority over the H-D account has been hinted at?

Schematically, it can be set out thus:

1. Some surprising, astonishing phenomena $p_1, p_2, p_3 \ldots$ are encountered.[3]
2. But $p_1, p_2, p_3 \ldots$ would not be surprising were an hypothesis of H's type to obtain. They would follow as a matter of course from something like H and would be explained by it.
3. Therefore there is good reason for elaborating an hypothesis of type H—for proposing it as a possible hypothesis from whose assumption $p_1, p_2, p_3 \ldots$ might be explained.

This is a free development of remarks in Aristotle (*Prior Analytics*, II, 25) and Peirce. (*Collected Papers*, Vol. I, 188. Peirce amplifies: "It must be remembered that retroduction, although it is very little hampered by logical rules, nevertheless is logical inference, asserting its conclusion only problematically, or conjecturally, it is true, but nevertheless having a perfectly definite logical form.")

How, then, would the discovery of universal gravitation fit this account?

1. The astonishing discovery that all planetary orbits are elliptical was made by Kepler.
2. But such an orbit would not be surprising if, in addition to other familiar laws, a law of 'gravitation', of the inverse square type, obtained. Kepler's First Law would follow as a matter of course; indeed that kind of hypothesis might even explain why (since the sun is in but one of the foci) the orbits are ellipses on which the planets travel with non-uniform velocity.

3. Therefore there is good reason for elaborating hypotheses of this kind further.

This says something about the rational context within which an hypothesis of H's type might come to be "caught" in the first place. It begins where all physics begins—with problematic phenomena requiring explanation. It suggests what might be done to particular hypotheses once proposed, viz., the H-D elaboration. And it points up how much philosophers have yet to learn about the kinds of reasons scientists might have for thinking that one kind of hypothesis may explain initial perplexities—why, e.g., an inverse square type of hypothesis may be preferred over others, *if* it throws initially perplexing data into patterns within which determinate modes of connection can be perceived. At least it appears that the ways in which scientists sometimes reason their way *towards* hypotheses, by eliminating those which are certifiably of the wrong type, may be as legitimate an area for conceptual inquiry as are the ways in which they reason their way *from* hypotheses.

Recently, in the Lord Portsmouth collection in the Cambridge University Library, a document was discovered which bears on our discussion. There, in "Additional manuscripts 3968, No. 41, bundle 2", is the following draft in Newton's own hand:

> ". . . And in the same year [1665, twenty years before the *Principia*] I began to think of gravity extending to ye orb of the Moon, and (having found out how to estimate the force with which a globe revolving within a sphere presses the surface of the sphere), from Kepler's rule . . . I deduced that the forces which keep the planets in their Orbs must be reciprocally as the squares of their distances from the centres about which they revolve"

This MS corroborates our argument. ("Deduce", in this passage, is used as when Newton speaks of deducing laws from phenomena—which is just what Aristotle and Peirce would call "retroduce".) Newton *knew* how to estimate the force of a small globe on the inner surface of a sphere. (To compare this with Halley's question and our pearl-within-eggshell reconstruction, note that a sphere can be regarded as a degenerate ellipsoid, i.e., where the foci superimpose.) From this, and from Kepler's rule $T^2 \propto r^3$, Newton determined that, whatever the final form of the law of gravitation, it would very probably be of the inverse-square type. These were the reasons which led Newton to think further about the details of universal gravitation. The reasons for accepting one such hypothesis of this type as a law are powerfully set out later in the *Principia* itself—and they are much more comprehensive than anything which occurred to him at this early age.

But without such preliminary reasoning Newton might have had no more grounds than Hooke or Wren for thinking the gravitation law to be of an inverse square type.

The morals of all this for our understanding of contemporary science are clear. With such a rich profusion of data and technique as we have, the arguments necessary for *eliminating* hypotheses of the wrong type become a central research inquiry. Such arguments are not always of the H-D type; but if, for that reason alone, we refuse to scrutinize the conceptual content of the arguments which precede the actual proposal of definite hypotheses, we will have a poorer understanding of scientific thought in our time. For our own sakes, we must attend as much to how scientific hypotheses are caught, as to how they are cooked.

Indiana University.

Notes

1. *Cf. De Motibus Stellae Martis* (Munich, pp. 250ff.).
2. Reichenbach writes that philosophy "cannot be concerned with the first, but only with the latter" (*Experience and Prediction,* p. 382).
3. The astonishment may consist in the fact that *p* is at variance with accepted *theories*—as, e.g., the discovery of discontinuous emission of radiation by hot black bodies, or the photoelectric effect, the Compton effect, and the continuous β-ray spectrum; or the orbital aberrations of Mercury, the refrangibility of white light, and the high velocities of Mars at 90°. What is important here is *that* the phenomena are encountered as anomalous, not *why* they are so regarded.

B
Force in Natural Philosophy

CHAPTER 17

Introduction to "Force in Natural Philosophy"

This section focuses on debates over the definition and characterization of "force" and "motion" in the analysis of causation in early modern natural philosophy. Much work in HOPOS focuses on the development and interaction of the major early modern systems of "natural philosophy." Seventeenth- and eighteenth-century natural philosophy is not the same as what, from the nineteenth century on, would be referred to with the term "science."

> in the early modern period the term "natural philosophy" denoted attempts to explain in a systematic way the nature of matter, the cosmological structuring of that matter, the principles of causation and the methodology for acquiring or justifying such natural knowledge.
>
> (Gaukroger et al. 2000, 2)

> In Descartes's day, "physics" or "natural philosophy" meant simply "the science of nature." "Nature" encompassed everything having a nature or essence (at least on the Earth), including human beings and human cognition. Aristotelian works on psychology [...] were classed within physics.
>
> (Hatfield 2013, chapter 9)

Natural philosophy is related to "science," but is not co-extensive with, say, modern physics.[1] As Hatfield emphasizes, it includes study of "human beings" and "human cognition," and many systems of natural philosophy include an account of cognition in explaining how we come to have knowledge of other natural phenomena.

Natural philosophy is neither what we would now call philosophy, nor what we would now call science. Another way to put it is that natural philosophy encompasses projects that we would call philosophical and scientific. The historical strategy of trying to divine "actor's categories" is thus particularly appropriate to the early modern period, from the HOPOS perspective. "Actor's categories" are the classifications that the historical figures might have used themselves, rather than the ones that would be assigned later.

Scholars who may have been considered natural philosophers in the early modern period included Francisco Suárez (1548–1617), Francis Bacon (1561–1626), William Harvey (1578–1657), Réne Descartes (1596–1650), Robert Boyle (1627–1691), Christiaan Huygens (1629–1695), Gottfried Leibniz (1646–1716), Isaac Newton (1642–1727), Joseph Priestley (1733–1804), and Antoine Lavoisier (1743–1794), among many others. Debates over the theory of matter and motion, the relationship between human (and divine) methods of cognition and knowledge, and the relationship between reason and observation are prevalent in the early modern period. But natural philosophy also encompassed medicine and physiology, chemistry, and biology, among other areas. The sources of some of the philosophical positions taken not only by Descartes, Leibniz, and Newton, but also by Hume and Kant, are illuminated by their positions on key questions in natural philosophy. The following is intended as a sketch of that relationship, with pointers to primary and secondary literature that can give a more nuanced picture.

The first part of this section focuses on a debate between Descartes, Leibniz, and Newton, on issues raised by Descartes's *Principles of Philosophy*. Newton responds to Descartes in *De Gravitatione*, reprinted almost entirely here, and Leibniz responds in his "Critical Remarks," also reprinted here. I have printed most of the sections of the *Principles of Philosophy* to which Newton and Leibniz respond directly, so that the reader can trace any criticism or remark by Newton or Leibniz back to its origin in Descartes.

A remark on the texts reprinted. They were chosen to demonstrate disagreements over specific questions between natural philosophers, which would be clear to the reader. But these texts, standing alone, do not provide a complete picture of the Cartesian, Leibnizian, or Newtonian systems. In Descartes's case, a student could read *The World*, the *Meditations, Objections, and Replies*, the *Discourse on the Method*, and more of the *Principles* to get a more rounded picture, supplemented by secondary literature such as

Cottingham 1992, Des Chêne 1996, Garber 1992, Gaukroger 2002, Gaukroger et al. 2000, Hattab 2007 and 2009, Ott 2013, and Schmaltz 2008.

In Leibniz's case, the *Monadology*, the *New Essays on Human Understanding*, the *Discourse on Metaphysics*, and for the scientifically inclined the *Specimen Dynamicum* would give a broader picture of his work, again supplemented by texts that might include Garber 2009, Ishiguro 1990, Jolley 1995, Look 1999 and 2011, Mercer 2001, Phemister 2005, Rutherford 1995, Rescher 1989, and Wilson 1989.

For Newton, the *De Gravitatione* is an early work, and not necessarily representative of his later positions. These are found in the *Mathematical Principles of Natural Philosophy* (*Principia*), in the *Optics*, and in other mature texts including those found in Newton 2004; commentary and analysis of these can be found in, for example, Buchwald and Cohen 2001, Cohen and Smith 2002, Domski 2013, Janiak 2009, Janiak and Schliesser 2012, Koyré 1965, the I. Bernard Cohen essay in Newton 1999/1687, Okruhlik and Brown 1985, Smith 2008, and Westfall 1971.

Descartes's *Principles* was a major work in the Cartesian system of natural philosophy. In it, Descartes dealt with several themes central to debates in natural philosophy, including the nature of motion, the motion of the planets, and the divisibility of matter. He argued for the impossibility of atoms and of the vacuum, and for extension as the "nature" or primary quality of body.

Descartes's positions on all of these questions were challenged by natural philosophers, and in at least two cases, by philosophers with rival systems: Gottfried Wilhelm Leibniz and Isaac Newton. Reprinted below are the key passages from Descartes's *Principles*, and then the responses to those passages by Leibniz and Newton.

In the case of the text reprinted from Newton, *De Gravitatione*, it is somewhat surprising that Descartes is Newton's special target. As Janiak puts it, "In the most general terms [. . .] *De Gravitatione* helps to underscore the centrality of Cartesian natural philosophy for understanding the development of Newton's own philosophical orientation, and his treatment of many significant questions in published texts" (Janiak 2009, §1).

Janiak observes that *De Gravitatione* has a significant historical function; to preserve the fact that Descartes was Newton's first target before the *Principia*, not Leibniz.

Since the scholarly consensus takes *De Gravitatione* to have been written before the appearance of the *Principia* in 1687, Newton's extensive attempt in *De Gravitatione* to refute Descartes's broadly relationalist conception of space and time suggests that the Scholium [to the *Principia*] should be read as providing a replacement for the *Cartesian* conception (see Stein 1970). Moreover, it is also unclear whether Newton was familiar with any of Leibniz's relevant writings when he wrote the Scholium. It may be thought a measure of Newton's success against his Cartesian predecessors that history records a debate between the *Leibnizians* and the Newtonians as influencing every subsequent discussion of space and time in the eighteenth century, including Kant's discussion in the *Critique of Pure Reason* in 1781.

(Janiak 2009, §1)

Leibniz and Newton alike challenge Descartes's view that extension is the primary, or essential, property of bodies. They also challenge the Cartesian positions on atoms and the void, on the basis of differing views on the divisibility and hardness of matter.

Leibniz's and Newton's discussions of Descartes's *Principles* illuminate their own philosophical and scientific positions. However, it is worth pointing out how central disputes over the divisibility of matter, the impenetrability of bodies, the nature of motion, and the nature of force were to natural philosophy at the time.

One text that brings this centrality sharply to the fore is the ninth chapter of Émilie du Châtelet's 1750 *Foundations of Physics*, "On the divisibility and subtlety of matter," newly translated for this volume. Du Châtelet was known as an associate of Voltaire and Maupertuis, but also as a shrewd interpreter of Newton. Her translation and commentary on Newton's *Principia*, which first appeared in 1756 and then again in 1759, was a significant influence on the reception of Newton in France. Chapter nine of the *Foundations of Physics* responds to debates in natural philosophy, and in particular to conflicts over the infinite divisibility of matter. Holden 2004 gives a detailed analysis of these debates, and gives an effective classification of the distinct positions taken on divisibility. Watkins 2000, Des Chêne 1996, Hesse 1961, Janiak 2000, and many others also analyze the issue of divisibility in natural philosophy.

Du Châtelet's text brings out what we would now call the philosophical aspects of these debates on positions in physics, e.g., positions on atomism, the divisibility of matter, and the nature of body and extension. As Iltis (1977) puts it, du Châtelet had broad-ranging philosophical sympathies, which led her to attempt to integrate Newtonian and Leibnizian principles and positions:

in March 1739 Samuel Koenig was brought to Cirey by Maupertuis as a tutor for [du Châtelet] and Voltaire in mathematics. By way of Koenig she became converted to Leibniz's philosophical views through their expression in the work of Christian Wolff. In the *Insti-*

tutions she states her indebtedness to Wolff's *Ontologie* of 1729 [. . .] As a result of Koenig's teachings she revised the philosophical chapters of the *Institutions de Physique* which had been written in secret and approved for publication by September 1738. Although Newtonian in its basic mechanical principles, the resulting work followed Leibniz on the subject of dynamics, while the natural philosophy of the early chapters presented an integration of elements from the thought of Leibniz, Descartes, and Newton. It is this integrative character of Madame du Châtelet's thought which sets her *Institutions* apart from other attempts to disseminate Newtonian mechanics. [. . .] Her attempts to integrate aspects of Cartesianism and Newtonianism with Leibnizian ideas reflected a need among natural philosophers of the 1740s to reconcile the conflicts among these systems. But these same integrative tendencies left her uncritical of logical inconsistencies in her account of the metaphysics underlying her system. (p. 31)

From our perspective, these inconsistencies are less interesting than the light this chapter sheds on the broader philosophical significance of these debates in natural philosophy. While the excerpt translated for this volume does not deal in depth with the Newtonian and Leibnizian systems (for more on this see du Châtelet 2009/various), it does bring out several intriguing philosophical and mathematical ramifications of these disputes in natural philosophy, including the Achilles paradox (see, e.g., Black 1950; Salmon 2001), and the relationship between mechanics and metaphysics (see, e.g., Detlefsen 2013; Hutton 2004; Iltis 1977).

Another debate in which du Châtelet took a prominent part, along with Jean Jacques d'Ortus de Mairan, Jean le Rond d'Alembert, Descartes, Leibniz, Newton, Immanuel Kant, and many others, was the so-called "vis viva" debate.

The question at issue was how to evaluate the force of a moving body, including cases in which the body is affected by gravity or engages in collisions. Descartes argued that this force was best evaluated as f = mv (mass times velocity), also called *vis mortua*, dead force, [. . .] or momentum, while Leibniz proposed *vis viva*, or living force, [. . .] f = mv². Descartes also argued that mv is conserved overall in the universe, which became a key issue in the dispute.

(Patton 2012)

While d'Alembert would argue later that the dispute was a "war of words," many natural philosophers took a position on vis viva, which in turn illuminated their views on force and body (Papineau 1977). Kant's pre-Critical text *Thoughts on the Estimation of True Living Forces* is a response to the vis viva debate, for instance (Kant 2012/various).

The chart below refers to the texts that are reprinted here, and shows in detail to which parts of Descartes's *Principles*, and to which themes in natural philosophy, Leibniz, Newton, and du Châtelet respond.

Responses to Descartes's Principles: Leibniz, Newton, du Châtelet

1. Primary truths and the cogito
   a. Descartes Part I, §7
   b. Leibniz "On Article 7"
2. The nature of motion; vis viva
   a. Descartes, Part II, §§13–15, §§21–26, 31
   b. Leibniz "On Articles 21, 22, and 23," "On Article 36"
   c. Newton, "Part II Article 25," "Part II Article 31," "Part II, Articles 10, 12, 18"
3. The motion of the planets
   a. Descartes, Part III, §28–29
   b. Newton, "Part II, Articles 13, 15, 29, 30"
4. The divisibility of matter; the impossibility of atoms and the vacuum
   a. Descartes Part II, §16, §20
   b. Leibniz "On Article 4," "On Articles 8–19," "On Article 45"
   c. Du Châtelet, entire text
5. Extension as the "nature" or primary quality of body
   a. Descartes Part I, §26, 27, Part II, §4, §6, §34
   b. Leibniz "On Article 4"
   c. Du Châtelet, §§169–171
   d. Newton, "Part II, Articles 4, 11," "Part I, Art. 26, 27; Part II, Art. 34"

Hume and Kant on Causality and Laws of Nature

At the beginning of *A Treatise of Human Nature*, Hume argues for an empiricist tenet usually called the "copy principle": that all our simple and complex ideas are copies of, or derive from, simple impressions and their combinations. Hume mounts his well-known skeptical argument against the idea of causality, or a necessary causal power or efficacy, on the basis of this principle. To understand the principle, it is helpful to review Hume's terminology and basic views.

First, Hume distinguishes between ideas and impressions (*Treatise* §1.1.1.1; *Enquiry* §2.1). Impressions are the "perception of the mind" at the moment that one is experiencing something. As Hume puts it memorably in the *Enquiry Concerning Human Understanding*:

Every one will readily allow, that there is a considerable difference between the perceptions of the mind, when a man feels the pain of excessive heat, or the pleasure of moderate warmth, and when he afterwards recalls to his memory the sensation, or anticipates it by his imagination. (§2.1)

As Owen observes:

In neither work does he make an attempt to explain what he means by the phrase, "perceptions of the mind," but it would have been obvious to any eighteenth century reader that he is using that expression much as Descartes and Locke had used the term "idea": for anything the mind is aware of or experiences. As he put it later in the *Treatise*: "To hate, to love, to think, to feel, to see; all this is nothing but to perceive" (§1.2.6.7).

(Owen 2009, 80)

Impressions, then, are perceptions at the time that one has an experience. Ideas are the product of memory or imagination, when one "afterwards recalls to his memory the sensation, or anticipates it by his imagination." Ideas and impressions, to Hume, are distinct. Impressions, as is implied by the passage from the *Enquiry* above, are more forceful, more vivid, and have a greater impact on our mind than ideas, which are fainter and weaker. As Owen goes on to observe:

The distinction between ideas and impressions is further characterized as "the difference betwixt feeling and thinking." Perceptions also differ in being either simple or complex. **Simple perceptions**, he says, "are such as admit of no distinction nor separation," a single colour or taste, for example. **Complex perceptions**, in contrast, are those that "may be distinguish'd into parts," for example, the several qualities (colour, taste, smell, etc.) "united together" in the perception of an apple (*Treatise* §1.1.1.1–2; cf. *Enquiry* §2.3).

(Owen 2009, 81, emphasis added)

The copy principle, then, may be parsed as the statement that all our complex ideas are derived from simple ideas, which in turn are the copies or faint images of simple impressions. No idea is sui generis.

With the copy principle in hand, Hume goes on to ask, which impression is the source of our idea of causality, or any of the synonyms of causality: force, necessity, (causal) efficacy, power, and the like? Hume argues that no simple impression corresponding exactly to these ideas can be found in our experience of objects and their relations.

However, in the section reprinted here, Hume argues that there is *something* that gives rise to our idea of cause and effect. When we experience a causal sequence over and over, our mind begins to expect the effect when we have an impression of the cause. If I repeatedly drop a glass, and the glass breaks each time, after several repetitions my mind will be moved by custom, and habit, to expect the glass to break, because of what Hume calls the "constant conjunction" of the effect with the cause. This movement or expectation of the mind is the source of my ideas of causality, force, and power. Watkins observes that this leads Hume to derive a second conclusion, about the source or foundation of our causal knowledge:

> Hume argues that the ultimate foundation of our knowledge of causality is not *a priori* reason, but rather custom, habit, or experience (where these terms indicate nothing more than repeated instances of certain kinds of impressions we have encountered in the past). For if we are presented with an object that we have never seen before, we are incapable of determining, on the basis of reason alone, what effects this object will have. [. . .] Hume then points out that since our knowledge of cause-effect relationships is based on various impressions from the past, we are not in a position to justify causal laws that would necessarily hold in the future. To do so would require establishing what Hume calls the principle of the uniformity of nature, which asserts that nature will continue to act in the future as it has in the past. However, since our empirical evidence is limited to the past and since no contradiction would seem to arise if nature were to change, [...] no inference to the future is warranted. In a final twist, Hume argues that, in spite of our lack of any rational or empirical justification, we still can and do expect the future to be like the past, given that it is simply part of human nature, whose fundamental basis is the passions, to make such inferences.
>
> (Watkins 2004, 450–451)

Hume's skeptical analysis of the notions of power and force is the basis of his criticism of "the Cartesians" in the passage from the *Treatise* printed below. In the *Principles* and in *The World*, Descartes argues that bodies are endowed by a deity with original motions. Some Cartesians argued on this basis for a continual intervention from a divine being, the position of "occasionalism." According to Hume, the Cartesians were forced into this view, because they conceded that we do not perceive any causal power or efficacy in the bodies themselves. Thus, "the Cartesians, proceeding upon their principle of innate ideas, have had recourse to a supreme spirit or deity, whom they consider as the only active being in the universe, and as the immediate cause of every alteration in matter." Hume had a rather different relationship to Newton; on this score see Schliesser 2008 and De Pierris 2006, among others.

According to Immanuel Kant, reading a brief translation of Hume's account of causality "interrupted" his "dogmatic slumber" (Ak. 4, 260; 10). His reading of Hume prompted Kant to realize that the status of other a priori principles, not just causality, were challenged by Hume's skepticism (see De Pierris and Friedman 2008, §1).

Kant's career is divided by scholars into the "pre-Critical" and "Critical" periods, the division of which is made, usually, by the date of publication of the first edition of the *Critique of Pure Reason* in 1781.[2] The *Prolegomena* is a Critical work, intended by Kant to illuminate the conclusions of the first *Critique*.

In the introduction to the *Prolegomena*, Kant lays out the basic structure of the work, and introduces some key distinctions. The "metaphysical knowledge" with which Kant deals is "knowledge lying beyond experience" (A265). Such knowledge cannot be a posteriori, that is, cannot be derived from experience. Instead, it must be a priori, meaning, at least, that its derivation is independent of experience (see Pereboom 1990, among others, for more on this distinction). Note that, for Kant, the metaphysics he is seeking is not just a priori, but pure a priori. Some a priori knowledge may begin with data given in experience, but then the a priori judgment made may be derived independently of that experience. This a priori knowledge has an empirical part, and is therefore not pure. The derivation of some physical laws is a good example of mixed empirical and a priori knowledge. Our knowledge of many physical laws begins with observation of phenomena in nature, such as spheres rolling down inclined planes, or gases expanding. But the laws themselves, perhaps Galileo's inverse square law or Boyle's gas law, are not derived directly from experience, nor do they apply only to the observed cases (see the *Critique of Pure Reason*, introduction to the B edition, for more on Kant's views here). They have a wider validity, that is, they apply to objects and to events not observed directly. Moreover, according to Kant, the laws have not only subjective validity, as stable descriptions of the observations made by a subject, but also objective validity, as applying to all objects of experience with similar properties in similar circumstances, observed by rational subjects. This objective validity, Kant argues, is due to the applicability of the law as an a priori principle (rule) of nature.

On the other hand, a priori knowledge derived without any empirical basis, that is, without any mixture or involvement of knowledge from experience, is called pure. The excerpt from the *Prolegomena* printed here sets out the foundations of such a system of *pure* a priori knowledge.

Kant divides the *Prolegomena* into four sections, on the basis of four fundamental questions:

1. *How is pure mathematics possible?*
2. *How is pure natural science possible?*
3. *How is metaphysics in general possible?*
4. *How is metaphysics as a science possible?*

Given our aims in studying the *Prolegomena*, I have reprinted the section dealing with the second question, *How is pure natural science possible?*

Kant's analysis of the role of causal concepts in determining temporal order is central to his account of causal judgments. Kant's response to Hume on causality cements Hume's attack on earlier accounts of "power," "force," and "necessity" as powers or efficacies with which objects are endowed. Kant does not think, any more than Hume does, that causal connections or forces are observed directly. As De Pierris and Friedman note:

> Kant agrees with Hume that neither the relation of cause and effect nor the idea of necessary connection is given in our sensory perceptions; both, in an important sense, are contributed by our mind. For Kant, however, the concepts of both causality and necessity arise from precisely the operations of our understanding—and, indeed, they arise entirely a priori as pure concepts or categories of the understanding. It is in precisely this way that Kant thinks that he has an answer to Hume's skeptical problem of induction: the problem, in Kant's terms, of grounding the transition from merely "comparative" to "strict universality" (A91–92/B123–124). Thus in §29 of the Prolegomena, as we have seen, Kant begins from a merely subjective "empirical rule" of constant conjunction or association among our perceptions (of heat following illumination by the sun), which is then transformed into a "necessary and universally valid law" by adding the a priori concept of cause. (2008, §2)

The key move, in the selection from the *Prolegomena* here, is the analysis of the relationship between subjective connections between perceptions ("judgments of perception") and objective laws that are valid of objects of experience ("judgments of experience").

There is a large amount of secondary literature on Kant. On the particular subject of causality, recommended literature includes (but is by no means limited to) Allison 2004; Buchdahl 1965; De Pierris and Friedman 2008; Friedman 1992; Guyer 1987; Van Cleve 1973; and Watkins 2004, 2005, and 2000.

Hume's arguments in the *Enquiry* and the *Treatise,* and Kant's arguments in the *Critique of Pure Reason,* the *Prolegomena,* and the *Metaphysical Foundations,* all go against the notion of force as an observable property of objects. Rather, in both cases, force is an *inference*; in Hume's case, a subjective inference as a result of custom or habit (in response to regularities in the phenomena), and in Kant's case, an inference on the basis of a priori causal concepts.

Suggestions for Further Research

The texts reprinted here were chosen for this volume partly because they are not widely available. As of printing, many of the primary texts I've mentioned above, including Newton's *Principia,* Leibniz's *Monadology* and *New Essays,* and Descartes's *Discourse* and *Meditations* are available online. Commentaries on and introductions to these texts also are available. The works reprinted here are thus intended to complement what is available freely, by giving a broader historical picture of the questions at issue in well-known debates between Leibniz, Descartes, Newton, and other natural philosophers.

As a result of the ready availability of public domain primary resources in natural philosophy, a number of options are available to the student choosing a paper topic or the professor seeking to supplement—or to go well beyond—the material in this textbook. I suggest a few options below; there are many more. As of printing, all the texts mentioned are available online, except where noted.

1. **A natural-philosophical revolution in medicine: Andreas Vesalius and William Harvey on the circulation of blood**
   "The History of Science," Thomas Kuhn, this volume.
   The Writings of Hippocrates and Galen, ed. John Coxe. Philadelphia: Lindsay and Blakiston, 1846.
   On the Fabric of the Human Body, Andreas Vesalius.
   The four hundredth anniversary celebration of the De humani corporis fabrica of Andreas Vesalius, ed. Yale University. New Haven: The Tuttle, Morehouse & Taylor Company, 1943.
   The Works of William Harvey, M.D. trans. Robert Willis. London: The Sydenham Society, 1847.
   The Discovery of the Circulation of the Blood, Charles Singer. London: Bell, 1922.

2. **The synthesis of water and the rejection of phlogiston: Joseph Priestley and Antoine Lavoisier**
   Elements of Chemistry, two volumes, Antoine Lavoisier, tr. Robert Kerr. Edinburgh: W. Creech.

"Experiments and Observations Relating to the Principle of Acidity, the Composition of Water, and Phlogiston," Joseph Priestley, *Philosophical Transactions of the Royal Society of London* 1788, 78: 147–157.

"Observations on the Doctrine of Phlogiston, and the Decomposition of Water," Joseph Priestley, Philadelphia: Printed by Thomas Dobson, 1797.

Many of Priestley's and Lavoisier's treatises and papers are available from archive.org.

Hasok Chang's *Is Water H2O? Evidence, Pluralism and Realism,* Boston Studies in the Philosophy of Science (Dordrecht: Springer, 2012) is a recent work, available in print, that provides a detailed philosophical analysis of this theme.

3. **The Scientific Revolution: Bacon, Boyle, Galileo, and their influence**

Advancement of Learning and *Novum Organum,* Francis Bacon. New York: The Colonial Press, 1900.

History of Scientific Ideas, volume 1. William Whewell. London: John W. Parker, 1858.

Francis Bacon of Verulam: Realistic Philosophy and its Age, trans. John Oxenford from the original by Kuno Fischer. London: Longman, Brown, Green, Longmans, & Roberts, 1857.

The Sceptical Chymist, Robert Boyle. London: J. Cadwell, 1661.

Dialogues Concerning Two New Sciences, Galileo Galilei, trans. Henry Crew and Alfonso de Salvio. New York: Macmillan, 1914.

The Sidereal Messenger of Galileo Galilei, and a Part of the Preface to Kepler's Dioptrics, Galileo Galilei, trans. Edward Carlos. London: Rivingtons, 1880.

See also the selection from *Leviathan and the Air Pump* in Part I; the rest of this book is not available freely but is in most academic libraries.

Notes

1. Grant 2007 and Des Chêne 1996, among others, provide historical and philosophical analysis of developments in natural philosophy.
2. Kant's philosophical education was in the Leibnizian school, filtered through Christian Wolff and his followers, including Martin Knutzen and Alexander Baumgarten (see Schönfeld 2000; Laywine 1993; Laywine 1998; Carl 1989).

References and Further Reading

Achinstein, Peter (ed.). 2004. *Science Rules.* Baltimore: Johns Hopkins University Press.

Allison, Henry. 2004/1983. *Kant's Transcendental Idealism: An Interpretation and Defense,* rev. expanded ed. New Haven, Conn.: Yale University Press.

Ariew, Roger. 2011. *Descartes Among the Scholastics*. Leiden, The Netherlands: Brill.

———. 1999. *Descartes and the Last Scholastics*. Ithaca, NY: Cornell University Press.

Ariew, Roger and Grene, Marjorie, eds. 1995. *Descartes and His Contemporaries: Meditations, Objections and Replies*. Chicago: University of Chicago Press.

Baumgarten, Alexander. 2013/1779. *Metaphysics: A Critical Translation with Kant's Elucidations, Selected Notes, and Related Materials*, trans. and ed. Courtney D. Fougate and John Hymers. London and New York: Bloomsbury Publishing.

Bertoloni Meli, Domenico. 1993. *Equivalence and Priority: Newton vs. Leibniz*. Oxford: Oxford University Press.

Black, Max. 1950. "Achilles and the Tortoise," *Analysis* 11: 91–101.

Bolton, Martha. 1996. "The Nominalist Argument of the *New Essays*," *The Leibniz Review*, 6: 1–24.

———. 1998. "Locke, Leibniz, and the Logic of Mechanism," *Journal of the History of Philosophy*, 36 (2): 189–213.

Boyle, Robert. 1661. *Skeptical Chymist*. London: Printed for J. Crooke.

Buchdahl, Gerd. 1965. "Causality, Causal Laws and Scientific Theory in the Philosophy of Kant," *British Journal for the Philosophy of Science* 16 (63): 187–208.

Buchwald, J. Z. and Cohen, I. B. (eds.). 2001. *Isaac Newton's Natural Philosophy*. Cambridge, Mass.: MIT Press.

Buckle, Stephen. 2004. *Hume's Enlightenment Tract*. Oxford: Oxford University Press.

Carl, Wolfgang. 1989. *Der schweigende Kant*. Göttingen: Vandenhoeck & Ruprecht.

Cohen, I. Bernard. 1980. *The Newtonian Revolution*. Cambridge: Cambridge University Press.

Cohen, I. Bernard and Smith, George (eds.). 2002. *The Cambridge Companion to Newton*. Cambridge: Cambridge University Press.

Cottingham, John (ed.). 1992. *The Cambridge Companion to Descartes*. Cambridge: Cambridge University Press.

Cover, J. A. and J. O'Leary-Hawthorne, 1999. *Substance and Individuation in Leibniz*. Cambridge: Cambridge University Press.

De Pierris, Graciela and Friedman, Michael. 2008. "Kant and Hume on Causality," *The Stanford Encyclopedia of Philosophy*, ed. Edward N. Zalta. URL = <http://plato.stanford.edu/archives/fall2008/entries/kant-hume-causality/>.

De Pierris, Graciela. 2006. "Hume and Locke on Scientific Methodology: The Newtonian Legacy," *Hume Studies* 32 (2): 277–330.

Descartes, René. 1976. *Oeuvres de Descartes*, C. Adams and P. Tannery, eds. Paris: J. Vrin.

———. 1979. *The World*, trans. M. S. Mahoney. New York: Abaris Books.

———. 1983. *Principles of Philosophy*, trans. V. R. Miller and R. P. Miller. Dordrecht: Kluwer Academic Publishers.

———. 1984a. *The Philosophical Writings of Descartes, Vol. 1*, trans. and eds. J. Cottingham, R. Stoothoff, and D. Murdoch. Cambridge: Cambridge University Press.

———. 1984b. *The Philosophical Writings of Descartes, Vol. 2*, trans. and eds. J. Cottingham, R. Stoothoff, and D. Murdoch. Cambridge: Cambridge University Press.

———. 1991. *The Philosophical Writings of Descartes, Vol. 3: The Correspondence*, trans. and eds. J. Cottingham, R. Stoothoff, and D. Murdoch. Cambridge: Cambridge University Press.

Des Chêne, Dennis. 1996. *Physiologia: Natural Philosophy in Late Aristotelian and Cartesian Thought*. Ithaca, NY: Cornell University Press.

Detlefsen, Karen. 2013. "Émilie du Châtelet," *The Stanford Encyclopedia of Philosophy* ed. Edward Zalta, URL = <http://plato.stanford.edu/archives/sum2013/entries/emilie-du-chatelet/>.

Domski, Mary. 2013. "Kant and Newton on the A Priori Necessity of Geometry," *Studies in the History and Philosophy of Science* 44 (3): 438–447.

Du Châtelet, Émilie. 2009/various. *Selected Philosophical and Scientific Writings*, ed. Judith Zinsser, trans. Isabelle Bour. Chicago: University of Chicago Press.

———. 1st edition, 1739; 2nd edition, 1744. *Dissertation sur la nature et la propagation du feu*. Paris: Chez Prault Fils.

———. 1st edition, 1740; 2nd edition, 1742. *Institutions de physique*. Paris: Chez Prault Fils.

———. 1st edition, 1756; 2nd edition, 1759. *Principes mathématiques de la philosophie naturelle, par feue Madame la Marquise du Châtelet*. Paris: Desaint & Saillant, Lambert.

———. 1741. *Réponse sur la question des forces vives*. Brussels: Foppens.

Euclid. 1956/ca. 300 BCE. *The Thirteen Books of Euclid's Elements*, second ed., trans. and ed. Thomas Heath. New York: Dover Publications. (Facsimile reprint of Heath, ed. Cambridge: Cambridge University Press, 1925).

Duchesneau, François. 1993. *Leibniz et la méthode de la science*. Paris: Presses Universitaires de France.

Fichant, Michel. 1998. *Science et métaphysique dans Descartes et Leibniz*. Paris: Presses Universitaires de France.

———. 2003. "Leibniz et les machines de la nature," *Studia Leibnitiana*, 35 (1): 1–28.

Friedman, Michael. 1992. *Kant and the Exact Sciences*. Cambridge, Mass.: Harvard University Press.

Galilei, Galileo and Drake, Stillman. 2000. *Two New Sciences and Drake's History of Free Fall*. Toronto, Ontario: Wall & Emerson, Inc.

Garber, Dan. 1992. *Descartes' Metaphysical Physics*. Chicago: University of Chicago Press.

———. 2009. *Leibniz: Body, Substance, Monad*. New York and Oxford: Oxford University Press.

Gaukroger, S., Schuster, J., and Sutton, J. (eds.). 2000. *Descartes' Natural Philosophy*. London: Routledge.

Gaukroger, Stephen. 2002. *Descartes' System of Natural Philosophy*. Cambridge: Cambridge University Press.

Grant, Edward. 2007. *A History of Natural Philosophy*. Cambridge: Cambridge University Press.

Guyer, Paul. 1987. *Kant and the Claims of Knowledge*. Cambridge: Cambridge University Press

Hatfield, Gary. 1979. "Force (God) in Descartes' Physics," *Studies in History and Philosophy of Science* 10: 113–140.

———. 2013. *Routledge Philosophy Guidebook to Descartes and the Meditations*. New York and London: Routledge.

Hattab, Helen. 2007. "Concurrence or Divergence? Reconciling Descartes's Metaphysics with His Physics," *Journal of the History of Philosophy*, 45: 49–78.

———. 2009. *Descartes on Forms and Mechanisms*. Cambridge: Cambridge University Press.

Hesse, Mary. 1961. *Forces and Fields: The Concept of Action at a Distance in the History of Physics*. London: Nelson.

Holden, Thomas. 2004. *The Architecture of Matter*. Oxford: Oxford University Press.

Hutton, Sarah. 2004. "Emilie du Châtelet's *Institutions de physique* as a Document in the History of French Newtonianism," *Studies in History and Philosophy of Science Part A* 35 (3): 515–531.

Huygens, Christiaan. 1888–1950. *Oeuvres Complètes de Christiaan Huygens*, 22 vols. The Hague: Martinus Nijhoff.

———. 1986/1673. *The Pendulum Clock or Geometrical Demonstrations Concerning the Motion of Pendula as Applied to Clocks*, trans. Richard J. Blackwell. Ames: Iowa State University. (Translation of *Horologium Oscillatorium*, Paris, vol. 18.)

Iltis, Carolyn. 1977. "Madame du Châtelet's Metaphysics and Mechanics," *Studies in History and Philosophy of Science Part A* 8 (1): 29–48.

Ishiguro, Hidé. 1990. *Leibniz's Philosophy of Logic and Language*, second ed. Cambridge: Cambridge University Press.

Jolley, Nicholas, 1984. *Leibniz and Locke: A Study of the "New Essays on Human Understanding."* Oxford: Clarendon Press.

Janiak, Andrew. 2009. "Newton's Philosophy," *The Stanford Encyclopedia of Philosophy*, ed. Edward N. Zalta. URL = <http://plato.stanford.edu/archives/win2009/entries/newton-philosophy/>.

———. 2000. "Space, Atoms and Mathematical Divisibility in Newton," *Studies in History and Philosophy of Science* 31: 203–230.

Janiak, Andrew and Schliesser, Eric. 2012. *Interpreting Newton: Critical Essays.* Cambridge: Cambridge University Press.

Jolley, Nicholas (ed.). 1995. *The Cambridge Companion to Leibniz.* Cambridge: Cambridge University Press.

Kant, Immanuel. 1999/1781–1785. *Critique of Pure Reason*, trans. Paul Guyer and Allan Wood. Cambridge: Cambridge University Press.

———. 1900 f.- *I. Kant, Gesammelte Werke.* Berlin: Ed. königlich preußische (später deutsche) Akademie der Wissenschaften.

———. 2004/1786. *Metaphysical Foundations of Natural Science*, trans. Michael Friedman. Cambridge: Cambridge University Press.

———. 2012/various. *Natural Science*, ed. Eric Watkins, trans. Beck, Edwards, Reinhardt, Schönfeld, and Watkins. Cambridge: Cambridge University Press.

———. 1977/1783. *Prolegomena to Any Future Metaphysics*, second ed. trans. James Ellington. Indianapolis/Cambridge: Hackett Publishing Co., Inc.

Keill, John. 1745. *An Introduction to Natural Philosophy.* Lectures originally given at Oxford in 1900. London: M. Senex, W. Innys, T. Longman, and T. Shewell.

Kepler, Johannes. 1952–55/1619. *Harmonice Mundi*, selections, vol. 16, *Great Books of the Western World.* Chicago: Encyclopaedia Britannica.

———. 1952–55/1617–21. *Epitome of Copernican Astronomy*, selections, vol. 16, *Great Books of the Western World.* Chicago: Encyclopaedia Britannica.

———. 1992/1609. *New Astronomy*, trans. William H. Donahue. Cambridge: Cambridge University Press.

Koyré, Alexandre. 1965. *Newtonian Studies.* Cambridge, Mass.: Harvard University Press.

Langton, Rae. 1998. *Kantian Humility: Our Ignorance of Things in Themselves.* Oxford: Oxford University Press.

Laywine, Alison. 1993. *Kant's Early Metaphysics and the Origins of the Critical Philosophy.* Atascadero: Ridgeview Publishing.

———. 1998. "Martin Knutzen," *Encyclopedia of Philosophy.* London: Routledge.

Leibniz, Gottfried Wilhelm. 1976/various. *Philosophical Papers and Letters*, ed. Leroy Loemker, 2nd ed. Dordrecht: Reidel.

———. 1996/1765. *New Essays on Human Understanding*, ed. Jon Bennett and Peter Remnant. Cambridge: Cambridge University Press.

———. 1976/1695. *Specimen Dynamicum*, pp. 435–452 in Loemker (ed.).

———. 1993/1689. "Tentamen de Motuum Coelestium Causis," *Acta Eruditorum*; translation in Bertoloni Meli 1993.

Locke, John. 1979/1689. *An Essay Concerning Human Understanding*, ed. Peter Nidditch. Clarendon Editions of the Works of John Locke. Oxford: Clarendon Press.

Longuenesse, Béatrice. 1998. *Kant and the Capacity to Judge.* Princeton, N.J.: Princeton University Press.

Look, Brandon. 1999. *Leibniz and the 'Vinculum Substantiale'.* Stuttgart: Steiner. *Studia Leibnitiana*, Supplement 30.

———. 2002. "On Monadic Domination in Leibniz's Metaphysics," *British Journal for the History of Philosophy* 10 (3): 379–399.

——. 2005. "Leibniz and the Shelf of Essence," *The Leibniz Review* 15: 27–47.

——. 2010. "Leibniz's Metaphysics and Metametaphysics: Idealism, Realism and the Nature of Substance," *Philosophy Compass*, 5 (11): 871–879.

Look, Brandon (ed.). 2011. *The Continuum Companion to Leibniz*. London: Continuum.

Mairan, Jean Jacques d'Ortus de. 1741/1728. *Dissertation sur l'estimation et la mésure des forces motrices des corps*, two volumes. Paris: Charles-Antoine Jombert.

——. 1741. "Lettre à Madame [du Chastelet] sur la question des forces vives," printed in the *Dissertation* vol. I.

Menn, Stephen. 1990. "Descartes and Some Predecessors on the Divine Conservation of Motion," *Synthese* 83 (2): 215–238.

Mercer, Christia. 2001. *Leibniz's Metaphysics: Its Origin and Development*. Cambridge: Cambridge University Press.

Nachtomy, Ohad. 2007. *Possibility, Agency, and Individuality in Leibniz's Metaphysics*. Dordrecht: Springer.

Newton, Isaac. 2004. *Philosophical Writings*, ed. A. Janiak. Cambridge: Cambridge University Press.

——. 1959–1984/various. *The Correspondence of Isaac Newton*, ed. H. W. Turnbull, J. F. Scott, A. R. Hall, and L. Tilling, 7 vols. Cambridge: Cambridge University Press.

——. 1974/1684–1691. *The Mathematical Papers of Isaac Newton*, ed. D. T. Whiteside, vol. 6. Cambridge: Cambridge University Press.

——. 1999/1687. *The Principia: Mathematical Principles of Natural Philosophy: A New Translation*, trans. I. B. Cohen and Anne Whitman, preceded by "A Guide to Newton's Principia" by I. B. Cohen. Berkeley: University of California Press.

Norton, David Fate and Taylor, Jacqueline Anne. 2009. *The Cambridge Companion to Hume*. Cambridge: Cambridge University Press.

Okruhlik, Kathleen, and James Brown (eds.). 1985. *The Natural Philosophy of Leibniz*. Dordrecht: D. Reidel.

Ott, Walter. 2013. *Causation and Laws of Nature in Early Modern Philosophy*. Oxford: Oxford University Press.

Owen, David. 2009. "Hume and the Mechanics of Mind," in Norton and Taylor 2009.

Papineau, David. 1977. "The *Vis Viva* Controversy: Do Meanings Matter?" *Studies in the History and Philosophy of Science* 8 (2): 111–141.

Patton, Lydia. 2012. "Review of *Natural Science*" (Kant 2012/various, ed. Watkins.) *Notre Dame Philosophical Reviews*.

Pereboom, Derk. 1990. "Kant on Justification in Transcendental Philosophy," *Synthese* 85 (1): 25–54.

Phemister, Pauline. 2005. *Leibniz and the Natural World: Activity, Passivity, and Corporeal Substances in Leibniz's Philosophy*. Dordrecht: Springer.

Proclus. 1992/unknown. *A Commentary on the First Book of Euclid's Elements*, ed. G. R. Morrow, foreword Ian Mueller. Princeton, N.J.: Princeton University Press.

Rescher, Nicholas (ed.). 1989. *Leibnizian Inquiries: A Group of Essays*. New York: University Press of America.

Rutherford, Donald. 1995. *Leibniz and the Rational Order of Nature*. Cambridge: Cambridge University Press.

Salmon, Wesley. 2001. *Zeno's Paradoxes*, 2nd ed. Indianapolis: Hackett Publishing Co.

Schliesser, Eric. 2008. "Hume's Newtonianism and Anti-Newtonianism," *The Stanford Encyclopedia of Philosophy*, Edward Zalta (ed.), URL = <http://plato.stanford.edu/archives/win2008/entries/hume-newton/>.

Schmaltz, Tad. 2008. *Descartes on Causation*. Oxford: Oxford University Press.

Schönfeld, Martin. 2000. *The Philosophy of the Young Kant*. Oxford: Oxford University Press.

Shea, William. 1991. *The Magic of Numbers and Motion: The Scientific Career of René Descartes*. Canton, Mass.: Science History Publications.

Simplicius. 1989/6th century BCE. *On Aristotle's Physics 6*, trans. David Konstan. London: Gerald Duckworth & Co. Ltd.

Slowik, Edward. 2002. *Cartesian Spacetime*. Dordrecht: Kluwer.

——. 2013. "Descartes' Physics," *Stanford Encyclopedia of Philosophy*, Edward N. Zalta (ed.), URL = <http://plato.stanford.edu/archives/fall2013/entries/descartes-physics/>.

Smith, George. 2008. "Newton's Philosophiae Naturalis Principia Mathematica," *The Stanford Encyclopedia of Philosophy* Edward N. Zalta (ed.), URL = <http://plato.stanford.edu/archives/win2008/entries/newton-principia/>.

Smith, Justin E. H. 2011. *Divine Machines: Leibniz and the Sciences of Life*. Princeton, N.L.: Princeton University Press.

——. 2011. *Machines of Nature and Corporeal Substances in Leibniz*. Dordrecht: Springer.

Southgate, Henry. 2013. "Kant's Critique of Leibniz's Rejection of Real Opposition," *HOPOS: The Journal of the International Society for the History of Philosophy of Science* 3 (1): 91–134.

Stan, Marius. 2013. "Kant's Third Law of Mechanics: The Long Shadow of Leibniz," *Studies in History and Philosophy of Science Part A* 44 (3): 493–504.

Stein, Howard. 1970. "Newtonian Space-Time," in *The Annus Mirabilis of Sir Isaac Newton, 1666–1966*, ed. R. Palter. Cambridge, Mass.: MIT Press.

——. 1990. "From the Phenomena of Motions to the Forces of Nature: Hypothesis or Deduction?" *PSA 1990, Proceedings of the 1990 Biennial Meeting of the Philosophy of Science Association*, vol. 2, East Lansing, Mich.: Philosophy of Science Association, pp. 209–222.

Strawson, Peter. 1990. *The Bounds of Sense: An Essay on Kant's Critique of Pure Reason*. New York and London: Routledge.

Suárez, Francisco. 1982/1597. *Francis Suarez on Individuation: Metaphysical Disputation V: Individual Unity and Its Principle*, trans. Jorge Gracia. Milwaukee: Marquette University Press.

Van Cleve, James. 1973. "Four Recent Interpretations of Kant's Second Analogy," *Kant-Studien* 64: 69–87.

Watkins, Eric. 2005. *Kant and the Metaphysics of Causality*. New York: Cambridge University Press.

——. 2004. "Kant's Model of Causality: Causal Powers, Laws, and Kant's Reply to Hume," *Journal of the History of Philosophy* 42 (4): 449–488.

——. 2000. *Kant and the Sciences*. Oxford: Oxford University Press.

Westfall, Richard. 1971. *Force in Newton's Physics: The Science of Dynamics in the Seventeenth Century*. London: Macdonald.

Wilson, Catherine. 1989. *Leibniz's Metaphysics: A Historical and Comparative Study*. Manchester: Manchester University Press.

Wilson, Jessica. 2007. "Newtonian Forces," *The British Journal for the Philosophy of Science*, 58 (2): 173–205.

Wolff, Christian. 1770. *Logic, or Rational Thoughts on the Powers of the Human Understanding with Their Use and Application in the Knowledge and Search of Truth*. London: Printed for L. Hawes, W. Clarke, and R. Collins.

——. 1963/1728. *Preliminary Discourse on Philosophy in General*, trans. Richard Blackwell. Indianapolis: The Bobbs-Merrill Company, Inc.

CHAPTER 18

Principles of Philosophy (selections)

RÉNE DESCARTES

Part I, §1–8, §13–14, §18–21, §26–8; Part II, §4–26, §32–45, §54–64.
Part III, §26–29, §119, §140, from *Principles of Philosophy*, in *The Philosophical Writings of Descartes*, vol. I, translated by Cottingham, Stoothoff and Murdoch. Cambridge University Press, 1985/1647.

PART ONE

The Principles of Human Knowledge

1. *The seeker after truth must, once in the course of his life, doubt everything,
   as far as is possible.*
Since we began life as infants, and made various judgements concerning the
things that can be perceived by the senses before we had the full use of our
reason, there are many preconceived opinions that keep us from knowledge
of the truth.[1] It seems that the only way of freeing ourselves from these
opinions is to make the effort, once in the course of our life, to doubt everything which we find to contain even the smallest suspicion of uncertainty.

2. *What is doubtful should even be considered as false.*
Indeed, it will even prove useful, once we have doubted these things, to consider them as false, so that our discovery of what is most certain and easy to
know may be all the clearer. [. . .]

4. *The reasons for doubt concerning the things that can be perceived by the senses.*

Given, then, that our efforts are directed solely to the search for truth, our initial doubts will be about the existence of the objects of sense-perception and imagination. The first reason for such doubts is that from time to time we have caught out the senses when they were in error, and it is prudent never to place too much trust in those who have deceived us even once. The second reason is that in our sleep we regularly seem to have sensory perception of, or to imagine, countless things which do not exist anywhere; and if our doubts are on the scale just outlined, there seem to be no marks by means of which we can with certainty distinguish being asleep from being awake.

5. *The reasons for doubting even mathematical demonstrations.*

Our doubt will also apply to other matters which we previously regarded as most certain – even the demonstrations of mathematics and even the principles which we hitherto considered to be self-evident. One reason for this is that we have sometimes seen people make mistakes in such matters and accept as most certain and self-evident things which seemed false to us. Secondly, and most importantly, we have been told that there is an omnipotent God who created us. Now we do not know whether he may have wished to make us beings of the sort who are always deceived even in those matters which seem to us supremely evident; for such constant deception seems no less a possibility than the occasional deception which, as we have noticed on previous occasions, does occur. We may of course suppose that our existence derives not from a supremely powerful God but either from ourselves or from some other source; but in that case, the less powerful we make the author of our coming into being, the more likely it will be that we are so imperfect as to be deceived all the time.

6. *We have free will, enabling us to withhold our assent in doubtful matters and hence avoid error.*

But whoever turns out to have created us, and however powerful and however deceitful he may be, in the meantime we nonetheless experience within us the kind of freedom which enables us always to refrain from believing things which are not completely certain and thoroughly examined. Hence we are able to take precautions against going wrong on any occasion.

7. *It is not possible for us to doubt that we exist while we are doubting; and this is the first thing we come to know when we philosophize in an orderly way.*

In rejecting – and even imagining to be false – everything which we can in any way doubt, it is easy for us to suppose that there is no God and no heaven, and that there are no bodies, and even that we ourselves have no hands or feet, or indeed any body at all. But we cannot for all that suppose that we, who are having such thoughts, are nothing. For it is a contradiction to suppose that what thinks does not, at the very time when it is thinking, exist. Accordingly, this piece of knowledge[2] – *I am thinking, therefore I exist* – is the first and most certain of all to occur to anyone who philosophizes in an orderly way.

8. *In this way we discover the distinction between soul and body, or between a thinking thing and a corporeal thing.*

This is the best way to discover the nature of the mind and the distinction between the mind and the body. For if we, who are supposing that everything which is distinct from us is false,[3] examine what we are, we see very clearly that neither extension nor shape nor local motion, nor anything of this kind which is attributable to a body, belongs to our nature, but that thought alone belongs to it. So our knowledge of our thought is prior to, and more certain than, our knowledge of any corporeal thing; for we have already perceived it, although we are still in doubt about other things.

9. *What is meant by 'thought'.*

By the term 'thought', I understand everything which we are aware of as happening within us, in so far as we have awareness of it. Hence, *thinking* is to be identified here not merely with understanding, willing and imagining, but also with sensory awareness. For if I say 'I am seeing, or I am walking, therefore I exist', and take this as applying to vision or walking as bodily activities, then the conclusion is not absolutely certain. This is because, as often happens during sleep, it is possible for me to think I am seeing or walking, though my eyes are closed and I am not moving about; such thoughts might even be possible if I had no body at all. But if I take 'seeing' or 'walking' to apply to the actual sense or awareness of seeing or walking, then the conclusion is quite certain, since it relates to the mind, which alone has the sensation or thought that it is seeing or walking. [. . .]

13. *The sense in which knowledge of all other things depends on the knowledge of God.*

The mind, then, knowing itself, but still in doubt about all other things, looks around in all directions in order to extend its knowledge further. First of all, it finds within itself ideas of many things; and so long as it merely contemplates these ideas and does not affirm or deny the existence outside itself of anything resembling them, it cannot be mistaken. Next, it finds certain common notions from which it constructs various proofs; and, for as long as it attends to them, it is completely convinced of their truth. For example, the mind has within itself ideas of numbers and shapes, and it also has such common notions as: *If you add equals to equals the results will be equal*; from these it is easy to demonstrate that the three angles of a triangle equal two right angles, and so on. And so the mind will be convinced of the truth of this and similar conclusions, so long as it attends to the premises from which it deduced them. But it cannot attend to them all the time; and subsequently,[4] recalling that it is still ignorant as to whether it may have been created with the kind of nature that makes it go wrong even in matters which appear most evident, the mind sees that it has just cause to doubt such conclusions, and that the possession of certain knowledge will not be possible until it has come to know the author of its being.

14. *The existence of God is validly inferred from the fact that necessary existence is included in our concept of God.*
The mind next considers the various ideas which it has within itself, and finds that there is one idea – the idea of a supremely intelligent, supremely powerful and supremely perfect being – which stands out from all the others. <And it readily judges from what it perceives in this idea, that God, who is the supremely perfect being, is, or exists. For although it has distinct ideas of many other things it does not observe anything in them to guarantee the existence of their object.> In this one idea the mind recognizes existence – not merely the possible and contingent existence which belongs to the ideas of all the other things which it distinctly perceives, but utterly necessary and eternal existence. [. . .]

18. *This gives us a second reason for concluding that God exists.*
Since, then, we have within us the idea of God, or a supreme being, we may rightly inquire into the cause of our possession of this idea. Now we find in the idea such immeasurable greatness that we are quite certain that it could have been placed in us only by something which truly possesses the sum of all perfections, that is, by a God who really exists. For it is very evident by the natural light not only that nothing comes from nothing but also that what is more perfect cannot be produced by – that is, cannot have as its efficient and total cause – what is less perfect. Furthermore, we cannot have within us the idea or image of anything without there being somewhere, either within us or outside us, an original which contains in reality all the perfections belonging to the idea. And since the supreme perfections of which we have an idea are in no way to be found in us, we rightly conclude that they reside in something distinct from ourselves, namely God – or certainly that they once did so, from which it most evidently follows that they are still there. [. . .]

20. *We did not make ourselves, but were made by God; and consequently he exists.*

However, this is something that not everyone takes note of. When people have an idea of some intricate machine, they generally know where they got the idea from; but we do not in the same way have a recollection of the idea of God being sent to us from God, since we have always possessed it. Accordingly, we should now go on to inquire into the source of our being, given that we have within us an idea of the supreme perfections of God. Now it is certainly very evident by the natural light that a thing which recognizes something more perfect than itself is not the source of its own being; for if so, it would have given itself all the perfections of which it has an idea. Hence, the source of its being can only be something which possesses within itself all these perfections – that is, God.

21. *The fact that our existence has duration is sufficient to demonstrate the existence of God.*

It will be impossible for anything to obscure the clarity of this proof, if we attend to the nature of time or of the duration of things. For the nature of time is such that its parts are not mutually dependent, and never coexist. Thus, from the fact that we now exist, it does not follow that we shall exist a moment from now, unless there is some cause – the same cause which originally produced us – which continually reproduces us, as it were, that is to say, which keeps us in existence. For we easily understand that there is no power in us enabling us to keep ourselves in existence. We also understand that he who has so great a power that he can keep us in existence, although we are distinct from him, must be all the more able to keep himself in existence; or rather, he requires no other being to keep him in existence, and hence, in short, is God. [. . .]

26. *We should never enter into arguments about the infinite. Things in which we observe no limits – such as the extension of the world, the division of the parts of matter, the number of the stars, and so on – should instead be regarded as indefinite.*

Thus we will never be involved in tiresome arguments about the infinite. For since we are finite, it would be absurd for us to determine anything concerning the infinite; for this would be to attempt to limit it and grasp it. So we shall not bother to reply to those who ask if half an infinite line would itself be infinite, or whether an infinite number is odd or even, and so on. It seems that nobody has any business to think about such matters unless he regards his own mind as infinite. For our part, in the case of anything in which, from some point of view, we are unable to discover a limit, we shall avoid asserting that it is infinite, and instead regard it as indefinite. There is, for example, no imaginable extension which is so great that we cannot understand the possibility of an even greater one; and so we shall describe the size of possible things as indefinite. Again, however many parts a body is divided into, each of the parts can still be understood to be divisible and so we shall hold that quantity is indefinitely divisible. Or again, no matter how great we imagine the number of stars to be, we still think that God could have created even more; and so we will suppose the number of stars to be indefinite. And the same will apply in other cases.

27. *The difference between the indefinite and the infinite.*

Our reason for using the term 'indefinite' rather than 'infinite' in these cases is, in the first place, so as to reserve the term 'infinite' for God alone. For in the case of God alone, not only do we fail to recognize any limits in any respect, but our understanding positively tells us that there are none. Secondly, in the case of other things, our understanding does not in the same way positively tell us that they lack limits in some respect; we merely acknowledge in a negative way that any limits which they may have cannot be discovered by us.

28. *It is not the final but the efficient causes of created things that we must inquire into.*

When dealing with natural things we will, then, never derive any explanations from the purposes which God or nature may have had in view when creating them <and we shall entirely banish from our philosophy the search for final causes>. For we should not be so arrogant as to suppose that we can share in God's plans. We should, instead, consider him as the efficient cause of all things and starting from the divine attributes which by God's will we have some knowledge of, we shall see, with the aid of our God-given natural light, what conclusions should be drawn concerning those effects which are apparent to our senses.[6] At the same time we should remember, as noted earlier, that the natural light is to be trusted only to the extent that it is compatible with divine revelation.

29. *God is not the cause of our errors.*

The first attribute of God that comes under consideration here is that he is supremely truthful and the giver of all light. So it is a complete contradiction to suppose that he might deceive us or be, in the strict and positive sense, the cause of the errors to which we know by experience that we are prone. For although the ability to deceive may perhaps be regarded among us men as a sign of intelligence, the will to deceive must undoubtedly always come from malice, or from fear and weakness, and so cannot belong to God.

30. *It follows that everything that we clearly perceive is true; and this removes the doubts mentioned earlier.*

It follows from this that the light of nature or faculty of knowledge which God gave us can never encompass any object which is not true in so far as it is indeed encompassed by this faculty, that is, in so far as it is clearly and distinctly perceived. For God would deserve to be called a deceiver if the faculty which he gave us was so distorted that it mistook the false for the true <even when we were using it properly>. This disposes of the most serious doubt which arose from our ignorance about whether our nature might not be such as to make us go wrong even in matters which seemed to us utterly evident. Indeed, this argument easily demolishes all the other reasons for doubt which were mentioned earlier. Mathematical truths should no longer be suspect, since they are utterly clear to us. And as for our senses, if we notice anything here that is clear and distinct, no matter whether we are awake or asleep, then provided we separate it from what is confused and obscure we will easily recognize – whatever the thing in question – which are the aspects that may be regarded as true. There is no need for me to expand on this point here, since I have already dealt with it in the *Meditations on Metaphysics*;[7] and a more precise explanation of the point requires knowledge of what I shall be saying later on.

PART TWO

4. *The nature of body consists not in weight, hardness, colour, or the like, but simply in extension.*

If we do this [refer to ideas "implanted in the intellect" only, cf. Section 3 of this Part], we shall perceive that the nature of matter, or body considered in general, consists not in its being something which is hard or heavy or coloured, or which affects the senses in any way, but simply in its being something which is extended in length, breadth and depth. For as regards hardness, our sensation tells us no more than that the parts of a hard body resist the motion of our hands when they come into contact with them. If, whenever our hands moved in a given direction, all the bodies in that area were to move away at the same speed as that of our approaching hands, we should never have any sensation of hardness. And since it is quite unintelligible to suppose that, if bodies did move away in this fashion, they would thereby lose their bodily nature, it follows that this nature cannot consist in hardness. By the same reasoning it can be shown that weight, colour, and all other such qualities that are perceived by the senses as being in corporeal matter, can be removed from it, while the matter itself remains intact; it thus follows that its nature does not depend on any of these qualities.

5. *This truth about the nature of body is obscured by preconceived opinions concerning rarefaction and empty space.*

But there are still two possible reasons for doubting that the true nature of body consists solely in extension. The first is the widespread belief that many bodies can be rarefied and condensed in such a way that when rarefied they possess more extension than when condensed. Indeed, the subtlety of some people goes so far that they distinguish the substance of a body from its quantity, and even its quantity from its extension.[8] The second reason is that if we understand there to be nothing in a given place but extension in length, breadth and depth, we generally say not that there is a body there, but simply that there is a space, or even an empty space; and almost everyone is convinced that this amounts to nothing at all.

6. *How rarefaction occurs.*

But with regard to rarefaction and condensation, anyone who attends to his own thoughts, and is willing to admit only what he clearly perceives, will not suppose that anything happens in these processes beyond a change of shape. Rarefied bodies, that is to say, are those which have many gaps between their parts – gaps which are occupied by other bodies; and they become denser simply in virtue of the parts coming together and reducing or completely closing the gaps. In this last eventuality a body becomes so dense that it would be a contradiction to suppose that it could be made any denser. Now in this condition, the extension of a body is no less than when it occupies more space in virtue of the mutual separation of its parts; for whatever extension is comprised in the pores or gaps left between the parts must be attributed not to the body itself but to the various other bodies which fill the gaps. In just the same way, when we see a sponge filled with water or some other liquid, we do not suppose that in terms of its own individual parts it has a greater extension than when it is squeezed dry; we simply suppose that its pores are open wider, so that it spreads over a greater space.

7. *This is the only intelligible way of explaining rarefaction.*

I really do not see what has prompted others to say that rarefaction occurs through an increase of quantity, in preference to explaining it by means of this example of the sponge.[9] It is true that when air or water is rarefied, we do not see any pores being made larger, or any new body coming to fill them up. But to invent something unintelligible so as to provide a purely verbal explanation of rarefaction is surely less rational than inferring the existence of pores or gaps which are made larger, and supposing that some new body comes and fills them. Admittedly, we do not perceive this new body with any of our senses; but there is no compelling reason to believe that all the bodies which exist must affect our senses. Moreover, it is very easy for us to see how rarefaction can occur in this way, but we cannot see how it could occur in any other way. Finally, it is a complete contradiction to suppose that something should be augmented by new quantity or new extension without new extended substance, i.e. a new body, being added to it at the same time. For any addition of extension or quantity is unintelligible without the addition of substance which has quantity and extension. This will become clearer from what follows.

8. *The distinction between quantity or number and the thing that has quantity or number is merely a conceptual distinction.*

There is no real difference between quantity and the extended substance; the difference is merely a conceptual one, like that between number and the thing which is numbered. We can, for example, consider the entire nature of the corporeal substance which occupies a space of ten feet without attending to the specific measurement; for we understand this nature to be exactly the same in any part of the space as in the whole space. And, conversely, we can think of the number ten, or the continuous quantity *ten feet*, without attending to this determinate substance. For the concept of the number ten is exactly the same irrespective of whether it is referred to this measurement of ten feet or to anything else; and as for the continuous quantity *ten feet*, although this is unintelligible without some extended substance of which it is the quantity, it can be understood apart from this determinate substance. In reality, however, it is impossible to take even the smallest fraction from the quantity or extension without also removing just as much from the substance; and conversely, it is impossible to remove the smallest amount from the substance without taking away just as much from the quantity or extension.

9. *If corporeal substance is distinguished from its quantity, it is conceived in a confused manner as something incorporeal.*

Others may disagree, but I do not think they have any alternative perception of the matter. When they make a distinction between substance and extension or quantity, either they do not understand anything by the term 'substance', or else they simply have a confused idea of incorporeal substance, which they falsely attach to corporeal substance; and they relegate the true idea of corporeal substance to the category of extension, which, however, they term an accident. There is thus no correspondence between their verbal expressions and what they grasp in their minds.

10. *What is meant by 'space', or 'internal place'.*

There is no real distinction between space, or internal place,[10] and the corporeal substance contained in it; the only difference lies in the way in which we are accustomed to conceive of them. For in reality the extension in length, breadth and depth which constitutes a space is exactly the same as that which constitutes a body. The difference arises as follows: in the case of a body, we regard the extension as something particular, and thus think of it as changing whenever there is a new body; but in the case of a space, we attribute to the extension only a generic unity, so that when a new body comes to occupy the space, the extension of the space is reckoned not to change but to remain one and the same, so long as it retains the same size and shape and keeps the same position relative to certain external bodies which we use to determine the space in question.

11. *There is no real difference between space and corporeal substance.*
It is easy for us to recognize that the extension constituting the nature of a body is exactly the same as that constituting the nature of a space. There is no more difference between them than there is between the nature of a genus or species and the nature of an individual. Suppose we attend to the idea we have of some body, for example a stone, and leave out everything we know to be non-essential to the nature of body: we will first of all exclude hardness, since if the stone is melted or pulverized it will lose its hardness without thereby ceasing to be a body; next we will exclude colour, since we have often seen stones so transparent as to lack colour; next we will exclude heaviness, since although fire is extremely light it is still thought of as being corporeal; and finally we will exclude cold and heat and all other such qualities, either because they are not thought of as being in the stone, or because if they change, the stone is not on that account reckoned to have lost its bodily nature. After all this, we will see that nothing remains in the idea of the stone except that it is something extended in length, breadth and depth. Yet this is just what is comprised in the idea of a space – not merely a space which is full of bodies, but even a space which is called 'empty'.[11]

12. *The difference between space and corporeal substance lies in our way of conceiving them.*
There is, however, a difference in the way in which we conceive of space and corporeal substance. For if a stone is removed from the space or place where it is, we think that its extension has also been removed from that place, since we regard the extension as something particular and inseparable from the stone. But at the same time we think that the extension of the place where the stone used to be remains, and is the same as before, although the place is now occupied by wood or water or air or some other body, or is even supposed to be empty. For we are now considering extension as something general, which is thought of as being the same, whether it is the extension of a stone or of wood, or of water or of air or of any other body – or even of a vacuum, if there is such a thing – provided only that it has the same size

and shape, and keeps the same position relative to the external bodies that determine the space in question.

13. *What is meant by 'external place'.*
The terms 'place' and 'space', then, do not signify anything different from the body which is said to be in a place; they merely refer to its size, shape and position relative to other bodies. To determine the position, we have to look at various other bodies which we regard as immobile; and in relation to different bodies we may say that the same thing is both changing and not changing its place at the same time. For example, when a ship is under way, a man sitting on the stern remains in one place relative to the other parts of the ship with respect to which his position is unchanged; but he is constantly changing his place relative to the neighbouring shores, since he is constantly receding from one shore and approaching another. Then again, if we believe the earth moves,[12] and suppose that it advances the same distance from west to east as the ship travels from east to west in the corresponding period of time, we shall again say that the man sitting on the stern is not changing his place; for we are now determining the place by means of certain fixed points in the heavens. Finally, if we suppose that there are no such genuinely fixed points to be found in the universe (a supposition which will be shown below to be probable[13]) we shall conclude that nothing has a permanent place, except as determined by our thought.

14. *The difference between place and space.*
The difference between the terms 'place' and 'space' is that the former designates more explicitly the position, as opposed to the size or shape, while it is the size and shape that we are concentrating on when we talk of space. For we often say that one thing leaves a given place and another thing arrives there, even though the second thing is not strictly of the same size and shape; but in this case we do not say it occupies the same space. By contrast, when something alters its position, we always say the place is changed, despite the fact that the size and shape remain unaltered. When we say that a thing is in a given place, all we mean is that it occupies such and such a position relative to other things; but when we go on to say that it fills up a given space or place, we mean in addition that it has precisely the size and shape of the space in question.

15. *How external place is rightly taken to be the surface of the surrounding body.*
Thus we always take a space to be an extension in length, breadth and depth. But with regard to place, we sometimes consider it as internal to the thing which is in the place in question, and sometimes as external to it. Now inter-

nal place is exactly the same as space; but external place may be taken as being the surface immediately surrounding what is in the place. It should be noted that 'surface' here does not mean any part of the surrounding body but merely the boundary between the surrounding and surrounded bodies, which is no more than a mode. Or rather what is meant is simply the common surface, which is not a part of one body rather than the other but is always reckoned to be the same, provided it keeps the same size and shape. For if there are two bodies, one surrounding the other, and the entire surrounding body changes, surface and all, the surrounded body is not therefore thought of as changing its place, provided that during this time it keeps the same position relative to the external bodies which are regarded as immobile. If, for example, we suppose that a ship on a river is being pulled equally in one direction by the current and in the opposite direction by the wind, so that it does not change its position relative to the banks, we will all readily admit that it stays in the same place, despite the complete change in the surrounding surface.

16. *It is a contradiction to suppose there is such a thing as a vacuum, i.e. that in which there is nothing whatsoever.*
The impossibility of a vacuum, in the philosophical sense of that in which there is no substance whatsoever, is clear from the fact that there is no difference between the extension of a space, or internal place, and the extension of a body. For a body's being extended in length, breadth and depth in itself warrants the conclusion that it is a substance, since it is a complete contradiction that a particular extension should belong to nothing; and the same conclusion must be drawn with respect to a space that is supposed to be a vacuum, namely that since there is extension in it, there must necessarily be substance in it as well.

17. *The ordinary use of the term 'empty' does not imply the total absence of bodies.*
In its ordinary use the term 'empty'[14] usually refers not to a place or space in which there is absolutely nothing at all, but simply to a place in which there is none of the things that we think ought to be there. Thus a pitcher made to hold water is called 'empty' when it is simply full of air; a fishpond is called 'empty', despite all the water in it, if it contains no fish; and a merchant ship is called 'empty' if it is loaded only with sand ballast. And similarly a space is called 'empty' if it contains nothing perceivable by the senses, despite the fact that it is full of created, self-subsistent matter; for normally the only things we give any thought to are those which are detected by our senses. But if we subsequently fail to keep in mind what ought to be understood by the terms 'empty' and 'nothing', we may suppose that a space we call empty contains not just nothing perceivable by the senses but nothing whatsoever;

that would be just as mistaken as thinking that the air in a jug is not a sub-sistent thing on the grounds that a jug is usually said to be empty when it contains nothing but air.

18. *How to correct our preconceived opinion regarding an absolute vacuum.*
Almost all of us fell into this error in our early childhood. Seeing no neces-sary connection between a vessel and the body contained in it, we reckoned there was nothing to stop God, at least, removing the body which filled the vessel, and preventing any other body from taking its place. But to correct this error we should consider that, although there is no connection between a vessel and this or that particular body contained in it, there is a very strong and wholly necessary connection between the concave shape of the vessel and the extension, taken in its general sense, which must be contained in the concave shape. Indeed, it is no less contradictory for us to conceive of a mountain without a valley than it is for us to think of the concavity apart from the extension contained within it, or the extension apart from the sub-stance which is extended; for, as I have often said, nothingness cannot possess any extension. Hence, if someone asks what would happen if God were to take away every single body contained in a vessel, without allowing any other body to take the place of what had been removed, the answer must be that the sides of the vessel would, in that case, have to be in contact. For when there is nothing between two bodies they must necessarily touch each other. And it is a manifest contradiction for them to be apart, or to have a distance between them, when the distance in question is nothing; for every distance is a mode of extension, and therefore cannot exist without an extended substance.

19. *The preceding conclusion confirms what we said regarding rarefaction.*
We have thus seen that the nature of corporeal substance consists simply in its being something extended; and its extension is no different from what is normally attributed to space, however 'empty'. From this we read-ily see that no one part of it can possibly occupy more space at one time than at another, and hence that rarefaction cannot occur except in the way explained earlier on.[15] Similarly, there cannot be more matter or corporeal substance in a vessel filled with lead or gold or any other body, no matter how heavy and hard, than there is when it contains only air and is thought of as empty. This is because the quantity of the parts of matter does not depend on their heaviness or hardness, but solely on their extension, which is always the same for a given vessel.

20. *The foregoing results also demonstrate the impossibility of atoms.*
We also know that it is impossible that there should exist atoms, that is, pieces of matter that are by their very nature indivisible <as some philoso-

phers have imagined>. For if there were any atoms, then no matter how small we imagined them to be, they would necessarily have to be extended; and hence we could in our thought divide each of them into two or more smaller parts, and hence recognize their divisibility. For anything we can divide in our thought must, for that very reason, be known to be divisible; so if we were to judge it to be indivisible, our judgement would conflict with our knowledge. Even if we imagine that God has chosen to bring it about that some particle of matter is incapable of being divided into smaller particles, it will still not be correct, strictly speaking, to call this particle indivisible. For, by making it indivisible by any of his creatures, God certainly could not thereby take away his own power of dividing it, since it is quite impossible for him to diminish his own power, as has been noted above.[16] Hence, strictly speaking, the particle will remain divisible, since it is divisible by its very nature.

21. *Similarly, the extension of the world is indefinite.*
What is more we recognize that this world, that is, the whole universe of corporeal substance, has no limits to its extension. For no matter where we imagine the boundaries to be, there are always some indefinitely extended spaces beyond them, which we not only imagine but also perceive to be imaginable in a true fashion, that is, real. And it follows that these spaces contain corporeal substance which is indefinitely extended. For, as has already been shown very fully, the idea of the extension which we conceive to be in a given space is exactly the same as the idea of corporeal substance.

22. *Similarly, the earth and the heavens are composed of one and the same matter; and there cannot be a plurality of worlds.*
It can also easily be gathered from this that celestial matter is no different from terrestrial matter.[17] And even if there were an infinite number of worlds, the matter of which they were composed would have to be identical; hence, there cannot in fact be a plurality of worlds, but only one. For we very clearly understand that the matter whose nature consists simply in its being an extended substance already occupies absolutely all the imaginable space in which the alleged additional worlds would have to be located; and we cannot find within us an idea of any other sort of matter.

23. *All the variety in matter, all the diversity of its forms, depends on motion.*
The matter existing in the entire universe is thus one and the same, and it is always recognized as matter simply in virtue of its being extended. All the properties which we clearly perceive in it are reducible to its divisibility and consequent mobility in respect of its parts, and its resulting capacity to be affected in all the ways which we perceive as being derivable from the

movement of the parts. If the division into parts occurs simply in our thought, there is no resulting change; any variation in matter or diversity in its many forms depends on motion. This seems to have been widely recognized by the philosophers, since they have stated that nature is the principle of motion and rest. And what they meant by 'nature' in this context is what causes all corporeal things to take on the characteristics of which we are aware in experience.

24. *What is meant by 'motion' in the ordinary sense of the term.*
Motion, in the ordinary sense of the term, is simply *the action by which a body travels from one place to another.* By 'motion', I mean local motion; for my thought encompasses no other kind, and hence I do not think that any other kind should be imagined to exist in nature.[18] Now I pointed out above that the same thing can be said to be changing and not changing its place at the same time;[19] and similarly the same thing can be said to be moving and not moving. For example, a man sitting on board a ship which is leaving port considers himself to be moving relative to the shore which he regards as fixed; but he does not think of himself as moving relative to the ship, since his position is unchanged relative to its parts. Indeed, since we commonly think all motion involves action, while rest consists in the cessation of action, the man sitting on deck is more properly said to be at rest than in motion, since he does not have any sensory awareness of action in himself.

25. *What is meant by 'motion' in the strict sense of the term.*
If, on the other hand, we consider what should be understood by *motion*, not in common usage but in accordance with the truth of the matter, and if our aim is to assign a determinate nature to it, we may say that *motion is the transfer of one piece of matter, or one body, from the vicinity of the other bodies which are in immediate contact with it, and which are regarded as being at rest, to the vicinity of other bodies.* By 'one body' or 'one piece of matter' I mean whatever is transferred at a given time, even though this may in fact consist of many parts which have different motions relative to each other. And I say 'the transfer' as opposed to the force or action which brings about the transfer, to show that motion is always in the moving body as opposed to the body which brings about the movement. The two are not normally distinguished with sufficient care; and I want to make it clear that the motion of something that moves is, like the lack of motion in a thing which is at rest, a mere mode of that thing and not itself a subsistent thing, just as shape is a mere mode of the thing which has shape.

26. *No more action is required for motion than for rest.*
It should be noted that in this connection we are in the grip of a strong preconceived opinion, namely the belief that more action is needed for motion

than for rest. We have been convinced of this since early childhood owing to the fact that our bodies move by our will, of which we have inner awareness, but remain at rest simply in virtue of sticking to the earth by gravity,[20] the force of which we do not perceive through the senses. And because gravity and many other causes of which we are unaware produce resistance when we try to move our limbs, and make us tired, we think that a greater action or force is needed to initiate a motion than to stop it; for we take *action* to be the effort we expend in moving our limbs and moving other bodies by the use of our limbs. We will easily get rid of this preconceived opinion if we consider that it takes an effort on our part not only to move external bodies, but also, quite often, to stop them, when gravity and other causes are insufficient to arrest their movement. For example, the action needed to move a boat which is at rest in still water is no greater than that needed to stop it suddenly when it is moving – or rather it is not much greater, for one must subtract the weight of the water displaced by the ship and the viscosity of the water, both of which could gradually bring it to a halt.

27. *Motion and rest are merely various modes of a body in motion.*
We are dealing here not with the action which is understood to exist in the body which produces or arrests the motion, but simply with the transfer of a body, and with the absence of a transfer, i.e. rest. So it is clear that this transfer cannot exist outside the body which is in motion, and that when there is a transfer of motion, the body is in a different state from when there is no transfer, i.e. when it is at rest. Thus motion and rest are nothing else but two different modes of a body. [. . .]

31. *How there may be countless different motions in the same body.*
Each body has only one proper motion, since it is understood to be moving away from only one set of bodies, which are contiguous with it and at rest. But it can also share in countless other motions, namely in cases where it is a part of other bodies which have other motions. For example, if someone walking on board ship has a watch in his pocket, the wheels of the watch have only one proper motion, but they also share in another motion because they are in contact with the man who is taking his walk, and they and he form a single piece of matter. They also share in an additional motion through being in contact with the ship tossing on the waves; they share in a further motion through contact with the sea itself; and lastly, they share in yet another motion through contact with the whole earth, if indeed the whole earth is in motion. Now all the motions will really exist in the wheels of the watch, but it is not easy to have an understanding of so many motions all at once, nor can we have knowledge of all of them. So it is enough to confine our attention to that single motion which is the proper motion of each body.

32. *How even the proper motion unique to each body may be considered as a plurality of motions.*

The single motion that is the proper motion of each body may also be considered as if it were made up of several motions. For example, we may distinguish two different motions in a carriage wheel – a circular motion about the axle and a rectilinear motion along the line of the road. But that these are not really distinct is clear from the fact that every single point on the moving object describes only one line. It does not matter that the line is often very twisted so that it seems to have been produced by many different motions; for we can imagine any line at all – even a straight line, which is the simplest of all – as arising from an infinite number of different motions. Thus if the line AB travels towards CD [see Figure 18.2], and at the same time the point A travels towards B, the straight line AD described by the point A will depend on two rectilinear motions, from A to B and from AB to CD, in just the same way as the curve described by any point of the wheel depends on a rectilinear motion and a circular motion. Although it is often useful to separate a single motion into several components in this way in order to facilitate our perception of it, nevertheless, absolutely speaking, there is only one motion that should be counted for any given body.

33. *How in every case of motion there is a complete circle of bodies moving together.*

I noted above[23] that every place is full of bodies, and that the same portion of matter always takes up the same amount of space, <so that it is impossible for it to fill a greater or lesser space, or for any other body to

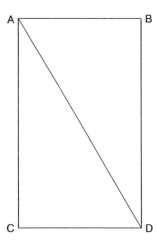

Figure 18.2 (redrawn)

occupy its place while it remains there>. It follows from this that each body can move only in a <complete> circle <of matter, or ring of bodies which all move together at the same time>: a body entering a given place expels another, and the expelled body moves on and expels another, and so on, until the body at the end of the sequence enters the place left by the first body at the precise moment when the first body is leaving it. We can easily understand this in the case of a perfect circle, since we see that no vacuum and no rarefaction or condensation is needed to enable part A of the circle [see Figure 18.3] to move towards B, provided that B simultaneously moves towards C, C towards D and D towards A. But the same thing is intelligible even in the case of an imperfect circle however irregular it may be, provided we notice how all the variations in the spaces can be compensated for by variations in speed. Thus all the matter contained in the space EFGH [see Figure 18.4] can move in a circle without the need for any condensation or vacuum, and the part that is around E can move towards G while the part that is around G simultaneously moves towards E, with this sole proviso: if the space in G is supposed to be four times as

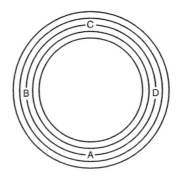

Figure 18.3 (redrawn)

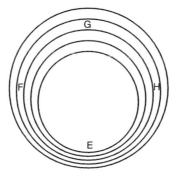

Figure 18.4 (redrawn)

wide as the space at E and twice as wide as the space at F and H, then the speed of the motion at E must be four times greater than that at G and twice as great as that at F or H; and at every other location an increase in speed must similarly compensate for a narrower space. In this way, the amount of matter passing through any given part of the circle in any given time will always be equal.

34. *From this it follows that the number of particles into which matter is divided is in fact indefinite, although it is beyond our power to grasp them all.*
It must, however, be admitted that in the case of this motion we come upon something the truth of which our mind perceives, while at the same time being unable to grasp exactly how it occurs. For what happens is an infinite, or indefinite,[24] division of the various particles of matter; and the resulting subdivisions are so numerous that however small we make a particle in our thought, we always understand that it is in fact divided into other still smaller particles. For it is impossible for the matter which now fills space G successively to fill all the spaces between G and E, which get gradually smaller by countless stages, unless some part of that matter adjusts its shape to the innumerable different volumes of those spaces. And for this to come about, it is necessary that all its imaginable particles, which are in fact innumerable, should shift their relative positions to some tiny extent. This minute shifting of position is a true case of division.

35. *How this division comes about; and the fact that it undoubtedly takes place, even though it is beyond our grasp.*
It should be noted, however, that I am not here speaking of the whole of this matter, but merely of some part of it. We may suppose that two or three of its parts at G are as wide as the space at E, and that there are also several smaller parts which remain undivided; but nevertheless we can still understand them to move in a circle towards E, provided they have mixed up with them various other particles which somehow bend and change shape in such a way as to join onto them. Now the former group do not change their own shape, but merely adapt their speed depending on the place they are to occupy, while the latter group exactly fill all the crevices which the former do not occupy. We cannot grasp in our thought how this indefinite division comes about, but we should not therefore doubt that it occurs. For we clearly perceive that it necessarily follows from what we <already> know most evidently of the nature of matter, and we perceive that it belongs to the class of things which are beyond the grasp of our finite minds.

36. *God is the primary cause of motion; and he always preserves the same quantity of motion in the universe.*

After this consideration of the nature of motion, we must look at its cause. This is in fact twofold: first, there is the universal and primary cause – the general cause of all the motions in the world; and second there is the particular cause which produces in an individual piece of matter some motion which it previously lacked. Now as far as the general cause is concerned, it seems clear to me that this is no other than God himself. In the beginning <in his omnipotence> he created matter, along with its motion and rest; and now, merely by his regular concurrence, he preserves the same amount of motion and rest in the material universe as he put there in the beginning. Admittedly motion is simply a mode of the matter which is moved. But nevertheless it has a certain determinate quantity; and this, we easily understand, may be constant in the universe as a whole while varying in any given part. Thus if one part of matter moves twice as fast as another which is twice as large, we must consider that there is the same quantity of motion in each part; and if one part slows down, we must suppose that some other part of equal size speeds up by the same amount. For we understand that God's perfection involves not only his being immutable in himself, but also his operating in a manner that is always utterly constant and immutable. Now there are some changes whose occurrence is guaranteed either by our own plain experience or by divine revelation, and either our perception or our faith shows us that these take place without any change in the creator; but apart from these we should not suppose that any other changes occur in God's works, in case this suggests some inconstancy in God. Thus, God imparted various motions to the parts of matter when he first created them, and he now preserves all this matter in the same way, and by the same process by which he originally created it;[25] and it follows from what we have said that this fact alone makes it most reasonable to think that God likewise always preserves the same quantity of motion in matter.

37. *The first law of nature: each and every thing, in so far as it can, always continues in the same state; and thus what is once in motion always continues to move.*

From God's immutability we can also know certain rules or laws of nature, which are the secondary and particular causes of the various motions we see in particular bodies. The first of these laws is that each thing, in so far as it is simple and undivided, always remains in the same state, as far as it can, and never changes except as a result of external causes. Thus, if a particular piece of matter is square, we can be sure without more ado that it will remain square for ever, unless something coming from outside changes its shape. If it is at rest, we hold that it will never begin to move unless it is pushed into motion by some cause. And if it moves, there is equally no reason for thinking it will ever lose this motion of its own accord and without being checked by something else. Hence we must conclude that what is in motion always, so far as it can, continues to move. But we live on the Earth, whose composition is such that all motions occurring near it are soon halted, often by causes undetectable by our senses. Hence from our earliest years we have often judged that such motions, which are in fact stopped by causes unknown to us, come to an end of their own accord. And we tend to believe that what we have apparently experienced in many cases holds good in all cases – namely that it is in the very nature of motion to come to an end, or to tend towards a state of rest. This, of course, <is a false preconceived opinion which> is utterly at variance with the laws of nature; for rest is the opposite of motion, and nothing can by its own nature tend towards its opposite, or towards its own destruction.

38. *The motion of projectiles.*

Indeed, our everyday experience of projectiles completely confirms this first rule of ours. For there is no other reason why a projectile should persist in motion for some time after it leaves the hand that throws it, except that what is once in motion continues to move until it is slowed down by bodies that are in the way.[26] And it is clear that projectiles are normally slowed down, little by little, by the air or other fluid bodies in which they are moving, and that this is why their motion cannot persist for long. The fact that air offers resistance to other moving bodies may be confirmed either by our own experience, through the sense of touch if we beat the air with a fan, or by the flight of birds. And in the case of any other fluid, the resistance offered to the motion of a projectile is even more obvious than in the case of air.

39. *The second law of nature: all motion is in itself rectilinear; and hence any body moving in a circle always tends to move away from the centre of the circle which it describes.*

The second law is that every piece of matter, considered in itself, always tends to continue moving, not in any oblique path but only in a straight line. This is true despite the fact that many particles are often forcibly deflected by the impact of other bodies; and, as I have said above,[27] in any motion the result of all the matter moving simultaneously is a kind of circle. The reason for this second rule is the same as the reason for the first rule, namely the immutability and simplicity of the operation by which God preserves motion in matter. For he always preserves the motion in the precise form in which it is occurring at the very moment when he preserves it, without taking any account of the motion which was occurring a little while earlier. It is true that no motion takes place in a single instant of time; but clearly whatever is in motion is determined, at the individual instants which can be specified as long as the motion lasts, to continue moving in a given direction along a straight line, and never in a curve . . .[28]

40. *The third law: if a body collides with another body that is stronger than itself, it loses none of its motion; but if it collides with a weaker body, it loses a quantity of motion equal to that which it imparts to the other body.*
The third law of nature is this: when a moving body collides with another, if its power of continuing in a straight line is less than the resistance of the other body, it is deflected so that, while the quantity of motion is retained, the direction is altered; but if its power of continuing is greater than the resistance of the other body, it carries that body along with it, and loses a quantity of motion equal to that which it imparts to the other body. Thus we find that when hard projectiles strike some other hard body, they do not stop, but rebound in the opposite direction; when, by contrast, they encounter a soft body, they are immediately halted because they readily transfer all their motion to it. All the particular causes of the changes which bodies undergo are covered by this third law – or at least the law covers all changes which are themselves corporeal. I am not here inquiring into the existence or nature of any power to move bodies which may be possessed by human minds, or the minds of angels, since I am reserving this topic for a treatise *On Man* <which I hope to produce>.[29]

41. *The proof of the first part of this rule.*
The first part of this law is proved by the fact that there is a difference between motion considered in itself <the motion of a thing> and its determination in a certain direction; for the determination of the direction can be altered, while the motion remains constant. As I have said above, everything that is not composite but simple, as motion is, always persists in being <as it is in itself and not in relation to other things>, so long as it is not destroyed by an external cause <by meeting another object>. Now if one body collides with a second, hard body <in its path which it is quite incapable of pushing>, there is an obvious reason why its motion should not remain fixed in the same direction, <namely the resistance of the body which deflects its path>; but there is no reason why its motion should be stopped or diminished, <since it is not removed by the other body or by any other cause, and> since one motion is not the opposite of another motion. Hence it follows that the motion in question ought not to diminish at all.

42. *The proof of the second part of this rule.*
The second part of the law is proved from the immutability of the workings of God, by means of which the world is continually preserved through an

action identical with its original act of creation. For the whole of space is filled with bodies, and the motion of every single body is rectilinear in tendency; hence it is clear that when he created the world in the beginning God did not only impart various motions to different parts of the world, but also produced all the reciprocal impulses and transfers of motion between the parts. Thus, since God preserves the world by the selfsame action and in accordance with the selfsame laws as when he created it, the motion which he preserves is not something permanently fixed in given pieces of matter, but something which is mutually transferred when collisions occur. The very fact that creation is in a continual state of change is thus evidence of the immutability of God.[30]

43. *The nature of the power which all bodies have to act on, or resist, other bodies.*

In this connection we must be careful to note what it is that constitutes the power of any given body to act on, or resist the action of, another body. This power consists simply in the fact that everything tends, so far as it can, to persist in the same state, as laid down in our first law. Thus what is joined to another thing has some power of resisting separation from it; and what is separated has some power of remaining separate. Again, what is at rest has some power of remaining at rest and consequently of resisting anything that may alter the state of rest; and what is in motion has some power of persisting in its motion, i.e. of continuing to move with the same speed and in the same direction. An estimate of this last power must depend firstly on the size of the body in question and the size of the surface which separates it from other bodies, and secondly on the speed of the motion, and on the various ways in which different bodies collide, and the degree of opposition involved.

44. *The opposite of motion is not some other motion but a state of rest; and the opposite of the determination of a motion in a given direction is its determination in the opposite direction.*

It should be noted that one motion is in no way contrary to another of equal speed. Strictly speaking there are only two sorts of opposition to be found here. One is the opposition between motion and rest, together with the opposition between swiftness and slowness of motion (in so far, that is, as such slowness shares something of the nature of rest). And the second sort is the opposition between the determination of motion in a given direction and an encounter somewhere in that direction with another body which is at rest or moving in another direction. The degree of this opposition varies in accordance with the direction in which a body is moving when it collides with another.

45. *How to determine how much the motion of a given body is altered by colli-*
    sion with other bodies. This is calculated by means of the following rules.
To enable us to determine, in the light of this, how individual bodies
increase or diminish their motions or change direction as a result of collision with other bodies, all that is necessary is to calculate the power of any
given body to produce or resist motion; we also need to lay it down as a firm
principle that the stronger power always produces its effects. Our calculation would be easy if there were only two bodies colliding, and these were
perfectly hard, and so isolated from all other bodies that no surrounding
bodies impeded or augmented their motions. In this case they would obey
the rules that follow.[31] [. . .]

54. *What hard bodies are, and what fluid bodies are.*
. . . If we go on to inquire how it comes about that some bodies readily abandon their place to other bodies, while others do not, we can easily see that a
body already in motion does not prevent another body occupying the place
which it is spontaneously leaving; a body at rest, on the other hand, cannot be expelled from its place except by some force <coming from outside,
which produces a change>. Hence we may infer that fluids are bodies made
up of numerous tiny particles which are agitated by a variety of mutually
distinct motions; while hard bodies are those whose particles are all at rest
relative to each other.

55. *There is no glue binding together the parts of hard bodies apart from the*
    simple fact that they are at rest <relative to each other>.
We certainly cannot think up any kind of glue which could fix together the
particles of two bodies any more firmly than is achieved simply by their
being at rest. For what could such a glue be? It could not be a substance,
for since the particles are themselves substances, there is no reason why
another substance should join them more effectively than they join themselves together. Nor could the 'glue' be any mode distinct from their being
at rest. For what mode could be more contrary to the motion that separates them than their being at rest? And we recognize no other categories of
things apart from substances and their modes.

56. *The particles of fluid bodies move with equal force in all directions. And if a*
    hard body is present in a fluid, the smallest force is able to set it in motion.
As far as fluids are concerned, even though we cannot observe through our
senses any motion of their particles, because they are too small, such motion
is easily inferred from their effects, especially in the case of air and water.
For air and water corrupt many other bodies; and no corporeal action – and
corruption is such an action – can occur without local motion [. . .]

64. *The only principles which I accept, or require, in physics are those of geometry and pure mathematics; these principles explain all natural phenomena, and enable us to provide quite certain demonstrations regarding them.*

I will not here add anything about shapes or about the countless different kinds of motions that can be derived from the infinite variety of different shapes. These matters will be quite clear in themselves when the time comes for me to deal with them. I am assuming that my readers know the basic elements of geometry already, or have sufficient mental aptitude to understand mathematical demonstrations. For I freely acknowledge that I recognize no matter in corporeal things apart from that which the geometers call quantity, and take as the object of their demonstrations, i.e. that to which every kind of division, shape and motion is applicable. Moreover, my consideration of such matter involves absolutely nothing apart from these divisions, shapes and motions; and even with regard to these, I will admit as true only what has been deduced from indubitable common notions so evidently that it is fit to be considered as a mathematical demonstration. And since all natural phenomena can be explained in this way, as will become clear in what follows, I do not think that any other principles are either admissible or desirable in physics.

PART THREE

26. *The earth is at rest in its own heaven, but nonetheless it is carried along by it.*

27. *The same view should be taken of all the planets.*

28. *Strictly speaking, the earth does not move, any more than the planets, although they are all carried along by the heaven.*

Here we must bear in mind what I said above about the nature of motion,[32] namely that if we use the term 'motion' in the strict sense and in accordance with the truth of things, then motion is simply the transfer of one body from the vicinity of the other bodies which are in immediate contact with it, and which are regarded as being at rest, to the vicinity of other bodies. But it often happens that, in accordance with ordinary usage, any action whereby a body travels from one place to another is called 'motion'; and in this sense it can be said that the same thing moves and does not move at the same time, depending on how we determine its location. It follows from this that in the strict sense there is no motion occurring in the case of the earth or even the other planets, since they are not transferred from the vicinity of those parts of the heaven with which they are in immediate contact, in so far as these parts are considered as being at rest. Such a transfer would require them to move away from all these parts at the same time, which does not occur. But since the celestial material is fluid, at any given time different groups of particles move away from the planet with which they are in contact, by a motion which should be attributed solely to the particles, not to the planet. In the same way, the partial transfers of water and air which occur on the surface of the earth are not normally attributed to the earth itself, but to the parts of water and air which are transferred.

29. *No motion should be attributed to the earth even if 'motion' is taken in the loose sense, in accordance with ordinary usage; but in this sense it is correct to say that the other planets move.*

But if we construe 'motion' in accordance with ordinary usage, then all the other planets, and even the sun and fixed stars should be said to move; but the same cannot without great awkwardness be said of the earth. For the common practice is to determine the position of the start from certain sites on the earth that are regarded as being immobile: the stars are deemed to move in so far as they pass these fixed spots. This is convenient for practical purposes, and so is quite reasonable. Indeed all of us from earliest years have reckoned the earth to be not a globe but a flat surface, such that 'up' and 'down' are everywhere the same, and the four directions, east, west, south and north, are the same for any point on the surface; and we have all used these directions for specifying the location of any other body. But what of a philosopher who realizes that the earth is a globe contained in a fluid and mobile heaven, and that the sun and the fixed stars always preserve the same positions relative to each other? If he takes these bodies as immobile for the purpose of determining the earth's location, and thus asserts that the earth itself moves, his way of talking is quite unreasonable. First of all, location in the philosophical sense must be determined not by means of very remote bodies like the stars, but with reference to bodies which are contiguous with the body which is said to move. Secondly, if we follow ordinary usage, there is no reason for considering that it is the stars which are at rest rather than the earth, unless we believe that there are no other bodies beyond them from which they are receding, and with reference to which it can be said that they move but the earth is at rest (in the same sense as the earth is said to move with reference to the stars). Yet to believe this is irrational. For since our mind is of such a nature as to recognize no limits in the universe, whoever considers the immensity of God and the weakness of our senses will conclude that it is much more reasonable to suspect that there may be other bodies beyond all the visible fixed stars; and that, with reference to these bodies, the earth may be said to be at rest, but all the stars may be said to be in simultaneous motion. This is surely more reasonable than supposing that there cannot possibly be any such bodies <because the creator's power is so imperfect; for this must be the supposition of those who maintain in this way that the earth moves. And if, later on, to conform to ordinary usage, we appear to attribute motion to the earth, it should be remembered that this is an improper way of speaking – rather like the way in which we may sometimes say that passengers asleep on a ferry 'move' from Calais to Dover, because the ship takes them there>. [. . .]

119. *How a fixed star is changed into a comet or a planet.*
120. *The direction in which such a star moves when it first ceases to be fixed.* [. . .]
140. *What sets a planet in motion.*

Notes

(*From the Cottingham, Stoothoff, and Murdoch edition of this text.*)
 1. Some examples of such preconceived opinions are given in art. 71.
 2. '. . . this inference' (French version).
 3. Lat. *falsum.* Descartes uses this term to refer not only to propositions which are false, but also to objects which are unreal, spurious or non-existent. The French version here reads: 'we who are now thinking that there is nothing outside of our thought which truly is or exists . . .'
 4. '. . . when it happens that it remembers a conclusion without attending to the sequence which enables it to be demonstrated' (added in French version).
 6. '. . . and we shall be assured that what we have once clearly and distinctly perceived to belong to the nature of these things has the perfection of being true' (added in French version, which also omits the last sentence of this article).
 7. Cf. *Med.* VI: vol. II, pp. 54ff.
 8. Cf. *The World,* above, p. 92.
 9. Scholastic philosophers explained rarefaction in terms of a given amount of matter occupying a larger quantity or volume of space: for Descartes, however, this is unintelligible, since there is no real distinction between the notions of 'quantity', 'matter' and 'space'. See below, art. 8–12.
10. The scholastics distinguished between *locus internus,* or 'internal place' (the space occupied by a body), and *locus externus,* or 'external space' (the external surface containing a body). Descartes employs the traditional terminology here and at art. 13, but puts it to his own use.
11. Lat. *vacuum.* See below, art. 16.
12. '. . . turns on its axis' (French version).
13. The French version has 'demonstrable' instead of 'probable'. Cf. Part 3, art. 29.
14. Lat. *vacuum,* from *vacuus,* 'void', 'unoccupied'; cf. art. 18.
15. See above, art. 6.
16. Cf. Part 1, art. 60.
17. Descartes here rejects the scholastic doctrine of a radical difference in kind between 'sublunary' or terrestrial phenomena and the incorruptible world of the heavens.
18. See note to Part 1, art. 69.
19. Above, art. 13.
20. Lat. *gravitas,* literally 'heaviness'. In scholastic physics this term was used to refer to the supposed inherent tendency of terrestrial bodies to downward motion. For Descartes'

own use of the term, and his purely mechanistic explanation of heaviness, see Part 4, art. 20–3. It should be remembered that neither for the scholastics nor for Descartes did the term 'gravity' have its modern (post-Newtonian) connotation of a universal attractive force.

23. Art. 18 and 19.
24. See above, Part 1, art. 26.
25. There is for Descartes no real distinction between God's action in creating the universe and his action in preserving it or maintaining it in existence. See below, art. 42 and Med. III.
26. *Cf. The World*, ch. 7.
27. Art. 33.
28. Descartes proceeds to illustrate the point by the example of a stone shot from a sling. Cf. Part 3, art. 57.
29. This treatise, originally planned to form Part 6 of the *Principles* [. . .], was never written. It is not to be confused with the earlier *Treatise on Man*.
30. '. . . is no way incompatible with the immutability of God, and may even serve as evidence to establish it' (French version).
31. Descartes' seven rules for calculating the speed and direction of bodies after impact cover seven ideal cases, which are, respectively: (1) where two bodies of equal size and speed collide head on; (2) as in case (1), but where one body is larger; (3) as in (1) but where one body is travelling faster; (4) where one body is at rest and larger; (5) where one body is at rest and smaller; (6) where one body is at rest and the bodies are equal in size; and (7) where two bodies collide when travelling in the same direction. The calculations in all seven rules presuppose that 'quantity of motion', measured as the product of size (extension) and speed, is preserved. For an English version of these articles, and other material omitted below, see V. R. and R. P. Miller, *Descartes, Principles of Philosophy* (Dordrecht: Reidel, 1983), pp. 64f.
32. See Part 2, art. 25.

CHAPTER **19**

Critical Thoughts on the General Part of the Principles of Descartes

GOTTFRIED LEIBNIZ

Pages 383–387, pages 391–398, and pages 403–412 in Gottfried Leibniz, *Philosophical Papers and Letters: A Selection*, translated and edited, and with an introduction by Leroy E. Loemker. Dordrecht: D. Reidel, 1969/1692.

The critical notes which Leibniz wrote to the first two parts of the Principia philosophiae *were intended to be the nucleus for a definitive judgment on Descartes's entire system from Leibniz's point of view and therefore summarized and united all the special criticisms of particular points which he had been making for the previous twenty years. In 1692 Leibniz sent the manuscript to Basnage de Beauval in Holland, with instructions to find a publisher (an assignment which Basnage was unable to carry out) and at the same time to submit the notes to Huygens, Bayle, and other scholars for their criticism. Basnage kept the document until 1697, when he sent it to John Bernoulli at Groningen.*

In the Principia philosophiae, *Part I, Descartes had restated the general metaphysical position of the* Meditations *more exactly, following them in Part II with his conception of the nature of body and of motion. Thus Leibniz's notes on Part I are valuable for their sharp formulation of his own views on knowledge and truth and for his careful analysis of the psychological and epistemological nature of error. Part II is one of the best statements available of the methodological use which Leibniz makes of the principle of continuity in physics. The text is that of Leibniz's revision. (Leroy Loemker)*

[G., IV, 354–92]

On Part I

On Article 1. Descartes's dictum that everything in which there is the least uncertainty is to be doubted might have been better and more exactly formulated in the precept that we must consider the degree of assent or dissent which a matter deserves or, more simply, that we must look into the reasons for every doctrine. This would end all the caviling about Cartesian doubt. But perhaps the author preferred to use paradoxes, in order to stimulate the sluggish reader through novelty. I wish, however, that he had remembered his own precept or rather, that he had understood its true force. We can best explain this matter and its application by the example of geometricians. They are agreed upon axioms and postulates, upon whose truth the rest depends. We accept these, both because they satisfy the mind immediately and because they are proved by countless experiences; nevertheless, it would be an aid to the perfection of science to prove them. This was attempted of old for certain axioms by Apollonius and Proclus and recently by Roberval. Euclid tried to prove that two sides of a triangle taken together are greater than the third (a fact which, as a certain old writer jokingly said, even a jackass knows when he goes to his feed in a straight line rather than a roundabout way), because it was his purpose to base geometric truths not on sensory images but on reason. In the same manner he might also have demonstrated that two straight lines (which do not coincide when extended) can have only one point in common if he had only had a good definition of a straight line.[1] I am convinced that the demonstration of the axioms is of great assistance to true analysis or the art of discovery. So if Descartes had wished to carry out what is best in his rule, he should have worked at the demonstration of scientific principles and thus achieved in philosophy what Proclus tried to do in geometry, where it is less necessary. But our author seems sometimes to have preferred applause rather than certainty. I should not blame him for being satisfied so often with verisimilitude, if he himself had not aroused expectations with so strong a profession of exactness. I blame Euclid much less for assuming certain things without proof, for he at least established the fact that if we assume a few hypotheses, we can be sure that what follows is equal in certainty, at least, to the hypotheses themselves. If Descartes or other philosophers had done something similar to this, we should not be in difficulty. Moreover, the skeptics, who despise the sciences on the pretext that they sometimes use undemonstrated principles, ought to regard this as said also to them. I hold, in contrast, that the geometricians should be praised because they have pinned down science with such pegs, as it were, and have discovered an art of advancing and of deriving so many things from a few. If they had tried to put off the discovery of theorems and problems until all the axioms and postulates had been proved, we should perhaps have no geometry today.

On Article 2. Furthermore, I do not see what good it does to consider what is doubtful as false. This would be not to lay aside prejudices but to change them. But if this is understood merely as a fiction, it should not have been abused, as for instance, in the fallacy which will be seen, below in Article 8, to arise where the difference between mind and body is discussed.

On Article 4. About sensible things we can know nothing more, nor ought we to desire to know more, than that they are consistent with each other as well as with rational principles that cannot be doubted, and hence that future events can to some extent be foreseen from past. To seek any other truth or reality than what this contains is vain, and skeptics ought not to demand any other, nor dogmatists promise it.[2]

On Article 5. There can be no doubt in mathematical demonstrations except insofar as we need to guard against error in our arithmetical calculations. For this there is no remedy except to re-examine the calculation frequently or to have it tested by others and also to add confirmatory proofs. This weakness of the human mind arises from a lack of attention and memory and cannot be completely overcome, and Descartes's mention of it, as if he knew a remedy, is in vain. It would be enough if the state of affairs in other fields were the same as that in mathematics; indeed, all reasoning, even the Cartesian, however convincing and accurate, is subject to this doubt, whatever may be said about some powerful deceiving spirit or about the distinction between dreams and waking.

On Article 6. We have a free will not in perceiving but in acting. Whether honey will seem sweet or bitter to me does not lie with my will, but neither does it lie with my will whether a proposed theorem will seem true or false to me; it is the business of consciousness merely to examine what appears to it. Whoever makes an affirmation of anything is conscious either of a present perception or reason or at least of a present memory bringing back a past perception or the perception of a past reason, although we are often deceived in this through unreliable memory or faulty attention. But consciousness of the present or past is in no way dependent on our will. This one thing we recognize to be within the power of will – to command attention and exertion. And so the will, though it does not bring about any opinion in us, can nevertheless contribute to it obliquely. Thus it happens that men often finally come to believe what they will to be true, after having accustomed the mind to attend most strongly to the things which they favor. In this way they finally succeed in making it satisfy not merely their will but also their consciousness.[3] See on Article 31.

On Article 7. I think, therefore I am. Descartes has well noted that this belongs to the first truths. But it would have been equitable not to neglect others of equal significance. In general, therefore, this can be expressed as follows. Truths are either of fact or of reason. The primary truth of reason is the principle of contradiction or, what amounts to the same thing, that of identity, as Aristotle has rightly observed. There are as many primary truths of fact as there are immediate perceptions or if I may say so, consciousnesses. However, I am conscious not only of myself thinking but also of my thoughts, and it is no more true and certain that I think than that this or that is thought by me. Hence the primary truths of fact can conveniently be reduced to these two: 'I think' and 'Various things are thought by me'. Whence it follows not only that I am, but that I am affected in various ways.

On Article 8. It is not valid to reason: 'I can assume or imagine that no corporeal body exists, but I cannot imagine that I do not exist or do not think. Therefore I am not corporeal, nor is thought a modification of the body.' I am amazed that so able a man could have based so much on so flimsy a sophism. Certainly he adds nothing more in this article; what he has added to it in the *Meditations* I shall examine in its proper place. No one who thinks that the soul is corporeal will admit that we can assume that nothing corporeal exists, but he will admit that we can doubt (as long as we are ignorant of the nature of the soul) whether anything corporeal exists or does not exist. And since we nevertheless see clearly that our soul exists, he will admit that only one thing follows from this: that we can still doubt that the soul is corporeal. And no amount of torture can extort anything more from this argument. But Descartes provided an opening for this fallacy above in Article 2 by taking the license of rejecting what is doubtful as false, so that it becomes possible to assume that there are no corporeal beings because we can doubt that they exist, a point which cannot be granted him. It would be different if we understood the nature of the soul as perfectly as we do its existence, for then it would be established that whatever does not appear in it is not in it.

On Article 13. I have already observed, on Article 5, that the errors which can arise from defective memory or attention and which can also occur in arithmetical calculations even after a perfect method has been found, as in numbers, have been mentioned here to no purpose, since no method can be devised in which such errors are not to be feared, especially when the reasoning is long drawn out. So one must resort to criteria. For the rest, God seems to be called in here merely as a kind of display or showpiece,[4] not to mention that strange fiction or doubt as to whether we are not led to err even in the most evident things, which should convince no one because the nature of evidence prevents it and the experiences and successes of the whole of life witness against it. And if this doubt could once be justly raised, it would be straightway insuperable; it would always confront Descartes himself and anyone else, however evident the assertions presented by them. Aside from this, we must recognize that this doubt cannot be established by denying God or removed by introducing him. For even if there were no

God, we should nonetheless be capable of truth, if only it remained possible for us to exist. And even if it be granted that God exists, it does not thence follow that there exists no created being that is entirely fallible and imperfect, especially since it is possible that its imperfection may not be native but superinduced, perhaps by some great sin, as the Christian theologians teach concerning original sin, so that this evil could not be imputed to God. But even though it does not seem appropriate to have introduced God here, I nevertheless think that the true knowledge of God is the principle of higher wisdom, though for other reasons. For God is the first cause no less than the ultimate reason of things, and there is no better knowledge of things than through their causes and reasons.

On Article 14. The argument for the existence of God taken from the concept of God was first discovered and stated, so far as is known, by Anselm, archbishop of Canterbury, in his book *Contra insipientem*, which still exists.[5] It was frequently examined by various authors of the Scholastic theology and by Thomas Aquinas himself, from whom Descartes, who was no stranger to this knowledge, having studied with the Jesuits of La Flèche, seems to have borrowed it. This reasoning contains something beautiful but is nevertheless imperfect. The argument reduces to this. Whatever can be demonstrated from the concept of a thing can be ascribed to that thing. Now from the concept of a most perfect or greatest being, its existence can be demonstrated. Therefore existence can be attributed to the most perfect being (God), or God exists. The minor premise is proved thus: The most perfect or greatest being contains all perfections, and therefore existence, which is undoubtedly one of the perfections, since it is more or greater to exist than not to exist. So much for the argument. But, by omitting perfection and greatness, the argument could have been formulated more fittingly and strictly in this way. A necessary being exists (that is, a being whose essence is existence or a being which exists of itself), as is clear from the terms. Now God is such a being, by definition of God. Therefore God exists. These arguments are valid, if only it is granted that a most perfect being or a necessary being is possible and implies no contradiction or, what amounts to the same thing, that an essence is possible from which existence follows. But as long as this possibility is not demonstrated, the existence of God can by no means be considered as perfectly demonstrated by such an argument. In general, we must recognize, as I have long since pointed out, that nothing can safely be inferred about a definite thing out of any given definition, as long as the definition is not known to express something possible. For if it should happen to imply some hidden contradiction, it would be possible for something absurd to be deduced from it.[6]

Meanwhile we do learn from this argument the admirable advantage of the divine nature, that if it is merely possible, by this fact itself it exists, an argument which does not suffice to prove the existence of other things. Therefore to have a geometric demonstration of the divine existence it remains only to demonstrate the possibility of God with an exactness adequate for geometric rigor. Meanwhile the existence of a thing which merely needs possibility thus acquires great credibility; and besides, that there is some necessary thing is established in another way, from the fact that contingent things exist.[7]

On Article 18. That we have an idea of a perfect being and that a cause of that idea therefore exists – that is, a perfect being exists (this is Descartes's second argument) – is more doubtful than the possibility of God and is denied by many of those who profess with the highest zeal not merely the possibility but also the existence of God. Nor is the remark valid which, as I recall, Descartes made somewhere to the effect that when we speak of something with an understanding of what we say, we have an idea of the thing.[8] For it often happens that we combine things that are incompatible, as when we think of a most rapid motion, which is certainly impossible, and hence not an idea; and yet we may speak of it, understanding what we mean. For I have elsewhere explained that we often think only confusedly of what we are talking about, and we are not conscious of the existence of an idea in our mind unless we understand the thing and analyze it sufficiently.[9]

On Article 20. The third argument suffers, among other things, from the same vice, namely, when it assumes that there is in us an idea of the highest perfection, God, and concludes from this that God exists because we who have this idea exist.

On Article 21. From the fact that we now are, it follows that we will be in the next moment, unless there exists a reason for change. And so, unless it were established in some other way that we cannot even exist without the beneficence of God, nothing is established about the existence of God from our own duration; as if one part of this duration were entirely independent of another, which we cannot admit.[10]

On Article 26. Even though we are finite, we can yet know many things about the infinite: for example, about asymptotic lines, or lines which approach each other continuously when infinitely produced but never meet; about spaces which are infinite in length but not greater in area than a given finite space; and about the sums of infinite series. Otherwise we should also know nothing with certainty about God. However, it is one thing to know something about a matter and another to comprehend the matter, that is, to have within our power all that is hidden in the matter.

On Article 28. As for the ends which God has proposed to himself, I am fully convinced both that they can be known and that it is of the highest value to investigate them; and that to disdain this inquiry is not without danger or suspicion. In general, whenever we see that anything is particularly useful, we may safely assert that one, among others, of the ends which God has proposed to himself in creating this thing is precisely that it render these services, since he both knew and planned this use of it. I have elsewhere pointed out, and shown by examples, that certain concealed physical truths of great importance can be discovered by considering final causes, which could not have been discovered as easily by efficient causes.[11]

On Article 30. Even if we admit that the perfect substance exists and that it is in no way the cause of imperfections, we shall not thereby remove the true or fictitious reasons for doubt which Descartes introduced, as I have already pointed out in Article 13.

On Articles 31, 35. I do not admit that errors are more dependent upon the will than upon the intellect. To give credence to what is true or to what is false – the former being to know, the latter to err – is nothing but the consciousness or memory of certain perceptions or reasons and so does not depend upon will except insofar as we may be brought by some oblique device to the point where we seem to see what we wish to see, even when we are actually ignorant. See Article 6. Hence we make judgments not because we will but because something appears. And when it is said that will reaches further than intellect, this is more ingenious than true; to put it bluntly, it is a bit of popular ornamentation. We will only what appears to the intellect. [...]

On Articles 71–74. In my comments on Articles 31 and 35, I have already made some remarks about the causes of error. These also provide reasons for the errors discussed here, for the prejudices of infancy belong to the class of unproved assumptions. Moreover, fatigue diminishes attention, and the ambiguity of words belongs with the abuse of signs and involves a formal fallacy. This is as if we put X in place of V in our calculation, to use a German proverb, or as if a druggist were to put sandarac into his prescription in place of dragon's blood [*sanguis draconis*]. [...]

On Part II

On Article 1. The argument by which Descartes tries to prove that material things exist is weak; it would have been better not to try. The gist of the argument is this: the reason why we perceive material things lies outside of us; therefore it is either in God or in someone else or in the material things themselves. It is not in God, for if no material things existed, he would be a deceiver; it is not in someone else – this he has forgotten to prove; therefore it is in the things themselves, and they therefore exist. To this we can reply that a sensation may come from some other being than God, who permits other evils for certain important reasons and who can also permit us to be deceived without having himself the character of a deceiver, especially since this involves no injury, since it would rather be disadvantageous for us to be undeceived. Besides, there is a further fallacy in that the argument neglects another possibility – that while our sensations may indeed be from God or from someone else, the judgment (as to whether the cause of the sensation lies in a real object outside of us), and hence the deception itself, may originate in us. A similar thing happens when colors and other things of this sort are held to be real objects. Through previous sins, moreover, souls may have deserved their condemnation to such a life full of deception, in which they snatch at shadows instead of things. The Platonists do not seem to have shrunk back from such an opinion, since this life seemed to them like a dream in the cave of Morpheus, the mind having lost its reason through the lethal drink before it came here, as the poets used to say.[16]

On Article 4. Descartes tries to prove that body consists in extension alone by enumerating the other attributes and eliminating them. But he should have shown that his enumeration is complete. Also, not all the attributes are correctly eliminated; in fact, those who hold to atoms, that is, to bodies of maximum hardness, will hold, not that hardness consists in a body not yielding to the pressure of the hands, but in it conserving its shape. And those who find the essence of body in antitypy or impenetrability derive the concept of it, not from our hands or from any senses, but from the fact that a body does not give place to another body homogeneous to it unless it can move elsewhere. Thus if we imagine a cube having six other cubes exactly similar to it converging upon it simultaneously and with equal velocities, so that a face of each one of them exactly coincides with one face of the confined cube, then it will be impossible for either the confined cube or any part of it to be moved from its position, whether it be thought of as flexible or rigid. But if this middle cube be thought of as penetrable extension or mere space, then the six concurrent cubes will oppose one another with their edges; but if they are flexible, nothing will prevent their middle parts from breaking into the confined cubical space. From this we may also understand what the difference is between hardness, which is a property of some bodies only, and impenetrability, which belongs to all. Descartes should have considered the latter as well as hardness.

On Articles 5, 6, and 7. Descartes has admirably explained that the rarefaction and condensation which we perceive by sense can occur without our having to admit either the existence of vacua interspersed within matter or a change of dimensions in the same part of matter.

On Articles 8–19. Many of those who defend a vacuum hold space to be a substance and hence cannot be refuted by Descartes's arguments. Other principles are needed to end this dispute. They will admit that quantity and number have no being outside the things to which they are attributed, but they will deny that space or place is the quantity of a body; they will assume rather that space has a quantity or capacity equal to that of the body contained in this space. Descartes should have shown that the space or internal place of a body is not different from its substance.[17] Those who hold the contrary will defend their view with the popular conception of mortals, according to which one body, in succeeding another, passes into the same place and the same space which the former body has deserted – a thing which cannot possibly be said if space coincides with the very substance of the body. Though it may be accidental to a body, however, to have a certain position or to be in a given place, these opponents will no more admit that place is itself an accident of the body than they will that since contact is an accident, the body contacted is also one. Indeed, it seems to me that Descartes does not so much offer sound reasons for his own opinions as he replies to opposing arguments, which he does very skilfully at this point. And he often uses this device instead of demonstration. But we expected something more, and if I am not mistaken, we were invited to expect it. It must be admitted that nothingness has no extension, and this is a fit retort to all who assume some imaginary – I know not what – sort of space. But those who consider space a substance are not touched by this argument; they would certainly be affected if Descartes had proved above what he assumes here, namely, that every extended substance is a body.[18]

On Article 20. The author's attack on atoms does not seem to be satisfactory. Those who defend them will admit that they can be divided in our thinking as well as by divine power. But the question which (to my amazement) Descartes does not even touch on here is whether bodies which have a firmness which natural forces cannot overcome (this is, according to them, the true concept of atoms) can exist naturally at all. Yet he declares here that he has destroyed the atoms and assumes it in the whole further course of his work. We shall say more about atoms below, on Article 54.[19]

On Articles 21, 22, and 23. That the world has no limits in extension and hence can be only one and that the whole of matter is everywhere homogeneous and therefore can be differentiated only through its motions and shapes – these are opinions which are here built upon the proposition that the extended and body are the same, though this is neither universally admitted nor demonstrated by the author.[20]

On Article 25. If motion is nothing but the change of contact or of immediate vicinity, it follows that we can never define which thing is moved. For just as the same phenomena may be interpreted by different hypotheses in astronomy, so it will always be possible to attribute the real motion to either one or the other of the two bodies which change their mutual vicinity or position. Hence, since one of them is arbitrarily chosen to be at rest or moving at a given rate in a given line, we may define geometrically what motion or rest is to be ascribed to the other, so as to produce the given phenomena.

Hence if there is nothing more in motion than this reciprocal change, it follows that there is no reason in nature to ascribe motion to one thing rather than to others. The consequence of this will be that there is no real motion. Thus, in order to say that something is moving, we will require not only that it change its position with respect to other things but also that there be within itself a cause of change, a force, an action.[21]

On Article 26. From what has been said in the preceding paragraph it follows that Descartes's assertion that no more action is required in a body for motion than for rest cannot be sustained. I admit that force is necessary for a body at rest to maintain its rest against colliding bodies. But this force is not in the body itself which is at rest, for the surrounding bodies themselves, opposing each other by their force of motion, cause the body at rest to preserve its given position.[22]

On Article 32. Archimedes was the first author transmitted to us who worked on the composition of motion, in his treatise on spirals. Kepler, in his *Optical paralimpomena*, was the first to apply this to an explanation of the equality of the angles of incidence and reflection, by dividing an oblique motion into a perpendicular and a parallel motion. In this Descartes followed him, as he did also in his *Dioptrics*. Galileo was the first to show the fullest use of the composition of motion in physics and mechanics.[23]

On Articles 33, 34, 35. What Descartes says here is most beautiful and worthy of his genius, namely, that every motion in filled space involves circulation and that matter must somewhere be actually divided into parts smaller than any given quantity. Yet he does not seem to have weighed sufficiently the importance of this last conclusion.

On Article 36. The most famous proposition of the Cartesians is that the same quantity of motion is conserved in things. They have given no demonstration of this, however, for no one can fail to see the weakness of their argument derived from the constancy of God. For although the constancy of God may be supreme, and he may change nothing except in accordance with the laws of the series already laid down, we must still ask what it is, after all, that he has decreed should be conserved in the series – whether the quantity of motion or something different, such as the quantity of force. I have proved that it is rather this latter which is conserved, that this is distinct from the quantity of motion, and that it often happens that the quantity of motion changes while the quantity of force remains permanent. The arguments by which I have shown this and defended it against objections may be read elsewhere.[24] But since the matter is of great importance, I shall give the heart of my conception in a brief example. Assume two bodies: A with a mass of 4 and a velocity of 1, and B with a mass of 1 and a velocity of 0, that is, at rest. Now imagine that the entire force of A is transferred to B, that is, that A is reduced to rest and B alone moves in its place. We ask what velocity B must assume. According to the Cartesians, the answer is that B should have a velocity of 4, since the original quantity of motion and the present quantity would then be equal, since mass 4 multiplied by velocity 1 is equal to mass 1 multiplied by velocity 4. Thus the increase in velocity is proportional to the decrease of the quantity of the body. But in my opinion the answer should be that B, whose mass is 1, will receive the velocity 2, in order to have only as much quantity of power as A, whose mass is 4 and whose velocity is 1. I shall explain my reason for this as briefly as possible, lest I appear to have proposed it without any reason. I say, then, that B will have only as much force as A had previously or that the present and the former force are equal, a thing which is worth proving. To go deeper, namely, and explain the true method of computation – which is the duty of any really universal mathematics, though it has not yet been carried out – it is clear, first of all, that force is doubled, tripled, or quadrupled when its simple quantity is repeated twice, three times, or four times, respectively. So two bodies of equal mass and velocity will have twice as much force as one of them. It does not follow, however, that one body with twice the velocity must have only twice the force of a body with simple velocity, for even though the degree of velocity may be doubled, the subject of this velocity is not itself duplicated, as it is when a body twice as great, or two bodies of the same velocity, are taken in place of one, so that they completely repeat the one in magnitude as well as motion. Similarly 2 pounds elevated to the height of 1 foot are exactly double in essence and power to one elevated the same distance, and two elastic bodies stretched equally are double one of them. But when the two bodies possessing this power are not fully homogeneous and cannot be compared with each other in this way, or reduced to a common measure of matter and force, an indirect comparison must be attempted by comparing their homogeneous effects or causes. Every cause whatever has a force equal to its total effect, or to the effect which it produces in using up its own force. Therefore, since the two bodies mentioned above, A with mass 4 and velocity 1, and B with mass 1 and velocity 2, are not exactly comparable, and no one quantity possessing force can be designated whose simple repetition will produce both, we must examine their effects. Let us assume, namely, that these two bodies are heavy and that A can change its direction and rise; then by virtue of its velocity of 1 it will rise to the height of 1 foot, while B, by virtue of its velocity of 2, will rise 4 feet, as Galileo and others have demonstrated. In each case the effect will entirely consume the force and so be equal to the cause which produces it. But these two effects are equal to each other in force or power, namely, the elevation of body A, 4 pounds, to 1 foot, and the elevation of body B, 1 pound, to 4 feet. Therefore the causes, too, are equal; that is, the body A of 4 pounds with velocity of 1 is equal in force or power to the body B of 1 pound with velocity 4, as was asserted. But if someone denies that the same power is needed to raise 4 pounds 1 foot and 1 pound 4 feet, or that these two effects are equivalent (though, unless I am mistaken, almost everyone will admit this), he can be convinced by the same principle. If we take a balance with unequal arms, 4 pounds can be raised exactly 1 foot by the descent of 1 pound for 4 feet, and no further work can be done; thus the effect exactly exhausts the power of the cause and is equal to it in force. Let me summarize, therefore. If the whole power of body A, 4, with a velocity of 1, is transferred to B, 1, B must receive a velocity of 2; or, what amounts to the same thing, if B is first at rest and A in motion, but A is then at rest and B has been placed in motion, other things remaining equal, the velocity of B must be double, since the mass of A was quadruple. If, as is popularly held, B should receive four times the velocity of A because it has one-fourth of its mass, we should have perpetual motion or an effect more powerful than its cause. For when A was moving, it could raise 4 pounds only 1 foot, or 1 pound 4 feet; but later, when B moved, it would be able to lift 1 pound 16 feet, for altitudes are as the square of the velocities by force of which bodies are lifted, and four times the velocity will raise a body sixteen times the altitude. With the aid of B not only could we thus once more raise A for 1 foot, after its descent had given it its original velocity, but we could do many other things besides and thus exhibit perpetual motion, since the original force is restored but there is still more left. Moreover, even though the assumption that the whole force of A is transmitted to B cannot actually be realized, this does not affect the matter, since we are here concerned with the true calculation, or with the question of how much force B would necessarily take on according to this hypothesis. Even if a part of the force is retained and only a part transmitted, the same absurdities would still arise, for if the quantity of motion is to be conserved, the quantity of forces can obviously not always be conserved, since the quantity of motion is known to be the product of mass and velocity, while the quantity of force is, as we have shown, the product of mass and the altitude to which it can be raised by force of its power, altitudes being proportional to the square of the velocities of ascent. Meanwhile this rule can be set up: The same quantity of force as well as of motion is conserved when bodies tend in the same directions both before and after their collision, as well as when the colliding bodies are equal.[25]

On Articles 37, 38. It is a very true and indubitable law of nature that the same thing, so far as in it lies, always persists in the same state – a law which both Galileo and Gassendi, and several others as well, have long held. It is surprising therefore that it has occurred to some men that a projectile owes the continuation of its motion to the air but that it has not occurred to them that by this same reasoning we should with equal right have to look for some new reason for the continued motion of the air itself. For air could not, as they hold, impel the stone forward unless it itself had the power to continue its received motion and found itself impeded in this by the resistance of the stone.

On Article 39. Not only did Kepler observe the very beautiful law of nature according to which bodies describing a circular or curved path strive to leave it in the line of the tangent straight line (others may have preceded him in this), but he already made that application of this law which I consider essential in making clear the cause of gravity. This is apparent from his *Epitome of the Copernican Astronomy*.[26] Descartes has rightly affirmed this law and brilliantly expounded it, but he has not demonstrated it, as one would have expected of him.

On Articles 40–44. In Articles 37 and 39 Descartes has presented two very true laws of nature which are clear in their own light. But the third seems to me to be so far, not merely from truth, but even from probability, that I wonder how it ever occurred to the mind of such a man. Yet he at once builds his laws of motion and of impact upon it and says that it contains all the causes of particular changes in bodies. He conceives it as follows: one body colliding with another stronger one loses none of its motion but merely changes its direction; however, it can receive some additional motion from the stronger body. But in colliding with a weaker body, it loses only as much motion as it transfers to the weaker body. In actual fact, however, it is only in the case of a collision of bodies moving in opposite directions that a body colliding with a stronger one loses no motion but either retains or increases its velocity. When a weaker but swifter body overtakes one that is stronger but slower, then the contrary occurs, and it is generally true, and can be observed in nature, that the velocity of the pursuing body is diminished by the impact. For if it continued its motion after the collision, it could not in any case proceed at its earlier velocity without giving this velocity to the body ahead as well, in which case the total power in the whole would be increased. If it came to rest after the collision, it would be clear in itself that its velocity had been diminished, and indeed destroyed, by the blow. This coming to rest, moreover, occurs in hard[27] bodies (which are here always to be understood) when the ratio of the excess of mass in the first body over that of the one overtaking it to the mass of the body overtaken[28] is double the ratio of the velocity of the first body to the one overtaking it.[29] Finally, if the body is thrown back after the collision, it is again clear that the motion of the repelled body is less than before. For otherwise the velocity of the forward body would necessarily be increased by the added impulsion of the body overtaking it, and whether we think of it as increasing its velocity after its rebound or merely retaining its earlier velocity, the aggregate of power would be increased, which is absurd.

If anyone were to defend Descartes by holding that this third law of his on the collision of bodies must be understood to deal only with the collision of bodies from opposite directions, I readily agree with it. But then it must be admitted that he has not provided for the collision of bodies moving in the same direction, though, as we have already seen, he himself claims that this law covers all particular cases. Also, if the demonstration which he attempts in Article 41 is correct, it includes all cases of colliding bodies, whether they move in the same or in opposite directions. But it does not seem to me to have even the semblance of a proof. I admit that it is correct to distinguish quantity and direction of motion and that one of these sometimes changes while the other remains constant. But it is also true that frequently both change together. In fact, both work together to preserve each other, and a body tends to preserve its determination or direction with its whole force and its whole quantity of motion. Whatever is taken away from the velocity when the direction remains constant is also lost from the determination, since a body proceeding more slowly in the same direction is less determined to conserve it. Besides, if a body A collides with a smaller body B at rest, it will continue in the same direction but with diminished motion; if it collides with a body B, at rest but equal to itself, it stops, so that, while it itself remains completely at rest, its motion is transferred to B; and finally, if it collides with B, which is at rest but greater than it, or is equal to it with an opposite movement, then A will simply be turned back.[30] Hence we can understand that a greater opposing force is necessary for A to be reflected in a direction opposite its original direction than for it merely to be brought to rest, a fact which directly contradicts Descartes's pronouncements. For the opposition must be greater, since the thing opposed is greater, or since the tendency opposing it is greater. But I hold his proposition that motion perseveres as a simple state until it is destroyed by an external cause, to be true not only for the quantity of motion but for its determination also. This determination of a moving body, or its tendency to advance, itself has its own quantity which is more easily diminished than reduced to nothing or to rest and can furthermore be more easily destroyed or reduced to rest than changed into a contrary or regressive motion, as we have just pointed out. Thus, even though one motion is not opposed to another in kind, the present motion opposes the present motion of a body which collides with it, and one advance opposes another contrary advance, since a smaller change and a smaller opposition are necessary, as we have shown, to diminish an advance than to destroy it entirely or to transform it into a retreat. So it seems to me that Descartes's reasoning is like trying to argue that when two bodies oppose each other, they ought never to break or fall apart but to bend each other so that their shapes are molded to each other, on the ground that matter is distinct from shape and that in this case matter is not opposed to matter, but shape to shape, and the quantity of matter can be conserved in the body while its shape is changed. Whence it would follow that the magnitude of a body can never change and that only the shape of the body can change. If Descartes had taken into consideration that every body which collides with another must, before it is repelled, first reduce its advance, then come to a stop, and only then be turned back, and must thus pass from one direction to the opposite, not by a leap but by degrees, he would have set up other rules of motion for us. We must recognize that, no matter how hard, every body is nevertheless flexible and elastic to some degree; like a ball inflated with air which gives way a little when it falls to the floor or is struck with a stone, until the impetus or advance of what strikes it is gradually broken and at last completely stopped, after which the ball resumes its shape and repels the stone, which now no longer resists, or until it rebounds by itself from the floor to which it had fallen. Experiments have taught us convincingly that something similar to this takes place in every rebound, even if the bending and restoration are not visible. But being too confident of posterity, Descartes in his letters very superciliously condemned this explanation of reflection by elastic force, which was first noted by Hobbes.[31] We do not need to examine anew the reasoning by which he tries, in Article 42, to demonstrate the last part of this law of nature which he wants to promulgate – the part, namely, which holds that whatever quantity of motion is lost to one colliding body is added to the other. For this assumes that the quantity of motion must remain constant, and we have already shown in Article 36 how great an error this is.

On Article 45. Before I turn to an examination of the special rules of motion given by the author, I shall set up a general criterion or touchstone, as it were, by which they can be examined. I usually call this the *law of continuity*. I have already explained this principle elsewhere, but it must be repeated and amplified here.[32] When two hypothetical conditions or two different data continuously approach each other until the one at last passes into the other, then the results sought for must also approach each other continuously until one at last passes over into the other, and vice versa. For example, if one focus of an ellipse remains fixed and the other recedes farther and farther away from it, while the *latus rectum* remains constant, the new ellipses which thus come into being continuously approach a parabola and finally pass over into it completely, namely, when the distance of the receding focus becomes immense. Therefore the properties of these ellipses must also approach more and more the properties of a parabola until at last they pass over into them, and the parabola can be considered as an ellipse whose second focus is infinitely distant. All the properties of an ellipse in general will thus be found in the parabola considered as such an ellipse. Geometry is full of examples of this kind, but nature, whose most wise Author uses the most perfect geometry, observes the same rule; otherwise it could not follow any orderly progress. Thus gradually decreasing motion finally disappears in rest, and gradually diminishing inequality passes into exact equality, so that rest can be considered as infinitely small motion or as infinite slowness, and equality as infinitely small inequality. Whatever is demonstrated about motion in general, or about inequality in general, must for this reason also be verifiable about rest or equality, if this interpretation is right. So the rules for rest or equality can in a sense be considered as special cases of the rules for motion or inequality. If this cannot be done, we may be certain that the proposed rules are inconsistent or wrongly conceived. Hence we shall also show, in Article 53, that to the continuous curve which represents the variations of the hypothetical conditions there must correspond a continuous curve representing the variations of the results but that the Cartesian rules of motion present these results by a figure which is absurd and incoherent.

On Article 46. Let us now examine the Cartesian rules of motion. We must understand the bodies involved to be hard and unimpeded by other conditions.

> Rule 1. If two equal bodies B and C, with equal velocities, collide directly, both will be deflected with the velocities of their approach.

This first rule is the only one of Descartes that is entirely true. It can be demonstrated in this way: since the properties of both bodies are equal, either both will continue in their motion and so penetrate each other, which is absurd; or both will come to rest, in which case power will be lost; or both will be repelled, and with their original velocity, because, if the velocity of one is diminished, the velocity of the other must also be diminished, because of the equality of their properties. But if the velocity of both is diminished, the whole will be diminished, which is impossible. [. . .]

On Articles 54, 55. I do not believe it to be entirely true, though there is some truth in it, that bodies are *fluid* if their particles are agitated by various motions in all directions; that they are *hard* if their adjacent parts are mutually at rest in relation to each other; and that matter is not held together by any other glue than the quiescence of one part in relation to another. Descartes infers therefore that hardness, or as I prefer to call it more generally, firmness (some of which there is even in soft bodies), arises from rest alone, because the glue or the cause of cohesion cannot be a body (for then the problem would repeat itself), and so it must be a mode of a body. But, he reasons further, rest is the only mode of a body which is fitting to explain this matter. Why so? Because rest is most contrary to motion. I marvel that so important a matter should be decided with so trifling and perfunctory, indeed, so sophistic a reason. The syllogism would be:

Rest is that mode of body which is most contrary to motion.
But that mode of body which is most contrary to motion is the cause
of firmness.
Therefore rest is the cause of firmness.

But both premises are false, though each makes some tenuous display of truth. It happens too often in Descartes that by assuming the most uncertain matters to be certain, he dismisses the careless reader by his dictatorial brevity; as when he concludes that extension constitutes matter, that thought is independent of matter, and that the same quantity of motion is conserved in nature – pronouncements based on authority rather than arguments. I am of the opinion that motion in the opposite direction is more contrary to motion than is rest and that a greater opposing force is required to reflect a body than merely to bring it to rest, as I have shown in Article 47. But the other premise would also have to be proved, namely, that what is most opposed to motion is the cause of firmness. Did the author by chance have in mind the following prosyllogism?

Firmness is most opposed to motion.
The cause of whatever is most opposed to motion is itself most opposed
to motion.
Therefore the cause of firmness is most opposed to motion.

But the premises of this prosyllogism are again both defective. Thus I deny that firmness is most opposed to motion; I admit that it is most opposed to the motion of one part without another, and it is the cause of this that he should have sought. Nor do I have any confidence in the axiom that the cause of whatever is opposed to a thing is itself also opposed to that thing. What is more opposed to death than life, but who would deny that death very often comes to an animal from a living being. No demonstration can be based upon such philosophical rules, which are entirely vague and not yet reduced to their proper limits.

There will be some who read this who will be offended at us for reducing such great philosophers to the limitations of Scholasticism by putting them in syllogisms; there will perhaps also be those who will condemn this as too trivial. But we have learned that these great philosophers, and indeed, often other men too, stumble on the most serious matters through neglect of this childish logic; in fact, they scarcely ever make mistakes in any other way. For what else does this logic contain than the most general dictates of supreme reason, expressed in rules that are easy to understand? It has seemed desirable to use this example, to show, for once, how useful such rules are for putting an argument into the prescribed form, so that the force of the argument may become apparent, especially in problems where the imagination does not come to the aid of reason as it does in mathematics, and where we are dealing with an author who puts great matters into precipitate arguments. Since Descartes does not help us with reasons in this problem, therefore, we must return to a consideration of the matters themselves.

In firmness, then, we must consider not so much rest as the force by which one part draws another along with it. Let two perfect cubes, A and B, be joined together and at rest with respect to each other, with their surfaces perfectly polished, and let cube B be placed to the left of cube A, with a surface of one congruent to a surface of the other, without any intervening space. Now let a small ball C strike the middle of cube A, in a direction parallel to the two congruent surfaces (Figure 27). Then the direction of the blow will not reach cube B unless it is assumed to adhere to cube A. Of course A resists the colliding body C by its rest and cannot be moved by it without diminishing the force of C, and so it is also true that in this case A by its rest resists being separated from B. But this is *per accidens*, not because it is unconnected with B, but because it has to absorb the force of the blow itself, just as it would if B were entirely absent. So, once it has received this force, it will begin its own path, abandoning B just as if B were entirely absent from its vicinity. It is therefore a sophism to try to conclude that because each thing perseveres in its own state as much as possible, it follows that two bodies at rest in relation to each other will mutually adhere and will have firmness from their mere state of rest. You might conclude with as much right that two bodies 10 feet apart are connected together and will strive to act so that they will always be 10 feet apart. A cause must therefore be found why two cubes A and B cohere sometimes and form a firm parallelopiped AB which moves as a whole when only the part A is impelled; or

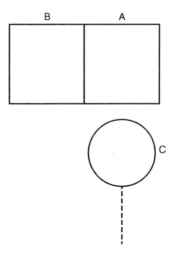

Figure 27 (redrawn)

a cause must be found why the cube *A*, when moved, draws the cube *B* with it. Thus we seek the cause of traction in nature. There are, it is true, learned men who affirm that perfect unity is itself the cause of firmness, and it seems that this opinion satisfies some advocates of atoms. For if the parallelopiped is taken as an atom which is conceptually divisible into two cubes *A* and *B*, but is not really so divisible, they say that the parallelopiped is also actually indivisible and will remain firm always. Many objections may be made to this; first of all, they offer no demonstration of their position. Let us assume that two atoms, *D* and *E*, which correspond to the cubes *A* and *B* with their anterior surfaces, strike simultaneously against the parallelopiped *AB* in directions parallel to the common surface of *A* and *B*, but with *D* coming from the back, from the direction of *F*, with its whole surface striking the entire congruent surface of *A*, and with *E* similarly coming from in front, in the direction from *G*, and striking *B* (Figure 28). We seek a cause for *A* not leaving *B* and being propelled toward *G*, and for *B* not leaving *A* and being propelled toward *F*. I find no reason for this in the atomists' doctrine. For what else is there in the statement that a unity is composed of the two cubes *A* and *B* than that they are not actually divided? But if you hold, as do certain thinkers, that there are no parts in the continuum before actual division, it follows either that this does not prevent separation, namely, when a further reason is added which tends to produce an actual division, and so determines and distinguishes the parts as it were (the reason being the impact

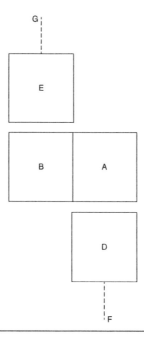

Figure 28 (redrawn)

of the bodies *D* and *E*), or that no continuum can ever be broken into parts. Suppose then that two cubical atoms *A* and *B*, formerly distinct, approach each other so that two of their surfaces coincide; will there, at the moment of contact, be no difference at all between them and the atomic parallelopiped described above? Then two atoms would be held by each other through simple contact as if by some kind of glue, and the same thing would happen even if only parts of their surfaces touched. By a natural progression it would further follow from this that atoms would continuously increase like snowballs rolled through the snow, and the outcome, finally, would be that everything would coalesce into a more than adamantine hardness, and congeal into eternal ice, since the cause of coalescence would subsist, but not that of dissolution. For those who hold these views there remains one escape – to say that there are no plane surfaces in nature, or that if there are any, they disappear in coalescence, but that all atoms terminate in curved surfaces, and these with the smallest possible areas of contact, as the case would be, for example, if all atoms were spherical, so that there would be no contact of any entire surfaces. But aside from the point that no adequate reason is given for excluding bodies with plane surfaces, or with other congruent surfaces, we again ask them to give us a reason why a continuum cannot be resolved into parts.

We have other strong arguments against atoms but do not propose to exhaust that matter here. There are those who explain the hardness of bodies by the same cause by which we see that two polished boards cannot be broken apart except with great force, the surrounding matter finding it impossible to penetrate so suddenly into the place to be relinquished by a separation of the boards. So they say that hardness arises from compression, which is right in most cases, but cannot be understood as a universal cause of hardness, because it again presupposes some hardness or firmness already to exist – that of the boards themselves. It is likewise irrelevant to say that the two cubes *A* and *B* are connected by some kind of glue, for some firmness is necessary in the glue itself by which its parts adhere both to each other and to the two bodies which they connect. But if anyone thinks that some kind of little projections pass over into *B* from *A*, penetrating into its small cavities, and also to *A* from *B*, and that this is the reason why the one cannot be moved without the other unless these points are broken, a new question arises: Whence the firmness of these points?

To pass over these theories, therefore, which either do not advance or do not solve the problem, I believe that the primary cause of cohesion is movement, namely, concurrent movement. (Evidently one must add impenetrability itself, when there is no place into which to give way, or when there is no reason for one body to give way rather than another; thus a perfect sphere which rotates in a uniform plenum at rest is prevented from throwing anything off by centrifugal force.) I believe that matter itself, which is homogeneous and equally divisible throughout, is differentiated by motion alone. We see that even fluids acquire a certain firmness when in motion. Thus a vigorous jet of water will prevent anything from breaking into its own path from without with more force than the same water at rest. For the irruption of new matter necessarily creates a strong disturbance in the co-ordinated motion, and force is necessary to produce this disturbance or to change the motion so greatly. If you touch a jet of water with your finger, you will see the little drops scatter in all directions with some violence, and you will also feel your finger repelled on entering the jet. We learn from the magnet, in an elegant experiment, that things which in themselves are separate and, so to speak, sand without lime, can acquire some firmness by motion alone. When iron filings are placed near a magnet, they suddenly become connected like a rope and form filaments, and the matter arranges itself in rows. It is no doubt also by some kind of magnetism, that is, by an internal co-ordinated motion, that other parts of certain bodies are linked together. This primary cause of consistency or cohesion therefore satisfies reason no less than the senses.

On Articles 56, 57. It is unnecessary to investigate the cause of fluidity, for matter is itself fluid except insofar as there are motions within it which are disturbed by the separation of certain parts. So it is not necessary for a fluid to be agitated by the varied motions of its particles. But since it is established on other grounds, by a general law of nature, that all bodies are agitated by internal motions, the conclusion is that bodies are firm insofar as these motions are concurrent, but remain fluid insofar as the motions are perturbed and not connected by any system. The result is that every body contains some degree of fluidity and some of firmness alike and that no body is so hard as not to have some flexibility, and the converse. Furthermore, this internal motion is insensible, since the parts which succeed each other continuously are not discernible by sense because of their smallness and similarity; moving rapidly, like a jet of water or the spokes of a wheel, they simulate one continuous solid. The internal motion of fluids is also confirmed by solutions of salt in water and by the corrosions which are made by acids and, indeed, also by heat in general. For when heat is great, it causes liquids to boil, when it is only moderate it produces only an agitation, but when the agitation arising from heat is weakened, as it is in winter, then the permanent internal motion of the parts of matter acting in harmony alone predominates in most liquids; hence they harden and sometimes freeze solid. Another crude example of this strangely perturbed agitation of fluids is offered in the dust particles revealed by the sun's rays in an otherwise dark place. Moreover, since the fluids which seem according to our sense perception to be at rest are really in equally unhindered motion everywhere, and in all directions, it follows that their perturbed motion is so equally distributed and compensated for within them, as it were, that when a solid is placed in such a fluid, it is assailed so equally on all sides by the blows and surges of the fluid that it is neither aided nor hindered in its own motion.

On Article 59. When a body is impelled by an external force in a fluid, the author thinks that this force, though in itself not enough to move the body, nevertheless does move it by concurring with particles of the fluid which support this motion and determines the remaining particles also to support it by retaining their own motion but changing its determination or direction. Add to this what the author says at the end of Article 56, and in his demonstration in Article 57. Hence he asserts that a hard body moving in a fluid does not obtain its whole motion from the hard body which impels it but also partly from the surrounding fluid. But soon he himself seems to demolish this in Article 60. In general, I believe that his remarks are to no purpose, because they rest upon a false principle (since he again speaks here as if rest were contrary to motion) and because they seem to have been thought up only to save the contradiction which phenomena show to the fourth rule of motion, in which our author has wrongly denied that a body at rest can be set in motion by a smaller one acting with any amount of velocity whatever (see the end of Article 61), whereas he himself was nevertheless forced to admit, in Article 56, that a hard body in a fluid is moved by the smallest force. So he makes use of this amazing comment to escape the difficulty and calls upon the particles of the fluid to help, but in vain, for since their motions from contrary directions compensate each other, they are of no avail. For if they had any effect, it would be too great and would give the movable object a motion greater than it can receive from the impelling force.

It is evident, however, that such a greater motion does not arise, and hence the movable body receives no more motion than if the fluid had not acted at all. Indeed, we must rather affirm the contrary – that far from motion being added by the fluid, some is rather subtracted by it, and the velocity of the moving body is diminished, partly because of the effect of a certain resistance but partly also from the mere fact that when a hard body enters into a fluid, a part of the fluid equal to the volume of the body must continually be displaced and excited to new motion, and a certain part of the moving power must be expended to do this. I have elsewhere reduced the quantity of both kinds of resistance to calculation; one part of it is absolute and always the same in the same fluid, the other is relative and increases with the velocity of motion.[35] [. . .]

On Article 64. The author closes his second part, which is a general part dealing with the principles of material things, with an observation which seems to me to need some restriction. He says, namely, that no other principles are necessary for the explanation of natural phenomena than those taken from abstract mathematics, or from the doctrine of size, figure, and motion, and that he recognizes no other matter than that which is the subject of geometry. I fully agree that all the particular phenomena of nature can be explained mechanically if we explore them enough and that we cannot understand the causes of material things on any other basis. But I hold, nevertheless, that we must also consider how these mechanical principles and general laws of nature themselves arise from higher principles and cannot be explained by quantitative and geometrical considerations alone; that there is rather something metaphysical in them, which is independent of the concepts which imagination offers, and which is to be referred to a substance devoid of extension. For in addition to extension and its variations, there is in matter a force or a power of action by which the transition is made from metaphysics to nature[36] and from material to immaterial things. This force has its own laws, which are derived from the principles not merely of absolute and, so to speak, brute necessity, as in mathematics, but from those of perfect reason.

Once these matters have been established in a general treatment, we may afterward, in accounting for natural phenomena, explain everything mechanistically, and it is as vain here to introduce the perceptions and appetites of an Archeus, operative ideas, substantial forms, and even minds, as it is to call upon a universal cause of all things, a *deus ex machina*, to move individual natural things by his simple will, as I recall the author of the *Philosophia mosaica* does by means of a fallacious interpretation of the words of Sacred Scripture.[37] Whoever considers these matters honestly will hold to the middle way in philosophy and do justice to theology as well as to physics. He will understand that the Scholastics sinned of old, not so much in holding to indivisible forms, as in applying them where they ought rather to have sought the modifications and instrumentalities of substance and its mode of action, that is, mechanism. Nature has, as it were, an empire within an empire, a double kingdom, so to speak, of reason and necessity, or of forms and of the particles of matter, for just as all things are full of souls, they are also full of organic bodies. These kingdoms are governed, each by its own law, with no confusion between them, and the cause of perception and appetite is no more to be sought in the modes of extension than is the cause of nutrition and of the other organic functions to be sought in the forms or souls. But that highest substance which is the universal cause of all brings it about, by its infinite wisdom and power, that two very different series in the same corporeal substance respond to each other and perfectly harmonize with each other, just as if one were ruled by the influence of the other. And if you observe the necessity of matter and the order of efficient causality, you will notice that nothing happens without a cause which satisfies our imagination, nothing which lies beyond the mathematical laws of mechanism; but if you contemplate the golden chain of ends and the circle of forms as an intelligible world, you will find that since the apex of metaphysics and that of ethics are united in one by reason of the perfections of their supreme author, nothing happens without the highest reason. For the same God is the supreme form and the first efficient cause and the end or ultimate reason of the universe. But it is our part to revere his traces in nature and to meditate not merely upon his instruments in operating and the mechanical efficiency of material things but also upon the more sublime uses of his admirable craftsmanship; to know God, not merely as the architect of bodies, but above all as the king of minds, whose intelligence orders all things for the best and constitutes the universe as the most perfect state under the dominion of the most powerful and most wise Monarch. By thus combining both types of interpretation, we shall serve, in the consideration of the individual phenomena of nature, both our welfare in life and our perfection of mind, and wisdom no less than piety.

Notes

(From the Loemker edition; page references are to the Loemker edition.)

1. Euclid's definition of a straight line is given on p. 247, note 5. For Leibniz's criticism of Euclid and his own definition see No. 27, II, p. 258, notes 8 and 10. See also the *New Essays*, Book IV, 12, 4.

2. This passage, somewhat anticipatory of Kant, involves the criteria for the reality of phenomena stated in No. 39. It can be reconciled with the monadology only when it is remembered that the rational principles upon which phenomena are well founded are themselves the metaphysical basis for God's creation of individuals.

3. *Conscientia.* The term is obviously used here in the sense of intellect or understanding.

4. The allusion is to Descartes's vicious circle in validating our general principles by a reference to God's moral nature.

5. See p. 234, note 5.

6. The first draft contained an example which was later deleted. "For example, assume a definitum *A* whose definition is 'an absolutely necessary beast.' *A* is shown to exist in this way: whatever is absolutely necessary exists, by an indubitable axiom. *A* is absolutely necessary, by definition. Therefore *A* exists. But this is absurd. The reply to it must be that definition or concept is impossible and hence cannot be admitted as an assumption."

7. See Nos. 51 and 67, Secs. 37–39.

8. To Mersenne, July 1, 1641 (*Correspondence*, ed. Adam and Tannery, III, 392). Leibniz had repeatedly refuted this view of ideas from the Paris periods on (see Nos. 21, 25, and 33 and *New Essays*, Book IV, 10, 7).

9. The first draft added: "Meanwhile it is very true that there is the idea of God in us, because it is most true that God is possible and hence existent and in both senses known by us. In a certain way all ideas are innate in us, nor can the senses do anything else but to turn the mind to them, as is shown elsewhere."

10. The comment is directed at Descartes's tendency to atomize the contents of consciousness (cf. *Meditations*, Part III; *Principles*, Part I, Art. 21), which led his school to interpret each state as a distinct divine creation. See also the beginnings of the correspondence with De Volder (No. 55).

11. See No. 50, below.

16. Leibniz had originally added: "I have elsewhere discussed wherein the reality of material things consists. See also Part I, Article 4."

17. Internal place is for Descartes the volume of a body; external space, the position of this volume within a larger whole. He then defines motion as change of this external position in relation to nearby bodies.

18. Addition in the first draft: "For the rest, I shall sometime make clear that a material mass is itself not a substance but an aggregate resulting from substances, and that space is nothing but the common order of all coexistents, as time is that of things which do not coexist."

19. Addition in the first draft: "We conclude that there are none on other grounds."

20. Addition in the first draft: "These opinions may be true, however, on other grounds."

21. The problem of the relativity of observed motion is thus, at this point in Leibniz's thought, an argument for an internal force in bodies. The point is further discussed in the correspondence with Huygens (No. 43) and in the *Specimen dynamicum* (No. 46).

22. Original addition: "although a perfectly quiescent body is in fact never found."

23. Leibniz's historical notes on the origins of the resolution of motions (and later, forces) are in general sound. Kepler's work is *Ad Vitellionem paralimpomena, quibus astronomiae pars optica traditur* (1604).

24. See No. 34. His argument will be often restated (cf. Nos. 46 and 55).

25. Leibniz's acknowledgment that the quantity of motion is conserved in the case of bodies which do not reverse their direction anticipates his own more general principle of the conservation of quantity of progress, which differs from Descartes's principle in considering the algebraic, not the arithmetic, sum of motions (see No. 46 and p. 451, note 9).

26. Kepler, *Epitome astronomiae Copernicanae* (1620), Book IV.

27. *Durus*. More exactly, Leibniz must mean bodies with perfect elasticity. Compare the discussion of Articles 56 and 57 below. See on the laws of impact, p. 302, note 9.

28. As BC. points out (I, 316), G's text is in error in reading *assequens* for *praecedens*.

29. In other words, if m is the mass and v the velocity of the first body before impact, and V its velocity after impact; and m_1 the mass and v_1 and V_1 the velocities before and after impact of the second body, which overtakes the first, then $V_1 = 0$ when $(m-m_1)/m = 2\, v/v_1$. Leibniz's formula can be derived from the principle of the conservation of vis *viva* and of momentum (or quantity of progress), both of which are true only in the case of perfectly elastic bodies:

$$mv^2 + m_1 v_1^2 = mV^2 + m_1 V_1^2; \tag{1}$$

and

$$mv + m_1\, v_1 = mV + m_1 V_1, \tag{2}$$

by dividing (1) by (2), solving for V in the quotient, substituting for V in (2), and then resolving the resulting equation for the case $V_1 = 0$.

30. These conclusions follow from the general formula for V_1 in note 29, for the cases, respectively, of $m < m_1$, $m = m_1$, and $m > m_1$, with $v = 0$.

31. See Descartes to Mersenne, January 21, February 7, and March 4, 1641 (*Correspondence*, ed. Adam and Tannery, III, 287ff., 300ff., 318ff.).

32. See No. 37.

35. 'Schediasma de resistentia medii et motu projectorum gravium in medio resistante', *Acta eruditorum*, January, 1689.

36. So G., IV, 391. BC. suggests reading "transitum a mathematica ad naturam". But G.'s reading is defensible in view of the narrower meaning of *naturalis* just below. The concept of force is metaphysical in Leibniz but appears in science in the form of derivative force.

37. Paracelsus was the source both of Van Helmont's *archeus* [. . .] and of the animistic views of Robert Fludd (1574–1637) in his *Philosophia mosaica* (1638).

CHAPTER 20

On the Gravity and Equilibrium
of Fluids

ISAAC NEWTON

Pages 121–148 in *Unpublished Scientific Papers of Isaac Newton*, ed. and trans. A. Rupert Hall and Marie Boas Hall, Cambridge: Cambridge University Press, 1978 [date unknown] [Material in the main text in brackets added by the editor(s)].

It is proper to treat the science of gravity and of the equilibrium of fluid and solid bodies in fluids by two methods. To the extent that it appertains to the mathematical sciences, it is reasonable that I largely abstract it from physical considerations. And for this reason I have undertaken to demonstrate its individual propositions from abstract principles, sufficiently well known to the student, strictly and geometrically. Since this doctrine may be judged to be somewhat akin to natural philosophy, in so far as it may be applied to making clear many of the phenomena of natural philosophy, and in order, moreover, that its usefulness may be particularly apparent and the certainty of its principles perhaps confirmed, I shall not be reluctant to illustrate the propositions abundantly from experiments as well, in such a way, however, that this freer method of discussion, disposed in scholia, may not be confused with the former which is treated in Lemmas, propositions and corollaries.

   a Axiom.
   b Corollarium 4.
   c Corollarium 2.
   d Contra Hypothesin.

The foundations from which this science may be demonstrated are either definitions of certain words; or axioms and postulates denied by none. And of these I treat directly.

Definitions

The terms *quantity, duration* and *space* are too well known to be susceptible of definition by other words.

> Def. 1. Place is a part of space which something fills evenly.[a]
> Def. 2. Body is that which fills place.[a]
> Def 3. Rest is remaining in the same place.
> Def. 4. Motion is change of place.[b]

Note. I said that a body fills place,[a] that is, it so completely fills it that it wholly excludes other things of the same kind or other bodies, as if it were an impenetrable being. Place could be said however to be a part of space in which a thing is evenly distributed; but as only bodies are here considered and not penetrable things, I have preferred to define [place] as the part of space that things fill.

Moreover, since body is here proposed for investigation not in so far as it is a physical substance endowed with sensible qualities but only in so far as it is extended, mobile and impenetrable, I have not defined it in a philosophical manner, but abstracting the sensible qualities (which Philosophers also should abstract, unless I am mistaken, and assign to the mind as various ways of thinking excited by the motions of bodies) I have postulated only the properties required for local motion. So that instead of physical bodies you may understand abstract figures in the same way that they are considered by Geometers when they assign motion to them, as is done in Euclid's *Elements*, Book 1, 4 and 8. And in the demonstration of the tenth definition, Book xi, this should be done; since it is mistakenly included among the definitions and ought rather to be demonstrated among the propositions, unless perhaps it should be taken as an axiom.

Moreover, I have defined motion as change of place,[b] because motion, transition, translation, migration and so forth seem to be synonymous words. If you prefer, let motion be transition or translation of a body from place to place.

For the rest, when I suppose in these definitions that space is distinct from body, and when I determine that motion is with respect to the parts of that space, and not with respect to the position of neighbouring bodies, lest this should be taken as being gratuitously contrary to the Cartesians, I shall venture to dispose of his fictions.

I can summarize his doctrine in the following three propositions:

(1) That from the truth of things only one particular motion fits each body (*Principia*, Part ii, Art. 28, 31, 32),[1] which is defined as being the translation of one part of matter or of one body from the neighbourhood of those bodies that immediately touch it, and which are regarded as being at rest, to the neighbourhood of others (*Principia*, Part ii, Art. 25; Part iii, Art. 28).[2]

(2) That by a body transferred in its particular motion according to this definition may be understood not only any particle of matter, or a body composed of parts relatively at rest, but all that is transferred at once, although this may, of course, consist of many parts which have different relative motions. (*Principia*, Part ii, Art. 25.)

(3) That besides this motion particular to each body there can arise in it innumerable other motions, through participation (or in so far as it is part of other bodies having other motions). (*Principia*, Part ii, Art. 31.) Which however are not motions in the philosophical sense and rationally speaking (Part iii, Art. 29),[3] and according to the truth of things (Part ii, Art. 25 and Part iii, Art. 28), but only improperly and according to common sense (Part ii, Art. 24, 25, 28, 31; Part iii, Art. 29).[4] That kind of motion he seems to describe (Part ii, Art. 24; Part iii, Art. 28) as the action by which any body migrates from one place to another.

[Part ii, Articles 13, 15, 29, 30][5]

And just as he formulates two types of motion, namely particular and derivative, so he assigns two types of place from which these motions proceed, and these are the surfaces of immediately surrounding bodies (Part ii, Art. 15),[6] and the position among any other bodies (Part ii, Art. 13; Part iii, Art. 29).[7]

Indeed, not only do its absurd consequences convince us how confused and incongruous with reason this doctrine is, but Descartes by contradicting himself seems to acknowledge the fact. For he says that speaking properly and according to philosophical sense the Earth and the other Planets do not move, and that he who declares it to be moved because of its translation with respect to the fixed stars speaks without reason and only in the vulgar fashion (Part iii, Art. 26, 27, 28, 29).[8] Yet later he attributes to the Earth and Planets a tendency to recede from the Sun as from a centre about which they are revolved, by which they are balanced at their [due] distances from the Sun by a similar tendency of the gyrating vortex (Part iii, Art. 140).[9] What then? Is this tendency to be derived from the (according to Descartes) true and philosophical rest of the planets, or rather from [their] common and non-philosophical motion? But Descartes says further that a Comet has a lesser tendency to recede from the Sun when it first enters the vortex, and keeping practically the same position among the fixed stars does not yet obey the impetus of the vortex, but with respect to it is transferred from the neighbourhood of the contiguous aether and so philosophically speaking whirls round the Sun, while afterwards the matter of the vortex carries the comet along with it and so renders it at rest, according to philosophical sense. (Part III, Art. 119, 120.)[10] The philosopher is hardly consistent who uses as the basis of Philosophy the motion of the vulgar which he had rejected a little before, and now rejects that motion as fit for nothing which alone was formerly said to be true and philosophical, according to the nature of things. And since the whirling of the comet around the Sun in his philosophic sense does not cause a tendency to recede from the centre, which a gyration in the vulgar sense can do, surely motion in the vulgar sense should be acknowledged, rather than the philosophical.

Secondly, he seems to contradict himself when he postulates that to each body corresponds a single motion, according to the nature of things; and yet he asserts that motion to be a product of our imagination, defining it as translation from the neighbourhood of bodies which are not at rest but only seem to be at rest even though they may instead be moving, as is more fully explained in Part II, Art. 29, 30.[11] And thus he thinks to avoid the difficulties concerning the mutual translation of bodies, namely, why one body is said to move rather than another, and why a boat on a flowing stream is said to be at rest when it does not change its position with respect to the banks (Part II, Art. 15). But that the contradiction may be evident, imagine that someone sees the matter of the vortex as at rest, and that the Earth, philosophically speaking, is at rest at the same time; imagine also that someone else simultaneously sees the same matter of the vortex as moving in a circle, and that the Earth, philosophically speaking, is not at rest. In the same way, a ship at sea will at the same time move and not move, and that without taking *motion* in the looser vulgar sense, by which there are innumerable motions for each body, but in his philosophical sense according to which, he says, there is but one in each body, and that one peculiar to it and corresponding to the nature of things and not to our imagination.

[Part II Article 31]

Thirdly, he seems hardly consistent when he supposes that a single motion corresponds to each body according to the truth of things, and yet (Part II, Art. 31) that there really are innumerable motions in each body. For the motions that really are in any body, are really natural motions, and thus motions in the philosophical sense and according to the truth of things, even though he contends that they are motions in the vulgar sense only. Add that when a whole thing moves, all the parts which constitute the whole and are translated together are really at rest; unless indeed it is conceded that they move by participating in the motion of the whole, and then indeed they have innumerable motions according to the truth of things.

But besides this, from its consequences we may see how absurd is this doctrine of Descartes. And first, just as he contends with heat that the Earth does not move because it is not translated from the neighbourhood of the contiguous aether, so from the same principles it follows that the internal particles of hard bodies, while they are not translated from the neighbourhood of immediately contiguous particles, do not have motion in the strict sense, but move only by participating in the motion of the external particles: it rather appears that the interior parts of the external particles do not move with their own motion because they are not translated from the neighbourhood of the internal parts: and thus that only the external surface of each body moves with its own motion and that the whole internal substance, that is the whole of the body, moves through participation in the motion of the external surface.[12] The fundamental definition of motion errs, therefore, that attributes to bodies that which only belongs to surfaces, and which denies that there can be any body at all which has a motion peculiar to itself.

[Part II Article 25]

Secondly, if we regard only Art. 25 of Part II, each body has not merely a unique proper motion but innumerable ones, provided that those things are said to be moved properly and according to the truth of things of which the whole is properly moved. And that is because he understands by the body whose motion he defines all that which is translated together, and yet this may consist of parts having other motions among themselves: suppose a vortex together with all the Planets, or a ship along with everything within it floating in the sea, or a man walking in a ship together with the things he carries with him, or the wheel of a clock together with its constituent metallic particles. For unless you say that the motion of the whole aggregate cannot be considered as proper motion and as belonging to the parts according to the truth of things, it will have to be admitted that all these motions of the wheels of the clock, of the man, of the ship, and of the vortex are truly and philosophically speaking in the particles of the wheels.

From both of these consequences it appears further that no one motion can be said to be true, absolute and proper in preference to others, but that all, whether with respect to contiguous bodies or remote ones, are equally philosophical—than which nothing more absurd can be imagined. For unless it is conceded that there can be a single physical motion of any body, and that the rest of its changes of relation and position with respect to other bodies are so many external designations, it follows that the Earth (for example) endeavours to recede from the centre of the Sun on account of a motion relative to the fixed stars, and endeavours the less to recede on account of a lesser motion relative to Saturn and the aetherial orb in which it is carried, and still less relative to Jupiter and the swirling aether which occasions its orbit, and also less relative to Mars and its aetherial orb, and much less relative to other orbs of aetherial matter which, although not bearing planets, are closer to the annual orbit of the Earth; and indeed relative to its own orb it has no endeavour, because it does not move in it. Since all these endeavours and non-endeavours cannot absolutely agree, it is rather to be said that only the motion which causes the Earth to endeavour to recede from the Sun is to be declared the Earth's natural and absolute motion. Its translations relative to external bodies are but external designations.

Thirdly. It follows from the Cartesian doctrine that motion can be generated where there is no force acting. For example, if God should suddenly cause the spinning of our vortex to stop, without applying any force to the Earth which could stop it at the same time, Descartes would say that the Earth is moving in a philosophical sense (on account of its translation from the neighbourhood of the contiguous fluid), whereas before he said it was resting, in the same philosophical sense.

Fourthly. It also follows from the same doctrine that God himself could not generate motion in some bodies even though he impelled them with the greatest force. For example, if God urged the starry heaven together with all the most remote part of creation with any very great force so as to cause it to revolve about the Earth (suppose with a diurnal motion): yet from this, according to Descartes, the Earth alone and not the sky would be truly said to move (Part III, Art. 38).[13] As if it would be the same whether, with a tremendous force, He should cause the skies to turn from east to west, or with a small force turn the Earth in the opposite direction. But who will imagine that the parts of the Earth endeavour to recede from its centre on account of a force impressed only upon the heavens? Or is it not more agreeable to reason that when a force imparted to the heavens makes them endeavour to recede from the centre of the revolution thus caused, they are for that reason the sole bodies properly and absolutely moved; and that when a force impressed upon the Earth makes its parts endeavour to recede from the centre of revolution thus caused, for that reason it is the sole body properly and absolutely moved, although there is the same relative motion of the bodies in both cases. And thus physical and absolute motion is to be defined from other considerations than translation, such translation being designated as merely external.

Fifthly. It seems repugnant to reason that bodies should change their relative distances and positions without physical motion; but Descartes says that the Earth and the other Planets and the fixed stars are properly speaking at rest, and nevertheless they change their relative positions.[14]

Sixthly. And on the other hand it seems not less repugnant to reason that of several bodies maintaining the same relative positions some one should move physically, while others are at rest. But if God should cause any Planet to stand still and make it continually keep the same position with respect to the fixed stars, would not Descartes say that although the stars are not moving, the planet now moves physically on account of its translation from the matter of the vortex?

Seventhly. I ask for what reason any body is properly said to move when other bodies from whose neighbourhood it is transported are not seen to be at rest, or rather when they cannot be seen to be at rest. For example, in what way can our own vortex be said to move circularly on account of the translation of matter near the circumference, from the neighbourhood of similar matter in other surrounding vortices, since the matter of surrounding vortices cannot be seen to be at rest, and this not only with respect to our vortex, but also in so far as those vortices are not at rest among themselves. For if the Philosopher refers this translation not to the numerical corporeal particles of the vortices, but to the generic space (as he calls it) in which those vortices exist, at last we do agree, for he admits that motion ought to be referred to space in so far as it is distinguished from bodies.

Lastly, that the absurdity of this position may be disclosed in full measure, I say that thence it follows that a moving body has no determinate velocity and no definite line in which it moves. And, what is worse, that the velocity of a body moving without resistance cannot be said to be uniform, nor the line said to be straight in which its motion is accomplished. On the contrary, there cannot be motion since there can be no motion without a certain velocity and determination.

But that this may be clear, it is first of all to be shown that when a certain motion is finished it is impossible, according to Descartes, to assign a place in which the body was at the beginning of the motion; it cannot be said whence the body moved. And the reason is that according to Descartes the place cannot be defined or assigned except by the position of the surrounding bodies, and after the completion of a certain motion the position of the surrounding bodies no longer stays the same as it was before. For example, if the place of the planet Jupiter a year ago be sought, by what reason, I ask, can the Cartesian philosopher define it? Not by the positions of the particles of the fluid matter, for the positions of these particles have greatly changed since a year ago. Nor can he define it by the positions of the Sun and fixed stars. For the unequal influx of subtle matter through the poles of the vortices towards the central stars (Part III, Art. 104), the undulation (Art. 114), inflation (Art. 111) and absorption of the vortices, and other more true causes, such as the rotation of the Sun and stars around their own centres, the generation of spots, and the passage of comets through the heavens, change both the magnitude and positions of the stars so much that perhaps they are only adequate to designate the place sought with an error of several miles; and still less can the place be accurately defined and determined by their help, as a Geometer would require.[15] Truly there are no bodies in the world whose relative positions remain unchanged with the passage of time, and certainly none which do not move in the Cartesian sense: that is, which are neither transported from the vicinity of contiguous bodies nor are parts of other bodies so transferred.[16] And thus there is no basis from which we can at the present pick out a place which was in the past, or say that such a place is any longer discoverable in nature. For since, according to Descartes, place is nothing but the surface of surrounding bodies or position among some other more distant bodies, it is impossible (according to his doctrine) that it should exist in nature any longer than those bodies maintain the same positions from which he takes the individual designation. And so, reasoning as in the question of Jupiter's position a year ago, it is clear that if one follows Cartesian doctrine, not even God himself could define the past position of any moving body accurately and geometrically now that a fresh state of things prevails, since in fact, due to the changed positions of the bodies, the place does not exist in nature any longer.

[Part II, Articles 10, 12, 18]

Now as it is impossible to pick out the place in which a motion began (that is, the beginning of the space passed over), for this place no longer exists after the motion is completed, so the space passed over, having no beginning, can have no length; and hence, since velocity depends upon the distance passed over in a given time, it follows that the moving body can have no velocity, just as I wished to prove at first. Moreover, what was said of the beginning of the space passed over should be applied to all intermediate points too; and thus as the space has no beginning nor intermediate parts it follows that there was no space passed over and thus no determinate motion, which was my second point. It follows indubitably that Cartesian motion is not motion, for it has no velocity, no definition, and there is no space or distance traversed by it. So it is necessary that the definition of places, and hence of local motion, be referred to some motionless thing such as extension alone or space in so far as it is seen to be truly distinct from bodies. And this the Cartesian philosopher may the more willingly allow, if only he notices that Descartes himself had an idea of extension as distinct from bodies, which he wished to distinguish from corporeal extension by calling it generic (*Principia*, Part II, Art. 10, 12, 18).[17] And also that the rotations of the vortices, from which he deduced the force of the aether in receding from their centres and thus the whole of his mechanical philosophy, are tacitly referred to generic extension.

[Part II, Articles 4, 11]

In addition, as Descartes in Part II, Art. 4 and 11 seems to have demonstrated that body does not differ at all from extension, abstracting hardness, colour, weight, cold, heat and the remaining qualities which body can lack, so that at last there remains only its extension in length, width and depth which hence alone appertain to its essence;[18] and as this has been taken as proved by many, and is in my view the only reason for having confidence in this opinion; lest any doubt should remain about the nature of motion, I shall reply to this argument by explaining what extension and body are, and how they differ from each other. For since the distinction of substances into thinking and extended [entities], or rather, into thoughts and extensions, is the principal foundation of Cartesian philosophy, which he contends to be even better known than mathematical demonstrations: I consider it most important to overthrow [that philosophy] as regards extension, in order to lay truer foundations of the mechanical sciences.

Perhaps now it may be expected that I should define extension as substance or accident or else nothing at all. But by no means, for it has its own manner of existence which fits neither substances nor accidents. It is not substance; on the one hand, because it is not absolute in itself, but is as it were an emanent effect of God, or a disposition of all being; on the other hand, because it is not among the proper dispositions that denote substance, namely actions, such as thoughts in the mind and motions in body. For although philosophers do not define substance as an entity that can act upon things, yet all tacitly understand this of substances, as follows from the fact that they would readily allow extension to be substance in the manner of body if only it were capable of motion and of sharing in the actions of body. And on the contrary they would hardly allow that body is substance if it could not move nor excite in the mind any sensation or perception whatever. Moreover, since we can clearly conceive extension existing without any subject, as when we may imagine spaces outside the world or places empty of body, and we believe [extension] to exist wherever we imagine there are no bodies, and we cannot believe that it would perish with the body if God should annihilate a body, it follows that [extension] does not exist as an accident inherent in some subject. And hence it is not an accident. And much less may it be said to be nothing, since it is rather something, than an accident, and approaches more nearly to the nature of substance. There is no idea of nothing, nor has nothing any properties, but we have an exceptionally clear idea of extension, abstracting the dispositions and properties of a body so that there remains only the uniform and unlimited stretching out of space in length, breadth and depth. And furthermore, many of its properties are associated with this idea; these I shall now enumerate not only to show that it is something, but what it is.

1. In all directions, space can be distinguished into parts whose common limits we usually call surfaces; and these surfaces can be distinguished in all directions into parts whose common limits we usually call lines; and

again these lines can be distinguished in all directions into parts which we call points. And hence surfaces do not have depth, nor lines breadth, nor points dimension, unless you say that coterminous spaces penetrate each other as far as the depth of the surface between them, namely what I have said to be the boundary of both or the common limit; and the same applies to lines and points. Furthermore spaces are everywhere contiguous to spaces, and extension is everywhere placed next to extension, and so there are everywhere common boundaries to contiguous parts; that is, there are everywhere surfaces acting as a boundary to solids on this side and that; and everywhere lines in which parts of the surfaces touch each other; and everywhere points in which the continuous parts of lines are joined together. And hence there are everywhere all kinds of figures, everywhere spheres, cubes, triangles, straight lines, everywhere circular, elliptical, parabolical and all other kinds of figures, and those of all shapes and sizes, even though they are not disclosed to sight. For the material delineation of any figure is not a new production of that figure with respect to space, but only a corporeal representation of it, so that what was formerly insensible in space now appears to the senses to exist. For thus we believe all those spaces to be spherical through which any sphere ever passes, being progressively moved from moment to moment, even though a sensible trace of the sphere no longer remains there. We firmly believe that the space was spherical before the sphere occupied it, so that it could contain the sphere; and hence as there are everywhere spaces that can adequately contain any material sphere, it is clear that space is everywhere spherical. And so of other figures. In the same way we see no material shapes in clear water, yet there are many in it which merely introducing some colour into its parts will cause to appear in many ways. However, if the colour were introduced, it would not constitute material shapes but only cause them to be visible.[19]

2. Space extends infinitely in all directions. For we cannot imagine any limit anywhere without at the same time imagining that there is space beyond it. And hence all straight lines, paraboloids, hyperboloids, and all cones and cylinders and other figures of the same kind continue to infinity and are bounded nowhere, even though they are crossed here and there by lines and surfaces of all kinds extending transversely, and with them form segments of figures in all directions. You may have in truth an instance of infinity; imagine any triangle whose base and one side are at rest and the other side so turns about the contiguous end of its base in the plane of the triangle that the triangle is by degrees opened at the vertex; and meanwhile take a mental note of the point where the two sides meet, if they are produced that far: it is obvious that all these points are found on the straight line along which the fixed side lies, and that they become perpetually more distant as the moving side turns further until the two sides become parallel

and can no longer meet anywhere. Now, I ask, what was the distance of the last point where the sides met? It was certainly greater than any assignable distance, or rather none of the points was the last, and so the straight line in which all those meeting-points lie is in fact greater than finite. Nor can anyone say that this is infinite only in imagination, and not in fact; for if a triangle is actually drawn, its sides are always, in fact, directed towards some common point, where both would meet if produced, and therefore there is always such an actual point where the produced sides would meet, although it may be imagined to fall outside the limits of the physical universe. And so the line traced by all these points will be real, though it extends beyond all distance.

If anyone now objects that we cannot imagine that there is infinite extension, I agree. But at the same time I contend that we can understand it. We can imagine a greater extension, and then a greater one, but we understand that there exists a greater extension than any we can imagine. And here, incidentally, the faculty of understanding is clearly distinguished from imagination.

Should it be further said that we do not understand what an infinite being is, save by negating the limitations of a finite being, and that this is a negative and faulty conception, I deny this. For the limit or boundary is the restriction or negation of greater reality or existence in the limited being, and the less we conceive any being to be constrained by limits, the more we observe something to be attributed to it, that is, the more positively we conceive it. And thus by negating all limits the conception becomes positive in the highest degree. 'End' [finis] is a word negative as to sense, and thus 'infinity' [not-end] as it is the negation of a negation (that is, of ends) will be a word positive in the highest degree with respect to our perception and comprehension, though it seems grammatically negative. Add that positive and finite quantities of many surfaces infinite in length are accurately known to Geometers. And so I can positively and accurately determine the solid quantities of many solids infinite in length and breadth and compare them to given finite solids. But this is irrelevant here.

[Part I, Art. 26, 27; Part II, Art. 34]

If Descartes now says that extension is not infinite but rather indefinite, he should be corrected by the grammarians. For the word 'indefinite' is never applied to that which actually is, but always relates to a future possibility signifying only something which is not yet determined and definite. Thus before God had decreed anything about the creation of the world (if there was ever a time when he had not), the quantity of matter, the number of the stars and all other things were indefinite; once the world was created they were defined. Thus matter is indefinitely divisible, but is always divided either finitely or infinitely (Part I, Art. 26; Part II, Art. 34).[20] Thus an indefinite line is one whose future length is still undetermined. And so an indefinite space is one whose future magnitude is not yet determined; for indeed that which actually is, is not to be defined, but does either have limits or not and so is either finite or infinite. Nor is it an objection that he takes space to be indefinite in relation to ourselves; that is, we simply do not know its limits and are not absolutely sure that there are none (Part I, Art. 27).[21] This is because although we are ignorant beings God at least understands that there are no limits not merely indefinitely but certainly and positively, and because although we negatively imagine it to transcend all limits, yet we positively and most certainly understand that it does so. But I see what Descartes feared, namely that if he should consider space Infinite, it would perhaps become God because of the perfection of infinity. But by no means, for infinity is not perfection except when it is an attribute of perfect things. Infinity of intellect, power, happiness and so forth is the height of perfection; but infinity of ignorance, impotence, wretchedness and so on is the height of imperfection; and infinity of extension is so far perfect as that which is extended.

3. The parts of space are motionless. If they moved, it would have to be said either that the motion of each part is a translation from the vicinity of other contiguous parts, as Descartes defined the motion of bodies; and that this is absurd has been sufficiently shown; or that it is a translation out of space into space, that is out of itself, unless perhaps it is said that two spaces everywhere coincide, a moving one and a motionless one. Moreover the immobility of space will be best exemplified by duration. For just as the parts of duration derive their individuality from their order, so that (for example) if yesterday could change places with today and become the later of the two, it would lose its individuality and would no longer be yesterday, but today; so the parts of space derive their character from their positions, so that if any two could change their positions, they would change their character at the same time and each would be converted numerically into the other. The parts of duration and space are only understood to be the same as they really are because of their mutual order and position; nor do they have any hint of individuality apart from that order and position which consequently cannot be altered.[22]

4. Space is a disposition of being *qua* being. No being exists or can exist which is not related to space in some way. God is everywhere, created minds are somewhere, and body is in the space that it occupies; and whatever is neither everywhere nor anywhere does not exist. And hence it follows that space is an effect arising from the first existence of being, because when any being is postulated, space is postulated. And the same may be asserted of duration: for certainly both are dispositions of being or attributes according to which we denominate quantitatively the presence and duration of any

existing individual thing. So the quantity of the existence of God was eternal, in relation to duration, and infinite in relation to the space in which he is present; and the quantity of the existence of a created thing was as great, in relation to duration, as the duration since the beginning of its existence, and in relation to the size of its presence as great as the space belonging to it.

Moreover, lest anyone should for this reason imagine God to be like a body, extended and made of divisible parts, it should be known that spaces themselves are not actually divisible, and furthermore, that any being has a manner proper to itself of being in spaces. For thus there is a very different relationship between space and body, and space and duration. For we do not ascribe various durations to the different parts of space, but say that all endure together. The moment of duration is the same at Rome and at London, on the Earth and on the stars, and throughout all the heavens. And just as we understand any moment of duration to be diffused throughout all spaces, according to its kind, without any thought of its parts, so it is no more contradictory that Mind also, according to its kind, can be diffused through space without any thought of its parts.

5. The positions, distances and local motions of bodies are to be referred to the parts of space. And this appears from the properties of space enumerated as 1. and 4. above, and will be more manifest if you conceive that there are vacuities scattered between the particles, or if you pay heed to what I have formerly said about motion. To that it may be further added that in space there is no force of any kind which might impede or assist or in any way change the motions of bodies. And hence projectiles describe straight lines with a uniform motion unless they meet with an impediment from some other source. But more of this later.

6. Lastly, space is eternal in duration and immutable in nature, and this because it is the emanent effect of an eternal and immutable being. If ever space had not existed, God at that time would have been nowhere; and hence he either created space later (in which he was not himself), or else, which is not less repugnant to reason, he created his own ubiquity. Next, although we can possibly imagine that there is nothing in space, yet we cannot think that space does not exist, just as we cannot think that there is no duration, even though it would be possible to suppose that nothing whatever endures. This is manifest from the spaces beyond the world, which we must suppose to exist (since we imagine the world to be finite), although they are neither revealed to us by God, nor known from the senses, nor does their existence depend upon that of the spaces within the world. But it is usually believed that these spaces are nothing; yet indeed they are true spaces. Although space may be empty of body, nevertheless it is not in itself a void; and *something* is there, because spaces are there, although nothing

more than that. Yet in truth it must be acknowledged that space is no more space where the world is, than where no world is, unless perchance you say that when God created the world in this space he at the same time created space in itself, or that if God should annihilate the world in this space, he would also annihilate the space in it. Whatever has more reality in one space than in another space must belong to body rather than to space; the same thing will appear more clearly if we lay aside that puerile and jejune prejudice according to which extension is inherent in bodies like an accident in a subject without which it cannot actually exist.

Now that extension has been described, it remains to give an explanation of the nature of body. Of this, however, the explanation must be more uncertain, for it does not exist necessarily but by divine will, because it is hardly given to us to know the limits of the divine power, that is to say whether matter could be created in one way only, or whether there are several ways by which different beings similar to bodies could be produced. And although it scarcely seems credible that God could create beings similar to bodies which display all their actions and exhibit all their phenomena and yet are not in essential and metaphysical constitution bodies; as I have no clear and distinct perception of this matter I should not dare to affirm the contrary, and hence I am reluctant to say positively what the nature of bodies is, but I rather describe a certain kind of being similar in every way to bodies, and whose creation we cannot deny to be within the power of God, so that we can hardly say that it is not body.

Since each man is conscious that he can move his body at will, and believes further that all men enjoy the same power of similarly moving their bodies by thought alone; the free power of moving bodies at will can by no means be denied to God, whose faculty of thought is infinitely greater and more swift. And by like argument it must be agreed that God, by the sole action of thinking and willing, can prevent a body from penetrating any space defined by certain limits.

If he should exercise this power, and cause some space projecting above the Earth, like a mountain or any other body, to be impervious to bodies and thus stop or reflect light and all impinging things, it seems impossible that we should not consider this space to be truly body from the evidence of our senses (which constitute our sole judges in this matter); for it will be tangible on account of its impenetrability, and visible, opaque and coloured on account of the reflection of light, and it will resonate when struck because the adjacent air will be moved by the blow.

Thus we may imagine that there are empty spaces scattered through the world, one of which, defined by certain limits, happens by divine power to be impervious to bodies, and *ex hypothesi* it is manifest that this would resist the motions of bodies and perhaps reflect them, and assume all the properties of a corporeal particle, except that it will be motionless. If we may further imagine that that impenetrability is not always maintained in the same part of space but can be transferred hither and thither according to certain laws, yet so that the amount and shape of that impenetrable space are not changed, there will be no property of body which this does not possess. It would have shape, be tangible and mobile, and be capable of reflecting and being reflected, and no less constitute a part of the structure of things than any other corpuscle, and I do not see that it would not equally operate upon our minds and in turn be operated upon, because it is nothing more than the product of the divine mind realized in a definite quantity of space. For it is certain that God can stimulate our perception by his own will, and thence apply such power to the effects of his will.

In the same way if several spaces of this kind should be impervious to bodies and to each other, they would all sustain the vicissitudes of corpuscles and exhibit the same phenomena. And so if all this world were constituted of this kind of being, it would seem hardly any different. And hence these beings will either be bodies or like bodies. If they are bodies, then we can define bodies as *determined quantities of extension which omnipresent God endows with certain conditions.* These conditions are, (1) that they be mobile; and therefore I did not say that they are numerical parts of space which are absolutely immobile, but only definite quantities which may be transferred from space to space; (2) that two of this kind cannot coincide anywhere; that is, that they may be impenetrable, and hence that when their motions cause them to meet they stop and are reflected in accord with certain laws; (3) that they can excite various perceptions of the senses and the fancy in created minds, and conversely be moved by them, nor is it surprising since the description of the origin [of things?] is founded in this.

Moreover, it will help to note the following points respecting the matters already explained.

1. That for the existence of these beings it is not necessary that we suppose some unintelligible substance to exist in which as subject there may be an inherent substantial form; extension and an act of the divine will are enough. Extension takes the place of the substantial subject in which the form of the body is conserved by the divine will; and that product of the divine will is the form or formal reason of the body denoting every dimension of space in which the body is to be produced.

2. These beings will not be less real than bodies, nor (I say) are they less able to be called substances. For whatever reality we attribute to bodies arises from their phenomena and sensible qualities. And hence we would judge these beings, since they can receive all qualities of this kind and can similarly exhibit all these phenomena, to be no less real, if they should exist.

Nor will they be less substance, since they will likewise subsist through God alone, and will acquire accidents.

3. Between extension and its impressed form there is almost the same analogy that the Aristotelians postulate between the *materia prima* and substantial forms, namely when they say that the same matter is capable of assuming all forms, and borrows the denomination of numerical body from its form. For so I suppose that any form may be transferred through any space, and everywhere denote the same body.

4. They differ, however, in that extension (since it is *what* and *how constituted* and *how much*) has more reality than *materia prima*, and also in that it can be understood, in the same way as the form that I assigned to bodies. For if there is any difficulty in this conception it is not in the form that God imparts to space, but in the manner by which he imparts it. But that is not to be regarded as a difficulty, since the same question arises with regard to the way we move our bodies, and nevertheless we do believe that we can move them. If that were known to us, by like reasoning we should also know how God can move bodies, and expel them from a certain space bounded in a given figure, and prevent the expelled bodies or any others from penetrating into it again, that is, cause that space to be impenetrable and assume the form of body.

5. Thus I have deduced a description of this corporeal nature from our faculty of moving our bodies, so that all the difficulties of the conception may at length be reduced to that; and further, so that God may appear (to our innermost consciousness) to have created the world solely by the act of will, just as we move our bodies by an act of will alone; and, besides, so that I might show that the analogy between the Divine faculties and our own is greater than has formerly been perceived by Philosophers. That we were created in God's image holy writ testifies. And his image would shine more clearly in us if only he simulated in the faculties granted to us the power of creation in the same degree as his other attributes; nor is it an objection that we ourselves are created beings and so a share of this attribute could not have been equally granted to us. For if for this reason the power of creating minds is not delineated in any faculty of created mind, nevertheless created mind (since it is the image of God) is of a far more noble nature than body, so that perhaps it may eminently contain [body] in itself. Moreover, in moving bodies we create nothing, nor can we create anything, but we only simulate the power of creation. For we cannot make any space impervious to bodies, but we only move bodies; and at that not any we choose, but only our own bodies, to which we are united not by our own will but by the divine constitution of things; nor can we move bodies in any way but only in accord with those laws which God has imposed on us. If anyone, however, prefers this our power to be called the finite and lowest level of the

power which makes God the Creator, this no more detracts from the divine power than it detracts from God's intellect that intellect in a finite degree belongs to us also; particularly since we do not move our bodies by our own independent power but through laws imposed on us by God. Rather, if any think it possible that God may produce some intellectual creature so perfect that he could, by divine accord, in turn produce creatures of a lower order, this so far from detracting from the divine power enhances it; for that power which can bring forth creatures not only directly but through the mediation of other creatures is exceedingly, not to say infinitely, greater. And so some may perhaps prefer to suppose that God imposes on the soul of the world, created by him, the task of endowing definite spaces with the properties of bodies, rather than to believe that this function is directly discharged by God. Therefore the world should not be called the creature of that soul but of God alone, who creates it by constituting the soul of such a nature that the world necessarily emanates [from it]. But I do not see why God himself does not directly inform space with bodies; so long as we distinguish between the formal reason of bodies and the act of divine will. For it is contradictory that it [body] should be the act of willing or anything other than the effect which that act produces in space. Which effect does not even differ less from that act than Cartesian space, or the substance of body according to the vulgar idea; if only we suppose that they are created, that is, that they borrow existence from the will, or that they are creatures of the divine reason.

Lastly, the usefulness of the idea of body that I have described is brought out by the fact that it clearly involves the chief truths of metaphysics and thoroughly confirms and explains them. For we cannot postulate bodies of this kind without at the same time supposing that God exists, and has created bodies in empty space out of nothing, and that they are beings distinct from created minds, but able to combine with minds. Say, if you can, which of the views, already well-known, elucidates any one of these truths or rather is not opposed to all of them, and obscures all of them. If we say with Descartes that extension is body, do we not manifestly offer a path to Atheism, both because extension is not created but has existed eternally, and because we have an absolute idea of it without any relationship to God, and so in some circumstances it would be possible for us to conceive of extension while imagining the non-existence of God? Nor is the distinction between mind and body in this philosophy intelligible, unless at the same time we say that mind has no extension at all, and so is not substantially present in any extension, that is, exists nowhere; which seems the same as denying the existence of mind, or at least renders its union with body totally unintelligible, not to say impossible. Moreover, if the distinction of substances between *thinking* and *extended* is legitimate and complete,

God does not eminently contain extension within himself and therefore cannot create it; but God and extension will be two substances separately complete, absolute, and having the same significance. But on the contrary if extension is eminently contained in God, or the highest thinking being, certainly the idea of extension will be eminently contained within the idea of thinking, and hence the distinction between these ideas will not be so great but that both may fit the same created substance, that is, but that a body may think, and a thinking being extend. But if we adopt the vulgar notion (or rather lack of it) of body, according to which there resides in bodies a certain unintelligible reality that they call substance, in which all the qualities of the bodies are inherent, this (apart from its unintelligibility) is exposed to the same inconveniences as the Cartesian view. For as it cannot be understood, it is impossible that its distinction from the substance of the mind should be understood. For the distinction drawn from substantial form or the attributes of substances is not enough: if bare substances do not have an essential difference, the same substantial forms or attributes can fit both, and render them by turns, if not at one and the same time, mind and body. And so if we do not understand that difference of substances deprived of attributes, we cannot knowingly assert that mind and body differ substantially. Or if they do differ, we cannot discover any basis for their union. Further, they attribute no less reality in concept (though less in words) to this corporeal substance regarded as being without qualities and forms, than they do to the substance of God, abstracted from his attributes. They conceive of both, when considered simply, in the same way; or rather they do not conceive of them, but confound them in some common idea of an unintelligible reality. And hence it is not surprising that Atheists arise ascribing that to corporeal substances which solely belongs to the divine. Indeed, however we cast about we find almost no other reason for atheism than this notion of bodies having, as it were, a complete, absolute and independent reality in themselves, such as almost all of us, through negligence, are accustomed to have in our minds from childhood (unless I am mistaken), so that it is only verbally that we call bodies created and dependent. And I believe that this preconceived idea explains why the same word, substance, is applied in the schools to God and to his creatures; and also why in forming an idea of body philosophers are brought to a stand and lose their drift, as when they try to form an independent idea of a thing dependent upon God. For certainly whatever cannot exist independently of God cannot be truly understood independently of the idea of God. God is no less present in his creatures than they are present in the accidents, so that created substance, whether you consider its degree of dependence or its degree of reality, is of an intermediate nature between God and accident. And hence the idea of it no less involves the concept of God than the idea of accident involves the concept of created substance. And so it ought to embrace no other reality in itself than a derivative and incomplete reality. Thus the preconception just mentioned must be laid aside, and substantial reality is rather to be ascribed to these kinds of attributes which are real and intelligible things in themselves and do not need to be inherent in a subject, than to the subject which we cannot conceive as dependent, much less form any idea of it. And this we can manage without difficulty if (besides the idea of body expounded above) we reflect that we can conceive of space existing without any subject when we think of a vacuum. And hence some substantial reality fits this. But if moreover the mobility of the parts (as Descartes imagined) should be involved in the idea of vacuum, everyone would freely concede that it is corporeal substance. In the same way, if we should have an idea of that attribute or power by which God, through the sole action of his will, can create beings, we should readily conceive of that attribute as subsisting by itself without any substantial subject and [thus as] involving the rest of his attributes. But while we cannot form an idea of this attribute nor even of our own power by which we move our bodies, it would be rash to say what may be the substantial basis of mind.

So much for the nature of bodies, in making which plain I judge that I have sufficiently proved that such a creation as I have expounded is most clearly the work of God; and that if this world were not constituted from that creation, at least another very like it could be constituted. And since there is no difference between the materials as regards their properties and nature, but only in the method by which God created one and the other, the distinction between body and extension is certainly brought to light from this. Because extension is eternal, infinite, uncreated, uniform throughout, not in the least mobile, nor capable of inducing change of motion in bodies or change of thought in the mind; whereas body is opposite in every respect, at least if God did not please to create it always and everywhere. For I should not dare to deny God that power. And if anyone thinks otherwise, let him say where he could have created the first matter, and whence the power of creating was granted to God. Or if there was no beginning to that power, but he had the same eternally that he has now, then he could have created from eternity. For it is the same to say that there never was in God an impotence to create, or that he always had the power to create and could have created, and that he could always create matter. In the same way, either a space may be assigned in which matter could not be created from, the beginning, or it must be conceded that God could have created it everywhere.

Moreover, so that I may respond more concisely to Descartes' argument: let us abstract from body (as he commands) gravity, hardness and all sensible qualities, so that nothing remains except what pertains to its essence. Will extension alone then remain? By no means. For we may also reject that faculty or power by which they [the qualities] stimulate the perceptions of thinking beings. For since there is so great a distinction between the ideas of thinking and of extension that it is impossible there should be any basis of connection or relation [between them] except that which is caused by divine power, the above faculty of bodies can be rejected without violating extension, but not without violating their corporeal nature. Clearly the changes which can be induced in bodies by natural causes are only accidental and they do not denote a true change of substance. But if any change is induced that transcends natural causes, it is more than accidental and radically affects the substance. And according to the sense of the demonstration, only those things are to be rejected which bodies can be deprived of, and made to lack, by the force of nature. But should anyone object that bodies not united to minds cannot directly arouse perceptions in minds, and that hence since there are bodies not united to minds, it follows that this power is not essential to them: it should be noticed that there is no question here of an actual union, but only of a faculty in bodies by which they are capable of a union through the forces of nature. From the fact that the parts of the brain, especially the more subtle ones to which the mind is united, are in a continual flux, new ones succeeding to those which fly away, it is manifest that that faculty is in all bodies. And, whether you consider divine action or corporeal nature, to remove this is no less than to remove that other faculty by which bodies are enabled to transfer mutual actions from one to another, that is, to reduce body into empty space.

However, as water offers less resistance to the motion of solid bodies through it than quicksilver does, and air much less than water, and aetherial spaces even less than air-filled ones, should we set aside altogether the force of resistance to the passage of bodies, we must also reject the corporeal nature [of the medium] utterly and completely. In the same way, if the subtle matter were deprived of all resistance to the motion of globules, I should no longer believe it to be subtle matter but a scattered vacuum. And so if there were any aerial or aetherial space of such a kind that it yielded without any resistance to the motions of comets or any other projectiles I should believe that it was utterly void. For it is impossible that a corporeal fluid should not impede the motion of bodies passing through it, assuming that (as I supposed before) it is not disposed to move at the same speed as the body (Part II, Epistle 96 to Mersenne).[23]

However it is manifest that all this force can be removed from space only if space and body differ from one another; and thence that they can exist apart is not to be denied before it has been proved that they do not differ, lest a mistake be made by *petitio principii*.

But lest any doubt remain, it should be observed from what was said earlier that there are empty spaces in the natural world. For if the aether were a corporeal fluid entirely without vacuous pores, however subtle its parts are made by division, it would be as dense as any other fluid, and it would yield to the motion of bodies through it with no less sluggishness; indeed with a much greater, if the projectile should be porous, because then the aether would enter its internal pores, and encounter and resist not only the whole of its external surface but also the surfaces of all the internal parts. Since the resistance of the aether is on the contrary so small when compared with the resistance of quicksilver as to be over ten or a hundred thousand times less, there is all the more reason for thinking that by far the largest part of the aetherial space is void, scattered between the aetherial particles. The same may also be conjectured from the various gravities of these fluids, for the descent of heavy bodies and the oscillations of pendulums show that these are in proportion to their densities, or as the quantities of matter contained in equal spaces. But this is not the place to go into this.

Thus you see how fallacious and unsound this Cartesian argument is, for when the accidents of bodies have been rejected, there remains not extension alone, as he imagined, but also the faculties by which they can stimulate perceptions in the mind and move other bodies. If we further reject these faculties and all power of moving so that there only remains a precise conception of uniform space, will Descartes fabricate any vortices, any world, from this extension? Surely not, unless he first invokes God, who alone can create bodies *de novo* in those spaces (by restoring those faculties, or the corporeal nature, as I explained above). And so in what has gone before I was correct in assigning the corporeal nature to the faculties already enumerated.

And thus at length since spaces are not the very bodies themselves but are only the places in which bodies exist and move, I think that what I laid down about local motion is sufficiently confirmed. Nor do I see what more could be desired in this respect unless perhaps I warn those to whom this is not satisfactory, that by the space whose parts I have defined as places filled by bodies, they should understand the Cartesian generic space in which spaces regarded singularly, or Cartesian bodies, are moved, and so they will find hardly anything to object to in our definitions.

I have already digressed enough; let us return to the main theme.

Definition 5. Force is the causal principle of motion and rest. And it is either an external one that generates or destroys or otherwise changes impressed motion in some body; or it is an internal principle by which existing motion or rest is conserved in a body, and by which any being endeavours to continue in its state and opposes resistance.

Definition 6. Conatus [endeavour] is resisted force, or force in so far as it is resisted.

Definition 7. Impetus is force in so far as it is impressed on a thing.

Definition 8. Inertia is force within a body, lest its state should be easily changed by an external exciting force.

Definition 9. Pressure is the endeavour of contiguous parts to penetrate into each others' dimensions. For if they could penetrate the pressure would cease. And pressure is only between contiguous parts, which in turn press upon others contiguous to them, until the pressure is transmitted to the most remote parts of any body, whether hard, soft or fluid. And upon this action is based the communication of motion by means of a point or surface of contact.

Definition 10. Gravity is a force in a body impelling it to descend. Here, however, by descent is not only meant a motion towards the centre of the Earth but also towards any point or region, or even from any point. In this way if the *conatus* of the aether whirling about the Sun to recede from its centre be taken for gravity, the aether in receding from the Sun could be said to descend. And so by analogy, the plane is called horizontal that is directly opposed to the direction of gravity or *conatus.*

Moreover, the quantity of these powers, namely motion, force, *conatus,* impetus, inertia, pressure and gravity may be reckoned in a double way: that is, according to either intension or extension.

Definition 11. The intension of any of the above-mentioned powers is the degree of its quality.

Definition 12. Its extension is the amount of space or time in which it operates.

Definition 13. Its absolute quantity is the product of its intension and its extension. So, if the quantity of intension is 2, and the quantity of extension 3, multiply the two together and you will have the absolute quantity 6.

Moreover, it will help to illustrate these definitions from individual powers. And thus motion is either more intense or more remiss, as the space traversed in the same time is greater or less, for which reason a body is usually said to move more swiftly or more slowly. Again, motion is more or less in extension as the body moved is greater or less, or as it is acting in a larger or smaller body. And the absolute quantity of motion is composed of both the velocity and the magnitude of the moving body. So force, *conatus,* impetus or inertia are more intense as they are greater in the same or an equivalent body: they have more extension when the body is larger, and their absolute quantity arises from both. So the intension of pressure is proportional to the increase of pressure upon the surface-area; its extension proportional to the surface pressed. And the absolute quantity results from the intension of the pressure and the area of the surface pressed. So, lastly, the intension of gravity is proportional to the specific gravity of the body; its extension is proportional to the size of the heavy body, and absolutely speaking the quantity of gravity is the product of the specific gravity and mass of the gravitating body. And whoever fails to distinguish these clearly, necessarily falls into many errors concerning the mechanical sciences.

In addition the quantity of these powers may sometimes be reckoned from the period of duration; for which reason there will be an absolute quantity which will be the product of intension, extension and duration. In this way if a body [of size] 2 is moved with a velocity 3 for a time 4 the whole motion will be $2 \times 3 \times 4$ or 12 [*sic*].

Definition 14. Velocity is the intension of motion, slowness is remission.

Definition 15. Bodies are denser when their inertia is more intense, and rarer when it is more remiss.

The rest of the above-mentioned powers have no names.

It is however to be noted that if, with Descartes or Epicurus, we suppose rarefaction and condensation to be accomplished in the manner of relaxed or compressed sponges, that is, by the dilation and contraction of pores which are either filled with some most subtle matter or empty of matter, then we ought to estimate the size of the whole body from the quantity of both its parts and its pores in Definition 15; so that one may consider inertia to be remitted by the increase of the pores and intensified by their diminution, as though the pores, which offer no inertial resistance to change, and whose mixtures with the truly corporeal parts give rise to all the various degrees of inertia, bear some ratio to the parts.

But in order that you may conceive of this composite body as a uniform one, imagine its parts to be infinitely divided and dispersed everywhere throughout the pores, so that in the whole composite body there is not the least particle of extension without an absolutely perfect mixture of parts and pores thus infinitely divided. Certainly it suits mathematicians to contemplate things in the light of such reasoning, or if you prefer in the Peripatetic manner; but in physics things seem otherwise.

Definition 16. An elastic body is one that can be condensed by pressure or compressed within narrower limits; and a non-elastic body is one that cannot be condensed by that force.

Definition 17. A hard body is one whose parts do not yield to pressure.

Definition 18. A fluid body is one whose parts yield to an overwhelming pressure.[24] Moreover, the pressures by which the fluid is driven in any direction whatsoever (whether these are exerted on the external surface alone, or on the internal parts by the action of gravity or any other cause), are said to be balanced when the fluid rests in equilibrium. This is asserted on the assumption that the pressure is exerted in some one direction and not towards all at once.

Definition 19. The limits defining the surface of the body (such as wood or glass) containing the fluid, or defining the surface of the external part of the same fluid containing some internal part, constitute the *vessel of fluid* [*vas fluidi*].

In these definitions, however, I refer only to absolutely hard or fluid bodies, for one cannot ratiocinate mathematically concerning ones partially so, on account of the innumerable circumstances affecting the figures, motions and contexture of the least particles. Thus I imagine that a fluid does not consist of hard particles, but that it is of such a kind that it has no small portion or particle which is not likewise fluid. And moreover, since the physical cause of fluidity is not to be examined here, I define the parts not as being in motion among themselves, but only as capable of motion, that is, as being everywhere so divided one from another that, although they may be supposed to be in contact and at rest with respect to one another, yet they do not cohere as though stuck together, but can be moved separately by any impressed force and can change the state of rest as easily as the state of motion if they move relatively. Indeed, I suppose that the parts of hard bodies do not merely touch each other and remain at relative rest, but that they do besides so strongly and firmly cohere, and are so bound together, as it were by glue, that no one of them can be moved without all the rest being drawn along with it; or rather that a hard body is not made up of conglomerate parts but is a single undivided and uniform body which preserves its shape most resolutely, whereas a fluid body is uniformly divided at all points.

And thus I have accommodated these definitions not to physical things but to mathematical reasoning, after the manner of the Geometers who do not accommodate their definitions of figures to the irregularities of physical bodies. And just as the dimensions of physical bodies are best determined from their geometry (as the measurement of a field from plane geometry, although a field is not a true plane; and the measurement of the Earth from the doctrine of the sphere even though the Earth is not precisely spherical) so the properties of physical fluids and solids are best known from this mathematical doctrine, even though they are not perhaps absolutely nor uniformly fluid or solid as I have defined them here. [. . .]

Notes

(*By A. Rupert Hall and Marie Boas Hall.*)

1. Descartes, *Principia Philosophia.* ii, 28: Motion, properly considered, cannot be referred to anything but the bodies contiguous to the moving one. ii, 31: How in the same body there can be innumerable diverse motions. ii, 32: How also motion properly considered, which is unique for each body, can be regarded as multiple.

2. ii, 25: What motion, properly considered, is. iii, 28: That the Earth, properly speaking, does not move, nor any of the Planets, although they are carried by the heaven.

3. iii, 29: That no motion is to be attributed to the Earth, although it is improperly taken as motion according to the vulgar usage, but then it is rightly said that the other planets move.

4. II, 24: What motion is, according to the vulgar understanding.

5. [These additions in square brackets are insertions by the editor to facilitate comparisons with Descartes's *Principles*, not features of the original text.]

6. II, 15: How external place is rightly considered as the surface surrounding the body.

7. II, 13: What external place is.

8. III, 26: That the Earth is at rest in its heaven, but nevertheless is borne by it. III, 27: That the same is to be understood of all the Planets.

9. III, 140: On the beginning of the motions of the Planets.

10. III, 119: How a fixed star is altered into a Comet or Planet, III, 120: How such a star is borne along when first it ceases to be fixed.

11. II, 29: Nor can it [motion] be referred to anything but to those contiguous bodies which appear to be at rest. II, 30: Why one of two contiguous bodies, which are separated from each other, is said to move rather than the other.

12. Compare with this whole passage the paragraph in the Scholium on space and time (*Principia*, 8), beginning 'Motus proprietas est, . . .'.

13. That according to Tycho's hypothesis, the Earth is said to move around its own centre.

14. See, e.g., Part III, §28 and §29.

15. III, 104: Why certain fixed stars disappear, or appear unexpectedly. III, 114: That the same star can alternately appear and disappear, III, 111: A description of the unexpected appearance of a star.

16. Cf. Principia, 7: 'Fieri enim potest ut nullum revera quiescat corpus, ad quod loca motusque referantur.'

17. II, 10: What space, or internal place is. II, 12: How it differs from [corporeal substance] in the way in which it is conceived, II, 18: How opinion about the vacuum, absolutely considered, is to be emended.

18. II, 4: That the nature of body does not consist of weight, hardness, colour or the like, but of extension alone. II, 11: How [space] does not differ in itself from corporeal substance.

19. Presumably Newton is thinking of dropping dye into water, which disturbances in the water cause to spread into swirls and 'shapes'; the boundaries to these disturbances exist before the dye reveals them.

20. I, 26: Infinity should never be discussed; those things in which we discern no limits should be taken as indefinite only. Such are the extension of the world, the divisibility of the parts of matter, the number of the stars, etc. III, 34: Hence it follows that the division of matter into particles is truly indefinite, although they are imperceptible to us.

21. What the difference between indefinite and infinite is.

22. Cf. Principia, 7: 'Ut partium Temporis ordo est immutabilis. . . .'

23. 'When I imagine that a body moves in a totally non-resistant medium, what I suppose is that all the particles of the fluid body which surround it have a tendency to move at precisely the same speed as it is doing, and no greater speed, whether yielding to it the place they occupied or going into that which it leaves; and thus there are no fluids which do not resist certain motions. But to suppose a kind of matter which does not at all resist the different motions of some body, one must imagine that God or an angel excites its particles more or less, so that the body which they surround moves more or less quickly.'

24. Newton originally wrote 'all whose parts are mobile among themselves', but crossed out these words.

CHAPTER 21

On the Divisibility and Subtlety of Matter

ÉMILIE DU CHÂTELET

Translation for this volume by Lydia Patton[1] of pages 179–200 (Chapter 9) of *Foundations of Physics*. Paris: Chez Prault Fils. Original publication date 1750, original title *Institutions de Physique*. Most remaining chapters of *Foundations of Physics* are translated in Du Châtelet 2009/1750.

§165 Extension can be conceived in terms of length, width, and depth; thus, the line AB is extended in length, the surface ABDE is extended in length and in width, and the cube ABCDEFGH is extended in length, width, and depth: these are the three dimensions of extension. (See Figures 7 and 10 below.)

§166 Every body has these three dimensions, or to speak with exactness, there are nothing but solids in nature. But since our mind[2] has the power to make abstractions, we can consider length without taking note of width nor depth, similarly, we can consider length and width only, without taking note of depth, and geometry is founded on these abstractions of our mind. Surfaces, lines, and points thus are not matter, but one conceives them in matter by abstraction.[3]

§167 However, to assist the imagination, and to form for oneself a distinct idea of the three dimensions of extension, one can imagine two points A and B at a distance from each other, and one can imagine that if the point A, in travelling to meet the point B, leaves a production of itself in each part of the interval that separates the points, it will form the line AB, which one supposes to be extended only in length.

Figure 21.7 (redrawn)

One can suppose further that the line AB, flowing[4] the length of the line AD, leaves a production of itself in the entire path that it traverses to travel from the point A to the point D, it will form the surface ABDE, which one supposes to be extended in length and in width.

Finally, if the surface ABCDE flows along the surface CDEF they will form the cube ABCDEFGH which has the three dimensions of nature, since it is extended in length, width, and depth.

§168 Most philosophers,[5] having confused the abstractions of our mind with physical body, have wanted to demonstrate the divisibility of matter to infinity by means of the reasoning of the geometers on the divisibility of lines that one pushes to infinity. This has given rise to the famous labyrinth of the divisibility of the continuum that has so embarrassed the philosophers. But they could have avoided all the difficulties that this divisibility involves if they had taken care never to apply the reasoning that one applies to the divisibility of the geometrical body to natural and physical bodies.

§169 The geometrical body is nothing but simple extension. The geometrical body has no determinate and actual parts, it contains nothing but simply possible parts, which one can increase to infinity as one wills; because the notion of extension contains nothing but parts coexisting and united, and the number of these parts is absolutely indeterminate, and does not enter into the notion of extension. Thus, one can determine this number as one wills, without harming the extension. That is to say that one

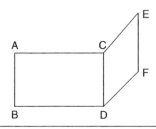

Figure 21.10 (redrawn)

can establish that an extension contains ten thousand, or a million, or ten million parts, and so on, depending on how one wishes to define some part or another as *one*. Thus, a line contains two parts, if one takes its half to be *one*, and it has either ten, or a thousand, if one takes the tenth or thousandth part as the unit. Hence, this unit is absolutely indeterminate, and depends on the will of whomever considers this extension.

§170 Each abstract and geometrical extension can thus be expressed by some number or other, but things are quite different in nature. All that exists naturally must be determined in every way, and it is not in our power to determine it otherwise.[6] A watch has its parts, for example, but these are not parts that are determinable merely by the imagination, these are real, actually existing parts. It is not open to say, *This watch has ten, one hundred, or a million parts*; for, as a watch, it has one number that constitutes its essence, and it cannot have more or fewer parts and remain a watch. It is the same with all the natural bodies, they are all machines which have determinate and dissimilar parts, which one is not permitted to express with just any number.

§171 It is in confusing geometrical extension with physical extension, and in supposing that physical extension is composed of an infinity of extended parts, that the Ancients formed those false and specious arguments against the possibility of movement.

The most famous of all the paradoxes[7] was the one that Zeno had called the *Achilles*, in recognition of Achilles' invincible force. He proposed Achilles running toward a tortoise, and, as Achilles ran ten times faster than the tortoise, he gave the tortoise a head start of one league, and he reasoned as follows: while Achilles traveled one tenth of a league, the tortoise would travel one hundredth of the league; thus, from tenth to tenth, the tortoise would always remain in front of Achilles, who would never catch up with it.[8]

First, if it were true that Achilles would never catch the tortoise, it would not follow from that that movement would be impossible. Since Achilles and the tortoise really move, Achilles always approaches the tortoise, who is assumed to always be ahead of him, albeit infinitely little.

But, secondly, since this ingenious paradox is founded on the divisibility of extension to infinity, the principle of sufficient reason provides a sufficient refutation. For it is proven by means of this principle that physical extension ultimately is composed of simple beings, and that consequently its divisions, even the possible ones, have positive and real limits.[9]

Entire treatises have been written to resolve Zeno's paradox. To refute it, perhaps it suffices to walk in his presence as Diogenes did; but failing such a de facto response, one can see whether it is easy to find one de jure.

Grégoire de Saint Vincent was the first to demonstrate its falsity, and who assigned the precise point at which Achilles would have to reach the tortoise, and this point was found, by means of infinite geometrical progressions, to be one and one ninth leagues. The sum of any infinite geometrical progression is finite, and this is because infinite being, and infinite extension, are two very different things. For any finitude whatsoever, a foot, for example, is a composite of finite and infinite: the foot is finite, insofar as it contains nothing but a certain number of simple beings, but I can suppose it to be divided into an infinity, or rather into a non-finite number[10] of parts, considering this foot as an abstract extension. Thus, if I were to consider half of the foot first, then half of what remains or a quarter of a foot, then half of that or an eighth of a foot, I would proceed thus, mentally, to infinity; always taking new and decreasing halves. These, all together, would never make up that foot, which would then become a geometrical body, because of all its properties I would only retain in my mind that of the extension on which my ideal division is performed.

Thus, the divisibility of extension to infinity is at the same time a geometrical truth and a physical error. And, thus, all the reasoning on the divisibility of matter drawn from the nature of asymptotes, from the incommensurability of the diagonal of a square, from infinite series, and from other geometrical considerations, are absolutely inapplicable to natural bodies, as are the theorems of M. Keill, by which he claims to prove that one could fill the entire universe with a grain of sand.[11] For one must not admit anything into physics besides actual parts, whose existence can be demonstrated by experience or by rigorous demonstration.

§172 One has seen above[12] that indivisible atoms, or parts, of matter are inadmissible, if one considers them as simple, irresolvable and primitive matters, because one cannot give a sufficient reason for their existence. But as long as one recognizes that they derive their origin from simple beings, one certainly can admit them. For it is very possible, and experience renders it very likely, that there is a certain determinate number of parts of matter in the universe which nature never resolves into their principle, which remain undivided in the present constitution of this universe, and that all the bodies that compose the universe result from the composition and the mixture of these solid particles, so that one can regard them as elements endowed with shapes and internal distinctions that result from their parts.

That nature halts in the analysis of matter at a certain fixed and determinate degree, is sufficiently probable, on the basis of the uniformity that rules nature's works, and on an infinity of experience.

1. If matter were resolvable[13] to infinity, it would be impossible that the same germs and the same seeds would consistently produce the same animals and the same plants, that plants and animals would always take exactly

the same time to grow, that they would always conserve the same properties, and that they would be the same at present that they were before. For if the substance that nourishes them were at times more subtle, and at times more coarse, it would be impossible that they would not be subject to perpetual variation. Because if the parts of the substance were more subtle, then it would take longer for the same body to develop than if the substance were more coarse, and consequently, the body would be more or less solid, and would develop in more or less time, depending on whether the parts of the substance that feeds it were more subtle or more coarse; and it follows that the form and the manner of being in composition would be subjected to a thousand changes, and the species of things would be blurred constantly.

But there is no such disturbance in the universe: plants, animals, fossils, each finally produces its likeness with its attributes, which constitute its essence: thus matter is not resolvable to infinity.[14]

2. If the parts of matter were resolvable to infinity, not only would species intermingle, but every day new ones would be formed; but no such new species is formed in nature, even monsters only produce their own. The hand of the creator has marked the limits of each being, and these limits will never be broached; however, if matter were divisible to infinity, they would be broached at each moment. The order that reigns in the universe, and the conservation of that order, thus appear to prove that there are solid particles in matter.

3. The dissolution of bodies has fixed limits just as their development does. The fire of a burning mirror, the most powerful solvent that we know, melts gold, pulverizes it, and then vitrifies it.[15] But the effects do not go beyond that, while, on the other hand, if matter were resolvable to infinity, the fire should destroy everything; and one could say neither why liquids only ever acquire a certain degree of heat, nor why the action of the fire on the body has such precise limits, if actual solidity and irresolvability were not bound up with the parts of matter, when they go beyond a certain littleness, and if those parts did not oppose an insurmountable barrier to this powerful agent with their solidity.

4. This irresolvability of the first bodies becomes indispensably necessary, if one adopts the system of germs, which the new discoveries that have been made with the microscope seem to demonstrate. All the world knows those discoveries of Hartsoëker, and every day it becomes more likely that nature does not act except through development.[16] Thus, if each grain of wheat contains the germ of all the grains of wheat it will produce, it is necessary that the actual divisions of matter have limits, even if those limits are unassignable by us.

It is thus strongly likely that there are particles of matter of a certain determinate littleness, which nature does not divide further.

§173 If one asks for the sufficient reason of this actual irresolvability of the little bodies of matter, it would be easy to find it in the mutual movements of its parts, for mutual movements are the cause of cohesion, according to Leibniz.

§174 Even though the actual divisions that matter can undergo have real limits, experience reveals to us a subtlety in the parts of natural bodies that stuns the imagination, and which one scarcely knows how to admire sufficiently. Wolff[17] has observed five hundred eggs within a grain of dust, from which are hatched animals similar to fish, and in which one finds an infinity of parts, as with the largest animals of the sea.

The same author reveals that a grain of barley can contain twenty-seven million living animals, each of which has twenty or twenty-four feet, and that the smallest grain of sand can be the habitat of two hundred ninety-four million organisms, which propagate their species, and who have nerves, veins, and fluids that compose them, and which no doubt have the same proportion to the bodies of these animals that the fluids of our bodies have to our mass.

The work of the drawers and beaters of gold provides fine proof of the subtlety of the parts of matter. Boyle reports that a single grain[18] of gold beaten into gold leaf fills a space of fifty square inches measured geometrically, but if one divides a side of one inch into two hundred parts, or of a line[19] in twenty, which still results in parts visible to the eye without a microscope, each square inch will have forty thousand parts of gold that one still can distinguish without a microscope, and consequently the entire leaf will have two million parts visible to the unaided eye. If one adds to this, that such a leaf is still divisible in thickness into at least six leaves, as one can conclude from the observations of Réaumur, who observed that the thickness of a leaf of gold is about $1/30000^{th}$ part of a line, and the thickness of silver wire is $1/175000^{th}$ part of a line, consequently the silver wire is about six times less thick than gold leaf.[20] Thus, this gold leaf, reduced to the thickness of silver leaf, would be divided in six parts, from which it can be seen that each grain of gold contains about twelve million parts visible to the unaided eye. Thus, since these parts are nothing but gold, and remain gold when one observes them using microscopes that magnify an object twenty or thirty thousand times, and which consequently reveal thirty thousand more parts in each of these twelve million parts that the unaided eye can distinguish in this grain of gold, one can conceive the point of fineness to which nature subdivides matter. For gold is a mixture of other, finer matters that are not gold, and it contains an infinity of pores that are filled with another matter besides the matter proper to it. Since one no longer distinguishes the constitutive parts of gold after this enormous division, nor the matter which passes through its pores, one can hardly expect to see the shapes and the movements of those parts of the mixture that must contain the immediate reason for the qualities that we notice in gold, and those parts themselves are composed again of simple beings.

§175 These considerations show us that the subtlety of the parts of matter is inexpressible, and that there is no one who can ever determine the number of parts of which a grain of sand is composed, since this number goes beyond our imagination and all that we can represent to ourselves. As reason shows us that this division has no limits, and that matter never ceases to be divisible as long as it is matter, one can say that, as far as we are concerned, matter is not only divisible, but divisible to infinity, although in reality the divisions have limits. For these divisions are so remote that they stretch to infinity for us, because for us, the infinite is a quantity that no number can express.

§176 Thus, it is evident that in nature there is an infinity of differently configured and differently altered matters, which escape our senses and our observations by their littleness, and which nonetheless produce the phenomena that we observe; and the first reasons of physical qualities are all found in these differently configured matters that are impossible for us to distinguish. We must conclude from this that an infinity of events could take place in the smallest space, just as in the entire world, but human attention could never perceive them, and it is a great deal for our understanding even to have comprehended their possibility. So it is a waste of time to attempt to divine these imperceptible mysteries, and we should limit ourselves to observing carefully the qualities that fall under our senses and the phenomena that result, which we can employ to make sense of other phenomena that depend on them.

§177 All bodies contain two types of matter, proper matter and foreign matter. Proper matter can be constant or variable. Constant matter is that without which the body could not subsist. Variable matter, like air and water, for example, is that which sometimes comes to rest in the largest pores, and which augments the weight of the body in introducing itself into and resting between its parts. All the proper matter of a body rests, moves, is weighed, and acts with it; but foreign matter doesn't always move with the body, but rather passes freely among its pores, as water does through a box pierced with several holes.

§178 The reality of the existence of the two matters is easily demonstrated by experience; for experience teaches us that bodies have different densities and different weights. Water, for example, weighs more than air, and gold is denser than wood, and weighs more.

All matters, including even gold, the densest of all, have pores that are not filled with the same matter as their proper matter. And, there being no point of absolute vacuum in the universe, it is necessary that these pores be filled with foreign matter that is not weighed with these bodies, and which does not enter into collisions with them if they encounter other bodies in their path, but which fills all their interstices, and which moves among them with as much liberty as air through a screen, or water through a net.

§179 This can be proven again by considering cohesion. For since the principle of sufficient reason rules out a vacuum between the parts of bodies, and shows that there will not be two parts of matter, one indiscernible from the other, in the universe, hence there cannot be any shape or diversity in nature except by means of motion. If all the parts of matter rested upon each other, it is evident that only a perfect, similar continuity, without any shape, would result. Thus, it is necessary, not only that all matter be in motion, but that its motion be varied to infinity in its speed and in its direction; so that the different qualities and all the internal differences of the parts of matter can emerge from it. Thus, if several parts of matter seem to be without force and at perfect rest, it must be the case that the movement of these parts tends in opposite directions with the same force; and that consequently they come to rest in the same place, which is what makes for cohesion. For we know that two bodies, strongly pressing against each other, cannot be separated except with difficulty, and appear to make up only one body. Thus, mutual movement is the origin of cohesion, according to Leibniz and his followers. So we have seen that the degree of speed with which a body moves, and the direction of its movement, are not determined by anything but the movement of several other bodies that contain its sufficient reason.[21] In this way, after the parts move in opposite directions with equal speeds, and after they cohere in this manner, it is necessary that the movement of an external matter, which does not cohere with these parts, must determine their direction and their speed. There are matters, very fine and very rapidly moving, which hide from our senses, and which produce many of the effects that we notice. Magnetic matter and electric matter are likely in this category, as well as the matter of fire, of cohesion, of elasticity, of gravity, and doubtless of an infinitude of others that are differently modified, and which cooperate in diverse ways to produce the sensible qualities of bodies.

§180 These reflections should be a warning against the impulsiveness of various philosophers, who, when they see phenomena like the fluids proposed by some, which cannot be explained at present, cut the tangle that they should unravel instead, and decide that no such fluid can exist, nor can it produce the effects that we observe.[22] For to make such a decision, it is necessary to know all the ways that matter can be moved, and all that can result from all its diverse motions, but we are still far from this.

§181 The only experiments with electricity show well enough which singular effects nature can produce by the motion of subtle matter, although the way in which she employs them to produce the effects is inexplicable for us.[23] For these matters are not sensibly perceived during the experiments with electricity, nonetheless, those who undertake to explain all the phenomena of electricity mechanically, by means of motion and a very subtle fluid, undertake a problem infinitely more difficult than that of the cause of planetary motion. For in planetary motion, a great regularity and a great uniformity rule, but the phenomena of electricity vary almost to infinity. Nonetheless, can one dare conclude that it is impossible that electrical phenomena would be brought about by fluids, because one has not yet discovered the way in which these phenomena are produced? Doubtless no; we should not be discouraged because we have not been able to divine all the secrets of nature up to the present. The first sources may elude our researches forever, but in trying to divine them, we will not fail to make discoveries as they fall along the path.

§182 In this way, however difficult it may be to apply mechanical principles to physical effects, one must never abandon this manner of philosophizing, which is the only good one, because it is the only one with which one can make sense of the phenomena in an intelligible fashion. Doubtless one must not abuse it, nor, in order to explain natural effects mechanically, invent motion and matter as one pleases (which ordinarily, even in explanation, does not produce the desired effect), nor certainly without taking pains to demonstrate the existence of these matters and these motions. But neither must one limit nature to the number of fluids that we believe are needed for the explication of the phenomena, as several philosophers have done, and in particular Hartsoëker, who has chosen two kinds of elements, the first perfectly fluid and the second perfectly solid, to make sense of the phenomena, and who believes the world to be composed of these two types of matter, which he supposes to be inalterable. But Leibniz made him see that these two matters or elements are nothing but a fiction, contrary to the principle of sufficient reason, for the principle is the touchstone which distinguishes truth from falsity. Those who know the diversity that reigns in nature, and the admirable mechanism employed there, will not fix by a rash hypothesis the number and quality of the sources that she employs, rather, they admit only those the existence of which is demonstrated by experience or by solid reasoning.

§183 The littleness of the individual parts of matter surpasses so strongly anything that our senses could discover, that there is no hope that we could ever know their qualities, motion, and shape, which makes us see how far we are from the simple beings from which solid parts are formed.

§184 In this way, one is in error if one believes oneself to be able to make sense of the phenomena that fall under our senses by simple shape and the size of their sensible parts, because we do not know how many mixtures of primitive and irresolvable parts of matter were necessary, before the parts that fall under our senses had resulted. For as much as the matter of a body is composed of other matters mixed together, one must determine the difference between the parts of this body by the matters that compose it, and by the proportion in which they are mixed. In this way, if someone wanted to explain the effects of cannon powder, for instance, they would have to begin by determining of how many types of matter it is composed, and the proportion of their mixture, before passing to the shape of its parts. For the mixed matters, and their proportion, must precede mechanical causes (that is, the determination of the shape and the size of the parts), of which it is not permitted to speak until one has arrived at the primitive matters: these physical qualities, which make up the effect of mechanical causes, must necessarily precede them in the explanation of the phenomena.

§185 But as we are left with little hope of discovering the simpler matters, the mixture of which results in sensible bodies, a physicist who does not wish to waste his time must content himself with discovering the closest reasons that human industry can perceive, and will admit only matters and motions the existence of which can be demonstrated.

Translator's Notes

1. These notes were added by the editor for clarity and references to outside sources, and are not features of the original text.
2. "Esprit", in the original.
3. See Chapter 7 (translated in du Châtelet 2009/1750) for her discussion of matter more generally.
4. Newton called the magnitudes evaluated by his calculus, that is, magnitudes that increase or decrease continuously, "fluxion" magnitudes. The Newtonian calculus of fluxions is a calculus of quantities that flow in this sense (see Newton 1974/1684–1691).
5. "Philosophes," in the original.
6. Du Châtelet is referring here to Leibniz's principle of sufficient reason. See du Châtelet 2009 / 1750, §8ff.; see also Iltis 1977.
7. Du Châtelet uses the term "sophisme." I have translated this "paradox," to follow the usual terminology, but also to avoid the pejorative meaning of "sophism."
8. See Salmon 2001, Black 1950, and Simplicius for discussion of this paradox.
9. See du Châtelet 2009/1750, §8ff. for this attempted proof.
10. "Quantité," in the original.
11. Du Châtelet probably is referring here to John Keill (1671–1721), a mathematician and Newtonian. The discussion of the properties of grains of sand is in Keill (1745), Lecture 5.
12. See Chapter 7, "The Elements of Matter," translated in du Châtelet 2009/1750.
13. It appears that du Châtelet means "actually resolvable" in this list, not "resolvable in principle." See §173 and §174 below.
14. Du Châtelet does not appear to mean that if a nourishing substance were divisible to infinity, then organisms that consume that substance would grow into larger or smaller members of their species, depending on how robust the nourishment was. Instead, she

appears to argue that organisms of the same species would develop differently and would become different species.

15. Converts it to glass.
16. Nicolaas Hartsoëker (1656–1725), the Dutch mathematician, inventor, and scientist.
17. "M. Volf" in the original. This is very likely a reference to Caspar Friedrich Wolff (1735–1794), a German physiologist.
18. Robert Boyle (1627–1691). A "graine" or grain is a pre-Revolutionary French unit of measurement equal to about 53 contemporary milligrams.
19. A "ligne" or line is a pre-Revolutionary French unit of measurement. A line is approximately 2.26 contemporary millimeters; there are 12 lines in a pre-Revolutionary, but not a contemporary, inch.
20. Réné Réaumur (1683–1756.)
21. §149, translated in du Châtelet 2009/1750.
22. Du Châtelet appears to refer here to hypotheses about the ether, which sometimes was supposed to be a fluid, and was employed in reasoning about electricity and magnetism.
23. Subtle matter is a term employed by Descartes, among others.

CHAPTER 22

Of the Idea of Necessary Connexion

DAVID HUME

Pages 155–172 of *A Treatise of Human Nature*. Book I, Section XIV.
Edition reprinted from the original ed. in three volumes and ed.,
with an analytical index, by A. Selby-Bigge. Oxford: Clarendon Press,
1888/1739.

Having thus explain'd the manner, *in which we reason beyond our immediate impressions, and conclude that such particular causes must have such
particular effects;* we must now return upon our footsteps to examine that
question, which[1] first occur'd to us, and which we dropt in our way, *viz.
What is our idea of necessity, when we say that two objects are necessarily
connected together.* Upon this head I repeat what I have often had occasion
to observe, that as we have no idea, that is not deriv'd from an impression,
we must find some impression, that gives rise to this idea of necessity, if we
assert we have really such an idea. In order to this I consider, in what objects
necessity is commonly suppos'd to lie; and finding that it is always ascrib'd
to causes and effects, I turn my eye to two objects suppos'd to be plac'd in
that relation; and examine them in all the situations, of which they are susceptible. I immediately perceive, that they are *contiguous* in time and place,
and that the object we call cause *precedes* the other we call effect. In no one
instance can I go any farther, nor is it possible for me to discover any third
relation betwixt these objects. I therefore enlarge my view to comprehend
several instances; where I find like objects always existing in like relations of
contiguity and succession. At first sight this seems to serve but little to my

purpose. The reflection on several instances only repeats the same objects; and therefore can never give rise to a new idea. But upon farther enquiry I find, that the repetition is not in every particular the same, but produces a new impression, and by that means the idea, which I at present examine. For after a frequent repetition, I find, that upon the appearance of one of the objects, the mind is *determin'd* by custom to consider its usual attendant, and to consider it in a stronger light upon account of its relation to the first object. 'Tis this impression, then, or *determination*, which affords me the idea of necessity.

I doubt not but these consequences will at first sight be receiv'd without difficulty, as being evident deductions from principles, which we have already establish'd, and which we have often employ'd in our reasonings. This evidence both in the first principles, and in the deductions, may seduce us unwarily into the conclusion, and make us imagine it contains nothing extraordinary, nor worthy of our curiosity. But tho' such an inadvertence may facilitate the reception of this reasoning, 'twill make it be the more easily forgot; for which reason I think it proper to give warning, that I have just now examin'd one of the most sublime questions in philosophy, *viz. that concerning the power and efficacy of causes;* where all the sciences seem so much interested. Such a warning will naturally rouze up the attention of the reader, and make him desire a more full account of my doctrine, as well as of the arguments, on which it is founded. This request is so reasonable, that I cannot refuse complying with it; especially as I am hopeful that these principles, the more they are examin'd, will acquire the more force and evidence.

There is no question, which on account of its importance, as well as difficulty, has caus'd more disputes both among antient and modern philosophers, than this concerning the efficacy of causes, or that quality which makes them be followed by their effects. But before they enter'd upon these disputes, methinks it wou'd not have been improper to have examin'd what idea we have of that efficacy, which is the subject of the controversy. This is what I find principally wanting in their reasonings, and what I shall here endeavour to supply.

I begin with observing that the terms of *efficacy, agency, power, force, energy, necessity, connexion,* and *productive quality,* are all nearly synonimous; and therefore 'tis an absurdity to employ any of them in defining the rest. By this observation we reject at once all the vulgar definitions, which philosophers have given of power and efficacy; and instead of searching for the idea in these definitions, must look for it in the impressions, from which it is originally deriv'd. If it be a compound idea, it must arise from compound impressions. If simple, from simple impressions.

I believe the most general and most popular explication of this matter, is to say,[2] that finding from experience, that there are several new productions in matter, such as the motions and variations of body, and concluding that there must somewhere be a power capable of producing them, we arrive at last by this reasoning at the idea of power and efficacy. But to be convinc'd that this explication is more popular than philosophical, we need but reflect on two very obvious principles. *First,* That reason alone can never give rise to any original idea, and *secondly,* that reason, as distinguish'd from experience, can never make us conclude, that a cause or productive quality is absolutely requisite to every beginning of existence. Both these considerations have been sufficiently explain'd; and therefore shall not at present be any farther insisted on.

I shall only infer from them, that since reason can never give rise to the idea of efficacy, that idea must be deriv'd from experience, and from some particular instances of this efficacy, which make their passage into the mind by the common channels of sensation or reflection. Ideas always represent their objects or impressions; and *vice versa,* there are some objects necessary to give rise to every idea. If we pretend, therefore, to have any just idea of this efficacy, we must produce some instance, wherein the efficacy is plainly discoverable to the mind, and its operations obvious to our consciousness or sensation. By the refusal of this, we acknowledge, that the idea is impossible and imaginary; since the principle of innate ideas, which alone can save us from this dilemma, has been already refuted, and is now almost universally rejected in the learned world. Our present business, then, must be to find some natural production, where the operation and efficacy of a cause can be clearly conceiv'd and comprehended by the mind, without any danger of obscurity or mistake.

In this research we meet with very little encouragement from that prodigious diversity, which is found in the opinions of those philosophers, who have pretended to explain the secret force and energy of causes.[3] There are some, who maintain, that bodies operate by their substantial form; others, by their accidents or qualities; several, by their matter and form; some, by their form and accidents; others, by certain virtues and faculties distinct from all this. All these sentiments again are mix'd and vary'd in a thousand different ways; and form a strong presumption, that none of them have any solidity or evidence, and that the supposition of an efficacy in any of the known qualities of matter is entirely without foundation. This presumption must encrease upon us, when we consider, that these principles of substantial forms, and accidents, and faculties, are not in reality any of the known properties of bodies, but are perfectly unintelligible and inexplicable. For 'tis evident philosophers wou'd never have had recourse to such obscure and uncertain principles had they met with any satisfaction in such as are clear and intelligible; especially in such an affair as this, which must be an object of the simplest understanding, if not of the senses. Upon the whole, we may conclude, that 'tis impossible in any one instance to shew the principle, in which the force and agency of a cause is plac'd; and that the most refin'd and most vulgar understandings are equally at a loss in this particular. If any one think proper to refute this assertion, he need not put himself to the trouble of inventing any long reasonings; but may at once shew us an instance of a cause, where we discover the power or operating principle. This defiance we are oblig'd frequently to make use of, as being almost the only means of proving a negative in philosophy.

The small success, which has been met with in all the attempts to fix this power, has at last oblig'd philosophers to conclude, that the ultimate force and efficacy of nature is perfectly unknown to us, and that 'tis in vain we search for it in all the known qualities of matter. In this opinion they are almost unanimous; and 'tis only in the inference they draw from it, that they discover any difference in their sentiments. For some of them, as the *Cartesians* in particular, having establish'd it as a principle, that we are perfectly acquainted with the essence of matter, have very naturally inferr'd, that it is endow'd with no efficacy, and that 'tis impossible for it of itself to communicate motion, or produce any of those effects, which we ascribe to it. As the essence of matter consists in extension, and as extension implies not actual motion, but only mobility; they conclude, that the energy, which produces the motion, cannot lie in the extension.

This conclusion leads them into another, which they regard as perfectly unavoidable. Matter, say they, is in itself entirely unactive, and depriv'd of any power, by which it may produce, or continue, or communicate motion: But since these effects are evident to our senses, and since the power, that produces them, must be plac'd somewhere, it must lie in the DEITY, or that divine being, who contains in his nature all excellency and perfection. 'Tis the deity, therefore, who is the prime mover of the universe, and who not only first created matter, and gave it its original impulse, but likewise by a continu'd exertion of omnipotence, supports its existence, and successively bestows on it all those motions, and configurations, and qualities, with which it is endow'd.

This opinion is certainly very curious, and well worth our attention; but 'twill appear superfluous to examine it in this place, if we reflect a moment on our present purpose in taking notice of it. We have establish'd it as a principle, that as all ideas are deriv'd from impressions, or some precedent *perceptions*, 'tis impossible we can have any idea of power and efficacy, unless some instances can be produc'd, wherein this power *is perceiv'd* to exert itself. Now as these instances can never be discover'd in body, the *Cartesians*, proceeding upon their principle of innate ideas, have had recourse to a supreme spirit or deity, whom they consider as the only active being in the universe, and as the immediate cause of every alteration in matter. But the principle of innate ideas being allow'd to be false, it follows, that the supposition of a deity can serve us in no stead, in accounting for that idea of agency, which we search for in vain in all the objects, which are presented to our senses, or which we are internally conscious of in our own minds. For if every idea be deriv'd from an impression, the idea of a deity proceeds from the same origin; and if no impression, either of sensation or reflection, implies any force or efficacy, 'tis equally impossible to discover or even imagine any such active principle in the deity. Since these philosophers, therefore, have concluded, that matter cannot be endow'd with any efficacious principle, because 'tis impossible to discover in it such a principle; the same course of reasoning shou'd determine them to exclude it from the supreme being. Or if they estem that opinion absurd and impious, as it really is, I shall tell them how they may avoid it; and that is, by concluding from the very first, that they have no adequate idea of power or efficacy in any object; since neither in body nor spirit, neither in superior nor inferior natures, are they able to discover one single instance of it.

The same conclusion is unavoidable upon the hypothesis of those, who maintain the efficacy of second causes, and attribute a derivative, but a real power and energy to matter. For as they confess, that this energy lies not in any of the known qualities of matter, the difficulty still remains concerning the origin of its idea. If we have really an idea of power, we may attribute power to an unknown quality: But as 'tis impossible, that that idea can be deriv'd from such a quality, and as there is nothing in known qualities, which can produce it; it follows that we deceive ourselves, when we imagine we are possest of any idea of this kind, after the manner we commonly understand it. All ideas are deriv'd from, and represent impressions. We never have any impression, that contains any power or efficacy. We never therefore have any idea of power.

It has been establish'd as a certain principle, that general or abstract ideas are nothing but individual ones taken in a certain light, and that, in reflecting on any object, 'tis as impossible to exclude from our thought all particular degrees of quantity and quality as from the real nature of things. If we be possest, therefore, of any idea of power in general, we must also be able to conceive some particular species of it; and as power cannot subsist alone, but is always regarded as an attribute of some being or existence, we must be able to place this power in some particular being, and conceive that being as endow'd with a real force and energy, by which such a particular effect necessarily results from its operation. We must distinctly and particularly conceive the connexion betwixt the cause and effect, and be able to pronounce, from a simple view of the one, that it must be follow'd or preceded by the other. This is the true manner of conceiving a particular power in a particular body: and a general idea being impossible without an individual; where the latter is impossible, 'tis certain the former can never exist. Now nothing is more evident, than that the human mind cannot form such an idea of two objects, as to conceive any connexion betwixt them, or comprehend distinctly that power or efficacy, by which they are united. Such a connexion wou'd amount to a demonstration, and wou'd imply the absolute impossibility for the one object not to follow, or to be conceiv'd not to follow upon the other: Which kind of connexion has already been rejected in all cases. If any one is of a contrary opinion, and thinks he has attain'd a notion of power in any particular object, I desire he may point out to me that object. But till I meet with such-a-one, which I despair of, I cannot forbear concluding, that since we can never distinctly conceive how any particular power can possibly reside in any particular object, we deceive ourselves in imagining we can form any such general idea.

Thus upon the whole we may infer, that when we talk of any being, whether of a superior or inferior nature, as endow'd with a power or force, proportion'd to any effect; when we speak of a necessary connexion betwixt objects, and suppose, that this connexion depends upon an efficacy or energy, with which any of these objects are endow'd; in all these expressions, *so apply'd*, we have really no distinct meaning, and make use only of common words, without any clear and determinate ideas. But as 'tis more probable, that these expressions do here lose their true meaning by being *wrong apply'd*, than that they never have any meaning; 'twill be proper to bestow another consideration on this subject, to see if possibly we can discover the nature and origin of those ideas, we annex to them.

Suppose two objects to be presented to us, of which the one is the cause and the other the effect; 'tis plain, that from the simple consideration of one, or both these objects we never shall perceive the tie, by which they are united, or be able certainly to pronounce, that there is a connexion betwixt them. 'Tis not, therefore, from any one instance, that we arrive at the idea of cause and effect, of a necessary connexion of power, of force, of energy, and of efficacy. Did we never see any but particular conjunctions of objects, entirely different from each other, we shou'd never be able to form any such ideas.

But again; suppose we observe several instances, in which the same objects are always conjoin'd together, we immediately conceive a connexion betwixt them, and begin to draw an inference from one to another. This multiplicity of resembling instances, therefore, constitutes the very essence of power or connexion, and is the source, from which the idea of it arises. In order, then, to understand the idea of power, we must consider that multiplicity; nor do I ask more to give a solution of that difficulty, which has so long perplex'd us. For thus I reason. The repetition of perfectly similar instances can never *alone* give rise to an original idea, different from what is to be found in any particular instance, as has been observ'd, and as evidently follows from our fundamental principle, *that all ideas are copy'd from impressions.* Since therefore the idea of power is a new original idea, not to be found in any one instance, and which yet arises from the repetition of several instances, it follows, that the repetition *alone* has not that effect, but must either *discover* or *produce* something new, which is the source of that idea. Did the repetition neither discover nor produce any thing new, our ideas might be multiply'd by it, but wou'd not be enlarg'd above what they are upon the observation of one single instance. Every enlargement, therefore, (such as the idea of power or connexion) which arises from the multiplicity of similar instances, is copy'd from some effects of the multiplicity, and will be perfectly understood by understanding these effects. Wherever we find any thing new to be discover'd or produc'd by the repetition, there we must place the power, and must never look for it in any other object.

But 'tis evident, in the first place, that the repetition of like objects in like relations of succession and contiguity *discovers* nothing new in any one of them; since we can draw no inference from it, nor make it a subject either of our demonstrative or probable reasonings;[4] as has been already prov'd. Nay suppose we cou'd draw an inference, 'twou'd be of no consequence in the present case; since no kind of reasoning can give rise to a new idea, such as this of power is; but wherever we reason, we must antecedently be possest of clear ideas, which may be the objects of our reasoning. The conception always precedes the understanding; and where the one is obscure, the other is uncertain; where the one fails, the other must fail also.

Secondly, 'Tis certain that this repetition of similar objects in similar situations *produces* nothing new either in these objects, or in any external body. For 'twill readily be allow'd, that the several instances we have of the conjunction of resembling causes and effects are in themselves entirely independent, and that the communication of motion, which I see result at present from the shock of two billiard-balls, is totally distinct from that which I saw result from such an impulse a twelve-month ago. These impulses have no influence on each other. They are entirely divided by time and place; and the one might have existed and communicated motion, tho' the other never had been in being.

There is, then, nothing new either discover'd or produc'd in any objects by their constant conjunction, and by the uninterrupted resemblance of their relation of succession and contiguity. But 'tis from this resemblance, that the ideas of necessity, of power, and of efficacy, are deriv'd. These ideas, therefore, represent not any thing, that does or can belong to the objects, which are constantly conjoin'd. This is an argument, which, in every view we can examine it, will be found perfectly unanswerable. Similar instances are still the first source of our idea of power or necessity; at the same time that they have no influence by their similarity either on each other, or on any external object. We must therefore, turn ourselves to some other quarter to seek the origin of that idea.

Tho' the several resembling instances, which give rise to the idea of power, have no influence on each other, and can never produce any new quality *in the object*, which can be the model of that idea, yet the *observation* of this resemblance produces a new impression *in the mind*, which is its real model. For after we have observ'd the resemblance in a sufficient number of instances, we immediately feel a determination of the mind to pass from one object to its usual attendant, and to conceive it in a stronger light upon account of that relation. This determination is the only effect of the resemblance; and therefore must be the same with power or efficacy, whose idea is deriv'd from the resemblance. The several instances of resembling conjunctions leads us into the notion of power and necessity. These instances are in themselves totally distinct from each other, and have no union but in the mind, which observes them, and collects their ideas. Necessity, then, is the effect of this observation, and is nothing but an internal impression of the mind, or a determination to carry our thoughts from one object to another. Without considering it in this view, we can never arrive at the most distant notion of it, or be able to attribute it either to external or internal objects, to spirit or body, to causes or effects.

The necessary connexion betwixt cause and effects is the foundation of our inference from one to the other. The foundation of our inference is the transition arising from the accustom'd union. These are, therefore, the same.

The idea of necessity arises from some impression. There is no impression convey'd by our senses, which can give rise to that idea. It must, therefore, be deriv'd from some internal impression, or impression of reflection. There is no internal impression, which has any relation to the present business, but that propensity, which custom produces, to pass from an object to the idea of its usual attendant. This therefore is the essence of necessity. Upon the whole, necessity is something, that exists in the mind, not in objects; nor is it possible for us ever to form the most distant idea of it, consider'd as a quality in bodies. Either we have no idea of necessity, or necessity is nothing but that determination of the thought to pass from causes to effects and from effects to causes, according to their experienc'd union.

Thus as the necessity, which makes two times two equal to four, or three angles of a triangle equal to two right ones, lies only in the act of the understanding, by which we consider and compare these ideas; in like manner the necessity or power, which unites causes and effects, lies in the determination of the mind to pass from the one to the other. The efficacy or energy of causes is neither plac'd in the causes themselves, nor in the deity, nor in the concurrence of these two principles; but belongs entirely to the soul, which considers the union of two or more objects in all past instances. 'Tis here that the real power of causes is plac'd, along with their connexion and necessity.

I am sensible, that of all the paradoxes, which I have had, or shall hereafter have occasion to advance in the course of this treatise, the present one is the most violent, and that 'tis merely by dint of solid proof and reasoning I can ever hope it will have admission, and overcome the inveterate prejudices of mankind. Before we are reconcil'd to this doctrine, how often must we repeat to ourselves, *that* the simple view of any two objects or actions, however related, can never give us any idea of power, or of a connexion betwixt them: *that* this idea arises from the repetition of their union: *that* the repetition neither discovers nor causes any thing in the objects, but has an influence only on the mind, by that customary transition it produces: *that* this customary transition is, therefore, the same with the power and necessity; which are consequently qualities of perceptions, not of objects, and are internally felt by the soul, and not perceiv'd externally in bodies? There is commonly an astonishment attending every thing extraordinary; and this astonishment changes immediately into the highest degree of esteem or contempt, according as we approve or disapprove of the subject. I am much afraid, that tho' the foregoing reasoning appears to me the shortest and most decisive imaginable; yet with the generality of readers the biass of the mind will prevail, and give them a prejudice against the present doctrine.

This contrary biass is easily accounted for. 'Tis a common observation, that the mind has a great propensity to spread itself on external objects, and to conjoin with them any internal impressions, which they occasion, and which always make their appearance at the same time that these objects discover themselves to the senses. Thus as certain sounds and smells are always found to attend certain visible objects, we naturally imagine a conjunction, even in place, betwixt the objects and qualities, tho' the qualities be of such a nature as to admit of no such conjunction, and really exist no where. But of this more fully[5] hereafter. Mean while 'tis sufficient to observe, that the same propensity is the reason, why we suppose necessity and power to lie in the objects we consider, not in our mind, that considers them; notwithstanding it is not possible for us to form the most distant idea of that quality, when it is not taken for the determination of the mind, to pass from the idea of an object to that of its usual attendant.

But tho' this be the only reasonable account we can give of necessity, the contrary notion is so riveted in the mind from the principles abovemention'd, that I doubt not but my sentiments will be treated by many as extravagant and ridiculous. What! the efficacy of causes lie in the determination of the mind! As if causes did not operate entirely independent of the mind, and wou'd not continue their operation, even tho' there was no mind existent to contemplate them, or reason concerning them. Thought may well depend on causes for its operation, but not causes on thought. This is to reverse the order of nature, and make that secondary, which is really primary. To every operation there is a power proportion'd; and this power must be plac'd on the body, that operates. If we remove the power from one cause, we must ascribe it to another: But to remove it from all causes, and bestow it on a being, that is no ways related to the cause or effect, but by perceiving them, is a gross absurdity, and contrary to the most certain principles of human reason.

I can only reply to all these arguments, that the case is here much the same, as if a blind man shou'd pretend to find a great many absurdities in the supposition, that the colour of scarlet is not the same with the sound of a trumpet, nor light the same with solidity. If we have really no idea of a power or efficacy in any object, or of any real connexion betwixt causes and effects, 'twill be to little purpose to prove, that an efficacy is necessary in all operations. We do not understand our own meaning in talking so, but ignorantly confound ideas, which are entirely distinct from each other. I am, indeed, ready to allow, that there may be several qualities both in material and immaterial objects, with which we are utterly unacquainted; and if we please to call these *power* or *efficacy*, 'twill be of little consequence to the world. But when, instead of meaning these unknown qualities, we make the terms of power and efficacy signify something, of which we have a clear idea, and which is incompatible with those objects, to which we apply it, obscurity and error begin then to take place, and we are led astray by a false philosophy. This is the case, when we transfer the determination of the thought to external objects, and suppose any real intelligible connexion betwixt them; that being a quality, which can only belong to the mind that considers them.

As to what may be said, that the operations of nature are independent of our thought and reasoning, I allow it; and accordingly have observ'd, that objects bear to each other the relations of contiguity and succession; that like objects may be observ'd in several instances to have like relations; and that all this is independent of, and antecedent to the operations of the understanding. But if we go any farther, and ascribe a power or necessary connexion to these objects; this is what we can never observe in them, but must draw the idea of it from what we feel internally in contemplating them. And this I carry so far, that I am ready to convert my present reasoning into an instance of it, by a subtility, which it will not be difficult to comprehend.

When any object is presented to us, it immediately conveys to the mind a lively idea of that object, which is usually found to attend it; and this determination of the mind forms the necessary connexion of these objects. But when we change the point of view, from the objects to the perceptions; in that case the impression is to be considered as the cause, and the lively idea as the effect; and their necessary connexion is that new determination, which we feel to pass from the idea of the one to that of the other. The uniting principle among our internal perceptions is as unintelligible as that among external objects, and is not known to us any other way than by experience. Now the nature and effects of experience have been already sufficiently examin'd and explain'd. It never gives us any insight into the internal structure or operating principle of objects, but only accustoms the mind to pass from one to another.

'Tis now time to collect all the different parts of this reasoning, and by joining them together form an exact definition of the relation of cause and effect, which makes the subject of the present enquiry. This order wou'd not have been excusable, of first examining our inference from the relation before we had explain'd the relation itself, had it been possible to proceed in a different method. But as the nature of the relation depends so much on that of the inference, we have been oblig'd to advance in this seemingly preposterous manner, and make use of terms before we were able exactly to define them, or fix their meaning. We shall now correct this fault by giving a precise definition of cause and effect.

There may two definitions be given of this relation, which are only different, by their presenting a different view of the same object, and making us consider it either as a *philosophical* or as a *natural* relation; either as a comparison of, two ideas, or as an association betwixt them. We may define a CAUSE to be 'An object precedent and contiguous to another, and where all the objects resembling the former are plac'd in like relation of precedency and contiguity to those objects, that resemble the latter.' If this definition be esteem'd defective, because drawn from objects foreign to the cause, we may substitute this other definition in its place, *viz.* 'A CAUSE is an object precedent and contiguous to another, and so united with it, that the idea of the one determines the mind to form the idea of the other, and the impression of the one to form a more lively idea of the other.' Shou'd this definition also be rejected for the same reason, I know no other remedy, than that the persons, who express this delicacy, should substitute a juster definition in its place. But for my part I must own my incapacity for such an undertaking. When I examine with the utmost accuracy those objects, which are commonly denominated causes and effects, I find, in considering a single instance, that the one object is precedent and contiguous to the other; and in inlarging my view to consider several instances, I find only, that like objects are constantly plac'd in like relations of succession and contiguity. Again, when I consider the influence of this constant conjunction,

I perceive, that such a relation can never be an object of reasoning, and can never operate upon the mind, but by means of custom, which determines the imagination to make a transition from the idea of one object to that of its usual attendant, and from the impression of one to a more lively idea of the other. However extraordinary these sentiments may appear, I think it fruitless to trouble myself with any farther enquiry or reasoning upon the subject, but shall repose myself on them as on establish'd maxims.

'Twill only be proper, before we leave this subject, to draw some corollaries from it, by which we may remove several prejudices and popular errors, that have very much prevail'd in philosophy. First, We may learn from the foregoing doctrine, that all causes are of the same kind, and that in particular there is no foundation for that distinction, which we sometimes make betwixt efficient causes, and causes *sine qua non*; or betwixt efficient causes, and formal, and material, and exemplary, and final causes. For as our idea of efficiency is deriv'd from the constant conjunction of two objects, wherever this is observ'd, the cause is efficient; and where it is not, there can never be a cause of any kind. For the same reason we must reject the distinction betwixt *cause* and *occasion*, when suppos'd to signify any thing essentially different from each other. If constant conjunction be imply'd in what we call occasion, 'tis a real cause. If not, 'tis no relation at all, and cannot give rise to any argument or reasoning.

Secondly, The same course of reasoning will make us conclude, that there is but one kind of *necessity*, as there is but one kind of cause, and that the common distinction betwixt *moral* and *physical* necessity is without any foundation in nature. This clearly appears from the precedent explication of necessity. 'Tis the constant conjunction of objects, along with the determination of the mind, which constitutes a physical necessity: And the removal of these is the same thing with *chance*. As objects must either be conjoin'd or not, and as the mind must either be determin'd or not to pass from one object to another, 'tis impossible to admit of any medium betwixt chance and an absolute necessity. In weakening this conjunction and determination you do not change the nature of the necessity; since even in the operation of bodies, these have different degrees of constancy and force, without producing a different species of that relation.

The distinction, which we often make betwixt *power* and the *exercise* of it, is equally without foundation.

Thirdly, We may now be able fully to overcome all that repugnance, which 'tis so natural for us to entertain against the foregoing reasoning, by which we endeavour'd to prove, that the necessity of a cause to every beginning of existence is not founded on any arguments either demonstrative or intuitive. Such an opinion will not appear strange after the foregoing definitions. If we define a cause to be *an object precedent and contiguous to another, and where all the objects resembling the former are plac'd in a like relation of priority and contiguity to those objects, that resemble the latter;* we may easily conceive, that there is no absolute nor metaphysical necessity, that every beginning of existence shou'd be attended with such an object. If we define a cause to be, *An object precedent and contiguous to another, and so united with it in the imagination, that the idea of the one determines the mind to form the idea of the other, and the impression of the one to form a more lively idea of the other;* we shall make still less difficulty of assenting to this opinion. Such an influence on the mind is in itself perfectly extraordinary and incomprehensible; nor can we be certain of its reality, but from experience and observation.

I shall add as a fourth corollary, that we can never have reason to believe that any object exists, of which we cannot form an idea. For as all our reasonings concerning existence are deriv'd from causation, and as all our reasonings concerning causation are deriv'd from the experienc'd conjunction of objects, not from any reasoning or reflection, the same experience must give us a notion of these objects, and must remove all mystery from our conclusions. This is so evident, that 'twou'd scarce have merited our attention, were it not to obviate certain objections of this kind, which might arise against the following reasonings concerning *matter* and *substance*. I need not observe, that a full knowledge of the object is not requisite, but only of those qualities of it, which we believe to exist.

Notes

1. Sect. 2.
2. See Mr. *Locke*; chapter on power [Editor's note: Locke 1979/1689, Chapter XXI of Book II.—LP].
3. See Father *Malebranche*, Book VI. Part ii. chap. 3, and the illustrations upon it [Editor's note: Probably *The Search After Truth*, written 1674–75. Translated by T. M. Lennon and P. J. Olscamp, 1997. Cambridge: Cambridge University Press.—LP].
4. Sect. 6.
5. Part IV. sect. 5.

CHAPTER 23

How Is Pure Natural Science Possible?

IMMANUEL KANT

Pages 38–64 of the *Prolegomena to any Future Metaphysics,* trans. Paul Carus, trans. revised James Ellington. Indianapolis: Hackett, 2001/1783.

§18. In the first place we must state that while all judgments of experience are empirical (i.e., have their ground in immediate sense-perception), yet conversely, all empirical judgments are not therefore judgments of experience; but, besides the empirical, and in general besides what is given to sensuous intuition, special concepts must yet be superadded—concepts which have their origin quite *a priori* in the pure understanding, and under which every perception must be first of all subsumed and then by their means changed into experience.

Empirical judgments, so far as they have objective validity, are *judgments of experience;* but those which are only subjectively valid I name mere *judgments of perception.* The latter require no pure concept of the understanding, but only the logical connection of perception in a thinking subject. But the former always require, besides the representation of the sensuous intuition, special *concepts originally generated in the understanding,* which make the judgment of experience objectively valid.

All our judgments are at first merely judgments of perception; they hold good only for us (i.e., for our subject), and we do not until afterward give them a new reference (to an object) and want that they shall always hold

356

good for us and in the same way for everybody else; for if a judgment agrees with an object, all judgments concerning the same object must likewise agree with one another, and thus the objective validity of the judgment of experience signifies nothing else than its necessary universal validity. And, conversely, if we have reason to hold a judgment to be necessarily universally valid (which never rests on perception, but on the pure concept of the understanding under which the perception is subsumed), we must consider it to be objective also, that is, that it expresses not merely a reference of our perception to a subject, but a quality of the object. For there would be no reason for the judgments of other men necessarily to agree with mine, if it were not the unity of the object to which they all refer and with which they accord; hence they must all agree with one another.

§19. Therefore objective validity and necessary universal validity (for everybody) are equivalent concepts, and though we do not know the object in itself, yet when we consider a judgment as universally valid, and hence necessary, we understand it thereby to have objective validity. By this judgment we cognize the object (though it remains unknown as it is in itself) by the universally valid and necessary connection of the given perceptions. As this is the case with all objects of sense, judgments of experience take their objective validity, not from the immediate cognition of the object (which is impossible), but merely from the condition of universal validity of empirical judgments, which, as already said, never rests upon empirical or, in short, sensuous conditions, but upon a pure concept of the understanding. The object in itself always remains unknown; but when by the concept of the understanding the connection of the representations of the object, which are given by the object to our sensibility, is determined as universally valid, the object is determined by this relation, and the judgment is objective.

To illustrate the matter: when we say, "The room is warm, sugar sweet, and wormwood nasty,"[1] we have only subjectively valid judgments. I do not at all expect that I or any other person shall always find it as I now do; each of these sentences only expresses a reference of two sensations to the same subject, i.e., myself, and that only in my present state of perception; consequently, they are not intended to be valid of the object. Such are judgments of perception. Judgments of experience are of quite a different nature. What experience teaches me under certain circumstances, it must always teach me and everybody; and its validity is not limited to the subject nor to its state at a particular time. Hence I pronounce all such judgments as being objectively valid. For instance, when I say the air is elastic, this judgment is as yet a judgment of perception only—I do nothing but refer two sensations in my senses to one another. But if I would have it called a judgment of experience, I require this connection to stand under a condition which makes it universally valid. I desire therefore that I and everybody else should always necessarily connect the same perceptions under the same circumstances.

§20. We must therefore analyze experience in general in order to see what is contained in this product of the senses and of the understanding, and how the judgment of experience itself is possible. The foundation is the intuition of which I become conscious, i.e., perception (*perceptio*), which pertains merely to the senses. But in the next place, there is judging (which belongs only to the understanding). But this judging may be twofold: first, I may merely compare perceptions and connect them in a consciousness of my state; or, secondly, I may connect them in consciousness in general. The former judgment is merely a judgment of perception and is of subjective validity only; it is merely a connection of perceptions in my mental state, without reference to the object. Hence it is not, as is commonly imagined, enough for experience to compare perceptions and connect them in consciousness through judgment; there arises no universal validity and necessity, by virtue of which alone consciousness can become objectively valid and be called experience.

Quite another judgment therefore is required before perception can become experience. The given intuition must be subsumed under a concept which determines the form of judging in general with regard to the intuition, connects the empirical consciousness of the intuition in consciousness in general, and thereby procures universal validity for empirical judgments. A concept of this nature is a pure *a priori* concept of the understanding, which does nothing but determine for an intuition the general way in which it can be used for judging. Let the concept be that of cause; then it determines the intuition which is subsumed under it, e.g., that of air, with regard to judging in general, viz., the concept of air as regards its expansion serves in the relation of antecedent to consequent in a hypothetical judgment. The concept of cause accordingly is a pure concept of the understanding, which is totally disparate from all possible perception and only serves to determine the representation contained under it with regard to judging in general, and so to make a universally valid judgment possible.

Before, therefore, a judgment of perception can become a judgment of experience, it is requisite that the perception should be subsumed under some such concept of the understanding; for instance, air belongs under the concept of cause, which determines our judgment about it with regard to its expansion as hypothetical.[2] Thereby the expansion of the air is represented, not as merely belonging to the perception of the air in my present state or in several states of mine, or in the state of perception of others, but as belonging to it necessarily. The judgment that air is elastic becomes universally valid and a judgment of experience only because certain judgments precede it which subsume the intuition of air under the concepts of cause and effect; and they thereby determine the perceptions, not merely as regards one another in me, but as regards the form of judging in general (which is here hypothetical), and in this way they render the empirical judgment universally valid.

If all our synthetic judgments are analyzed so far as they are objectively valid, it will be found that they never consist of mere intuitions connected only (as is commonly supposed) by comparison into a judgment; but that they would be impossible were not a pure concept of the understanding superadded to the concepts abstracted from intuition, under which pure concept these latter concepts are subsumed and in this manner only combined into an objectively valid judgment. Even the judgments of pure mathematics in their simplest axioms are not exempt from this condition. The principle that a straight line is the shortest distance between two points presupposes that the line is subsumed under the concept of quantity, which certainly is no mere intuition but has its seat in the understanding alone and serves to determine the intuition (of the line) with regard to the judgments which may be made about it in respect to the quantity, that is, to plurality (as *judica plurativa*).[3] For under them it is understood that in a given intuition there is contained a plurality of homogeneous parts.

§21. To prove, then, the possibility of experience so far as it rests upon pure *a priori* concepts of the understanding, we must first represent what belongs to judgments in general and the various moments (functions) of the understanding in them, in a complete table. For the pure concepts of the understanding must run parallel to these moments, inasmuch as such concepts are nothing more than concepts of intuitions in general, so far as these are determined by one or other of these moments of judging, in themselves, i.e., necessarily and universally. Hereby also the *a priori* principles of the possibility of all experience, as objectively valid empirical cognition, will be precisely determined. For they are nothing but propositions which subsume all perception (conformably to certain universal conditions of intuition) under those pure concepts of the understanding.

Logical Table of Judgments

1	2
As to Quantity	*As to Quality*
Universal	Affirmative
Particular	Negative
Singular	Infinite
3	4
As to Relation	*As to Modality*
Categorical	Problematic
Hypothetical	Assertoric
Disjunctive	Apodeictic

TRANSCENDENTAL TABLE OF THE CONCEPTS OF THE UNDERSTANDING

1	2
As to Quantity	*As to Quality*
Unity (Measure)	Reality
Plurality (Quantity)	Negation
Totality (Whole)	Limitation

3	4
As to Relation	*As to Modality*
Substance	Possibility
Cause	Existence
Community	Necessity

PURE PHYSIOLOGICAL[4] TABLE OF THE UNIVERSAL PRINCIPLES OF NATURAL SCIENCE

1	2
Axioms of Intuition	Anticipations of Perception

3	4
Analogies of Experience	Postulates of Empirical Thought in General

§ 21a. In order to comprise the whole matter in one idea, it is first necessary to remind the reader that we are discussing, not the origin of experience, but what lies in experience. The former pertains to empirical psychology and would even then never be adequately developed without the latter, which belongs to the critique of cognition, and particularly of the understanding.

Experience consists of intuitions, which belong to the sensibility, and of judgments, which are entirely a work of the understanding. But the judgments which the understanding makes entirely out of sensuous intuitions are far from being judgments of experience. For in the one case the judgment connects only the perceptions as they are given in sensuous intuition, while in the other the judgments must express what experience in general and not what the mere perception (which possesses only subjective validity) contains. The judgment of experience must therefore add to the sensuous intuition and its logical connection in a judgment (after it has been rendered universal by comparison) something that determines the synthetic judgment as necessary and therefore as universally valid. This can be nothing but that concept which represents the intuition as determined in itself with regard to one form of judgment rather than another, viz., a concept of that synthetic unity of intuitions which can only be represented by a given logical function of judgments.

§ 22. The sum of the matter is this: the business of the senses is to intuit, that of the understanding is to think. But thinking is uniting representations in a consciousness. This unification originates either merely relative to the subject and is contingent and subjective, or it happens absolutely and is necessary or objective. The uniting of representations in a consciousness is judgment. Thinking therefore is the same as judging, or referring representations to judgments in general. Hence judgments are either merely subjective when representations are referred to a consciousness in one subject only and are united in it, or they are objective when they are united in a consciousness in general, that is, necessarily. The logical moments of all judgments are so many possible ways of uniting representations in consciousness. But if they serve as concepts, they are concepts of the necessary unification of representations in a consciousness and so are principles of objectively valid judgments. This uniting in a consciousness is either analytic by identity, or synthetic by the combination and addition of various representations one to another. Experience consists in the synthetic connection of appearances (perceptions) in consciousness, so far as this connection is necessary. Hence the pure concepts of the understanding are those under which all perceptions must first be subsumed before they can serve for judgments of experience, in which the synthetic unity of the perceptions is represented as necessary and universally valid.[5]

§ 23. Judgments, when considered merely as the condition of the unification of given representations in a consciousness, are rules. These rules, so far as they represent the unification as necessary, are rules *a priori*, and so far as they cannot be deduced from higher rules, are principles. But in regard to the possibility of all experience, merely in relation to the form of thinking in it, no conditions of judgments of experience are higher than those which bring the phenomena, according to the different form of their intuition, under pure concepts of the understanding, and render the empirical judgments objectively valid. These are therefore the *a priori* principles of possible experience.

The principles of possible experience are then at the same time universal laws of nature, which can be cognized *a priori*. And thus the problem in our second question: How is pure natural science possible? is solved. For the systematization which is required for the form of a science is to be met with in perfection here, because, beyond the above-mentioned formal conditions of all judgments in general and of all rules in general, that are offered in logic, no others are possible, and these constitute a logical system. The concepts grounded thereupon, which contain the *a priori* conditions of all synthetic and necessary judgments, accordingly constitute a transcendental system. Finally, the principles by means of which all appearances are subsumed under these concepts constitute a physiological system, that is, a system of nature, which precedes all empirical cognition of nature, first makes it possible, and hence may in strictness be called the universal and pure natural science.

§ 24. The first[6] of the physiological principles[7] subsumes all appearances, as intuitions in space and time, under the concept of *quantity*, and is so far a principle of the application of mathematics to experience. The second[8] subsumes the strictly empirical element, viz., sensation, which denotes the real in intuitions, not indeed directly under the concept of *quantity*, because sensation is not an intuition that *contains* either space or time, though it puts the object corresponding to sensation in both space and time. But still there is between reality (sense-representation) and zero, or total lack of intuition in time, a difference which has a quantity. For between any given degree of light and darkness, between any degree of heat and complete cold, between any degree of weight and absolute lightness, between any degree of occupied space and of totally empty space, ever smaller degrees can be thought, just as even between consciousness and total unconsciousness (psychological darkness) ever smaller degrees obtain. Hence there is no perception that can show an absolute absence; for instance, no psychological darkness that cannot be regarded as a consciousness only surpassed by a stronger consciousness. This occurs in all cases of sensation; and so the understanding can anticipate sensations, which constitute the peculiar quality of empirical representations (appearances), by means of the principle that they all have a degree, consequently, that what is real in all appearance has a degree. Here is the second application of mathematics (*mathesis intensorum*) to natural science.

§ 25. As regards the relation of appearances merely with a view to their existence, the determination is not mathematical but dynamical, and can never be objectively valid and fit for experience, if it does not come under *a priori* principles[9] by which the cognition of experience relative to appearances first becomes possible. Hence appearances must be subsumed under the concept of substance, which as a concept of the thing itself is the foundation of all determination of existence; or, secondly—so far as a succession is found among appearances, that is, an event—under the concept of an effect with reference to cause; or, lastly—so far as coexistence is to be known objectively, that is, by a judgment of experience—under the concept of community (action and reaction). Thus *a priori* principles form the basis of objectively valid, though empirical, judgments—that is, of the possibility of experience so far as it must connect objects as existing in nature. These principles are properly the laws of nature, which may be called dynamical.

Finally[10] the cognition of the agreement and connection, not only of appearances among themselves in experience, but of their relation to experience in general, belongs to judgments of experience. This relation contains either their agreement with the formal conditions which the understanding cognizes, or their coherence with the material of the senses and of perception, or combines both into one concept and consequently contains possibility, actuality, and necessity according to universal laws of nature. This would constitute the physiological doctrine of method (distinction between truth and hypotheses, and the bounds of the reliability of the latter).

§ 26. The third table of principles drawn from the nature of the understanding itself according to the critical method shows an inherent perfection, which raises it far above every other table which has hitherto, though in vain, been tried or may yet be tried by analyzing the objects themselves dogmatically. It exhibits all synthetic *a priori* principles completely and according to one principle, viz., the faculty of judging in general, which constitutes the essence of experience as regards the understanding, so that we can be certain that there are no more such principles. This affords a satisfaction such as can never be attained by the dogmatic method. Yet this is not all; there is a still greater merit in it.

We must carefully bear in mind the ground of proof which shows the possibility of this cognition *a priori* and, at the same time, limits all such principles to a condition which must never be lost sight of, if they are not to be misunderstood and extended in use beyond what is allowed by the original sense which the understanding places in them. This limit is that they contain nothing but the conditions of possible experience in general so far as it is subjected to laws *a priori*. Consequently, I do not say that things *in themselves* possess a quantity, that their reality possesses a degree, their existence a connection of accidents in a substance, etc. This nobody can prove, because such a synthetic connection from mere concepts, without any reference to sensuous intuition on the one side or connection of such intuition in a possible experience on the other, is absolutely impossible. The essential limitation of the concepts in these principles, then, is that all things stand necessarily *a priori* under the aforementioned conditions only *as objects of experience*.

Hence there follows, secondly, a specifically peculiar mode of proof of these principles; they are not directly referred to appearances and to their relation, but to the possibility of experience, of which appearances constitute the matter only, not the form. Thus they are referred to objectively and universally valid synthetic propositions, in which we distinguish judgments of experience from those of perception. This takes place because appearances, as mere intuitions *occupying a part of space and time*, come under the concept of quantity, which synthetically unites their multiplicity *a priori* according to rules. Again, insofar as the perception contains, besides intuition, sensation, and between the latter and nothing (i.e., the total disappearance of sensation), there is an ever-decreasing transition, it is apparent

that the real in appearances must have a degree, so far as it (viz., the sensation) *does not itself occupy any part of space or of time.*[11] Still the transition to
sensation from empty time or empty space is only possible in time. Consequently, although sensation, as the quality of empirical intuition in respect
of its specific difference from other sensations, can never be cognized *a priori*, yet it can, in a possible experience in general, as a quantity of perception
be intensively distinguished from every other similar perception. Hence the
application of mathematics to nature, as regards the sensuous intuition by
which nature is given to us, is first made possible and determined.

Above all, the reader must pay attention to the mode of proof of the
principles which occur under the title of Analogies of Experience. For these
do not refer to the generation of intuitions, as do the principles of applying mathematics to natural science in general, but to the connection of
their existence in experience; and this can be nothing but the determination of their existence in time according to necessary laws, under which
alone the connection is objectively valid and thus becomes experience. The
proof, therefore, does not turn on the synthetic unity in the connection of
things in themselves, but merely of perceptions, and of these, not in regard
to their content, but to the determination of time and of the relation of
their existence in it according to universal laws. If the empirical determination in relative time is indeed to be objectively valid (i.e., experience), these
universal laws thus contain the necessity of the determination of existence
in time generally (viz., according to a rule of the understanding *a priori*).
Since these are prolegomena I cannot further descant on the subject, but
my reader (who has probably long been accustomed to consider experience
as a mere empirical synthesis of perceptions, and hence has not considered
that it goes much beyond them since it imparts to empirical judgments universal validity, and for that purpose requires a pure and *a priori* unity of the
understanding) is recommended to pay special attention to this distinction
of experience from a mere aggregate of perceptions and to judge the mode
of proof from this point of view.

§ 27. Now we are prepared to remove Hume's doubt. He justly maintains that we cannot comprehend by reason the possibility of causality,
that is, of the reference of the existence of one thing to the existence of
another which is necessitated by the former. I add that we comprehend just
as little the concept of subsistence, that is, the necessity that at the foundation of the existence of things there lies a subject which cannot itself be a
predicate of any other thing; nay, we cannot even form a concept of the
possibility of such a thing (though we can point out examples of its use in
experience). The very same incomprehensibility affects the community of
things, as we cannot comprehend how from the state of one thing an inference to the state of quite another thing beyond it, and *vice versa*, can be

drawn, and how substances which have each their own separate existence should depend upon one another necessarily. But I am very far from holding these concepts to be derived merely from experience, and the necessity represented in them to be fictitious and a mere illusion produced in us by long habit. On the contrary, I have amply shown that they and the principles derived from them are firmly established *a priori* before all experience and have their undoubted objective rightness, though only with regard to experience.

§ 28. Though I have no conception of such a connection of things in themselves, how they can either exist as substances, or act as causes, or stand in community with others (as parts of a real whole) and I can just as little think such properties in appearances as such (because those concepts contain nothing that lies in the appearances, but only what the understanding alone must think), we have yet a concept of such a connection of representations in our understanding and in judgments generally. This is the concept that representations belong in one sort of judgments as subject in relation to predicates; in another as ground in relation to consequent; and, in a third, as parts which constitute together a total possible cognition. Further we know *a priori* that without considering the representation of an object as determined with regard to one or the other of these moments, we can have no valid cognition of the object; and, if we should occupy ourselves with the object in itself, there is not a single possible attribute by which I could know that it is determined with regard to one or the other of these moments, that is, belonged under the concept of substance, or of cause, or (in relation to other substances) of community, for I have no conception of the possibility of such a connection of existence. But the question is not how things in themselves but how the empirical cognition of things is determined, as regards the above moments of judgments in general, that is, how things, as objects of experience, can and must be subsumed under these concepts of the understanding. And then it is clear that I completely comprehend, not only the possibility, but also the necessity, of subsuming all appearances under these concepts, that is, of using them as principles of the possibility of experience.

§ 29. In order to put to a test Hume's problematic concept (his *crux metaphysicorum*), the concept of cause, we have, in the first place, given *a priori* by means of logic the form of a conditional judgment in general, i.e., we have one given cognition as antecedent and another as consequent. But it is possible that in perception we may meet with a rule of relation which runs thus: that a certain appearance is constantly followed by another (though not conversely); and this is a case for me to use the hypothetical judgment and, for instance, to say that if the sun shines long enough upon a body it grows warm. Here there is indeed as yet no necessity of connection, or concept of cause. But I proceed and say that if this proposition, which is merely a subjective connection of perceptions, is to be a judgment of experience, it must be regarded as necessary and universally valid. Such a proposition would be that the sun is by its light the cause of heat. The empirical rule is now considered as a law, and as valid not merely of appearances but valid of them for the purposes of a possible experience which requires universal and therefore necessarily valid rules. I therefore easily comprehend the concept of cause as a concept necessarily belonging to the mere form of experience, and its possibility as a synthetic unification of perceptions in a consciousness in general; but I do not at all comprehend the possibility of a thing in general as a cause, inasmuch as the concept of cause denotes a condition not at all belonging to things, but to experience. For experience can only be an objectively valid cognition of appearances and of their succession, only so far as the antecedent appearances can be conjoined with the consequent ones according to the rule of hypothetical judgments.

§ 30. Hence if the pure concepts of the understanding try to go beyond objects of experience and be referred to things in themselves (*noumena*), they have no meaning whatever. They serve, as it were, only to spell out appearances, so that we may be able to read them as experience. The principles which arise from their reference to the sensible world only serve our understanding for use in experience. Beyond this they are arbitrary combinations without objective reality; and we can neither cognize their possibility *a priori*, nor verify their reference to objects, let alone make such reference understandable, by any example, because examples can only be borrowed from some possible experience, and consequently the objects of these concepts can be found nowhere but in a possible experience.

This complete (though to its originator unexpected) solution of Hume's problem rescues for the pure concepts of the understanding their *a priori* origin and for the universal laws of nature their validity as laws of the understanding, yet in such a way as to limit their use to experience, because their possibility depends solely on the reference of the understanding to experience, but with a completely reversed mode of connection which never occurred to Hume: they are not derived from experience, but experience is derived from them.

This is, therefore, the result of all our foregoing inquiries: "All synthetic principles *a priori* are nothing more than principles of possible experience" and can never be referred to things in themselves, but only to appearances as objects of experience. And hence pure mathematics as well as pure natural science can never be referred to anything more than mere appearances, and can only represent either that which makes experience in general possible, or else that which, as it is derived from these principles, must always be capable of being represented in some possible experience.

§ 31. And thus we have at last something determinate upon which to depend in all metaphysical enterprises, which have hitherto, boldly enough but always at random, attempted everything without discrimination. That the goal of their exertions should be set up so close struck neither the dogmatic thinkers nor those who, confident in their supposed sound common sense, started with concepts and principles of pure reason (which were legitimate and natural, but destined for mere empirical use) in search of insights for which they neither knew nor could know any determinate bounds, because they had never reflected nor were able to reflect on the nature or even on the possibility of such a pure understanding.

Many a naturalist of pure reason (by which I mean the man who believes he can decide in matters of metaphysics without any science) may pretend that he, long ago, by the prophetic spirit of his sound sense, not only suspected but knew and comprehended what is here propounded with so much ado, or, if he likes, with prolix and pedantic pomp: "that with all our reason we can never reach beyond the field of experience." But when he is questioned about his rational principles individually, he must grant that there are many of them which he has not taken from experience and which are therefore independent of it and valid *a priori.* How then and on what grounds will he restrain both himself and the dogmatist, who makes use of these concepts and principles beyond all possible experience because they are recognized to be independent of it? And even he, this adept in sound sense, in spite of all his assumed and cheaply acquired wisdom, is not exempt from wandering inadvertently beyond objects of experience into the field of chimeras. He is often deeply enough involved in them, though in announcing everything as mere probability, rational conjecture, or analogy, he gives by his popular language a color to his groundless pretensions.

§ 32. Since the oldest days of philosophy, inquirers into pure reason have thought that, besides the things of sense, or appearances (*phenomena*), which make up the sensible world, there were certain beings of the understanding (*noumena*), which should constitute an intelligible world. And as appearance and illusion were by those men identified (a thing which we may well excuse in an undeveloped epoch) actuality was only conceded to the beings of the understanding.

And we indeed, rightly considering objects of sense as mere appearances, confess thereby that they are based upon a thing in itself, though we know not this thing as it is in itself but only know its appearances, viz., the way in which our senses are affected by this unknown something. The understanding therefore, by assuming appearances, grants also the existence of things in themselves, and thus far we may say that the representation of such things as are the basis of appearances, consequently of mere beings of the understanding, is not only admissible but unavoidable.

Our critical deduction by no means excludes things of that sort (*noumena*), but rather limits the principles of the Aesthetic[12] in such a way that they shall not extend to all things (as everything would then be turned into mere appearance) but that they shall hold good only of objects of possible experience. Hereby, then, beings of the understanding are admitted, but with the inculcation of this rule which admits of no exception: that we neither know nor can know anything determinate whatever about these pure beings of the understanding, because our pure concepts of the understanding as well as our pure intuitions extend to nothing but objects of possible experience, consequently to mere things of sense; and as soon as we leave this sphere, these concepts retain no meaning whatever.

§ 33. There is indeed something seductive in our pure concepts of the understanding which tempts us to a transcendent use—a use which transcends all possible experience. Not only are our concepts of substance, of power, of action, of reality, and others, quite independent of experience, containing nothing of sense appearance, and so apparently applicable to things in themselves (*noumena*), but, what strengthens this conjecture, they contain a necessity of determination in themselves, which experience never attains. The concept of cause contains a rule according to which one state follows another necessarily; but experience can only show us that one state of things often or, at most, commonly follows another, and therefore affords neither strict universality nor necessity,

Hence concepts of the understanding seem to have a deeper meaning and content than can be exhausted by their merely empirical use, and so the understanding inadvertently adds for itself to the house of experience a much more extensive wing which it fills with nothing but beings of thought, without ever observing that it has transgressed with its otherwise legitimate concepts the bounds of their use.

§ 34. Two important and even indispensable, though very dry, investigations therefore became indispensable in the *Critique of Pure Reason* [viz., the two chapters "The Schematism of the Pure Concepts of the Understanding" and "The Ground of the Distinction of All Objects in General into Phenomena and Noumena"]. In the former there is shown that the senses furnish, not the pure concepts of the understanding *in concreto*, but only the schema for their use, and that the object conformable to it occurs only in experience (as the product of the understanding from materials of sensibility). In the latter there is shown that, although our pure concepts of the understanding and our principles are independent of experience, and despite the apparently greater sphere of their use, still nothing whatever can be thought by them beyond the field of experience, because they can do nothing but merely determine the logical form of the judgment with regard to given intuitions. But as there is no intuition at all beyond the field of sensibility, these pure concepts, since they cannot possibly be exhibited in *concreto*, are void of all meaning; consequently all these *noumena*, together with their sum total, the intelligible world,[13] are nothing but representations of a problem, the object of which in itself is quite possible but the solution, from the nature of our understanding, totally impossible. For our understanding is not a faculty of intuition but of the connection of given intuitions in an experience. Experience must therefore contain all the objects for our concepts; but beyond it no concepts have any meaning, since no intuition can be subsumed under them.

§ 35. The imagination may perhaps be forgiven for occasional vagaries and for not keeping carefully within the limits of experience, since it gains life and vigor by such flights and since it is always easier to moderate its boldness than to stimulate its languor. But the understanding which ought to *think* can never be forgiven for indulging in vagaries; for we depend upon it alone for assistance to set bounds, when necessary, to the vagaries of the imagination.

But the understanding begins its aberrations very innocently and modestly. It first discerns the elementary cognitions which inhere in it prior to all experience, but yet must always have their application in experience. It gradually drops these limits; and what is there to prevent it, inasmuch as it has quite freely derived its principles from itself? And then it proceeds first to newly thought out forces in nature, then to beings outside nature—in short, to a world for whose construction the materials cannot be wanting, because fertile fiction furnishes them abundantly, and though not confirmed is yet never refuted by experience. This is the reason why young thinkers are so partial to metaphysics in the truly dogmatical manner, and often sacrifice to it their time and their talents, which might be otherwise better employed.

But there is no use in trying to moderate these fruitless endeavors of pure reason by all manner of cautions as to the difficulties of solving questions so occult, by complaints of the limits of our reason, and by degrading our assertions into mere conjectures. For if their impossibility is not distinctly shown, and reason's knowledge of itself does not become a true science, in which the field of its right use is distinguished, so to say, with geometrical certainty from that of its worthless and idle use, these fruitless efforts will never be entirely abandoned.

§ 36. *How is nature itself possible?* This question—the highest point that transcendental philosophy can ever reach, and to which, as its boundary and completion, it must proceed—properly contains two questions.

First: How is nature possible in general in the *material* sense, i.e., according to intuition, as the totality of appearances; how are space, time, and that which fills both—the object of sensation—possible in general? The answer is: by means of the constitution of our sensibility, according to which it is in its special way affected by objects which are in themselves unknown to it and totally distinct from those appearances. This answer is given in the *Critique* itself in the Transcendental Aesthetic, and in these *Prolegomena* by the solution of the first main question.

Secondly: How is nature possible in the *formal* sense, as the totality of the rules under which all appearances must come in order to be thought as connected in an experience? The answer must be this: it is only possible by means of the constitution of our understanding, according to which all those representations of sensibility are necessarily referred to a consciousness, and by which the peculiar way in which we think (viz., by rules) and hence experience also are possible, but must be clearly distinguished from an insight into the objects in themselves. This answer is given in the *Critique* itself in the Transcendental Logic, and in these *Prolegomena* in the course of the solution of the second main question.

But how this peculiar property of our sensibility itself is possible, or that of our understanding and of the apperception which is necessarily its basis and also that of all thinking, cannot be further analyzed or answered, because it is of them that we are in need for all our answers and for all our thinking about objects.

There are many laws of nature which we can only know by means of experience; but conformity to law in the connection of appearances, i.e., nature in general, we cannot discover by any experience, because experience itself requires laws which are *a priori* at the basis of its possibility.

The possibility of experience in general is therefore at the same time the universal law of nature, and the principles of experience are the very laws of nature. For we know nature as nothing but the totality of appearances, i.e., of representations in us; and hence we can only derive the law of their connection from the principles of their connection in us, that is, from the conditions of their necessary unification in a consciousness, which constitutes the possibility of experience.

Even the main proposition expounded throughout this section—that universal laws of nature can be cognized *a priori*—leads of itself to the proposition that the highest legislation of nature must lie in ourselves, i.e., in our understanding; and that we must not seek the universal laws of nature in nature by means of experience, but conversely must seek nature, as to its universal conformity to law, in the conditions of the possibility of experience, which lie in our sensibility and in our understanding. For how would it otherwise be possible to know *a priori* these laws, as they are not rules of analytic cognition but truly synthetic extensions of it? Such a necessary agreement of the principles of possible experience with the laws of the possibility of nature can only proceed from one of two reasons: either these laws are drawn from nature by means of experience, or conversely nature is derived from the laws of the possibility of experience in general and is quite the same as the mere universal conformity to law of the latter. The former is self-contradictory, for the universal laws of nature can and must be cognized *a priori* (that is, independent of all experience) and be the foundation of all empirical use of the understanding; the latter alternative therefore alone remains.[14]

But we must distinguish the empirical laws of nature, which always presuppose particular perceptions, from the pure or universal laws of nature, which, without being based on particular perceptions, contain merely the conditions of their necessary unification in experience. With regard to the latter, nature and possible experience are quite the same, and as the conformity to law in the latter depends upon the necessary connection of appearances in experience (without which we cannot cognize any object whatever in the sensible world), consequently upon the original laws of the understanding, it seems at first strange, but is not the less certain, to say: *the understanding does not derive its laws (a priori) from, but prescribes them to, nature.*

§ 37. We shall illustrate this seemingly bold proposition by an example, which will show that laws which we discover in objects of sensuous intuition (especially when these laws are cognized as necessary) are already held by us to be such as have been placed there by the understanding, in spite of their being similar in all points to the laws of nature which we ascribe to experience.

§ 38. If we consider the properties of the circle, by which this figure at once combines into a universal rule so many arbitrary determinations of the space in it, we cannot avoid attributing a nature to this geometrical thing. Two lines, for example, which intersect each other and the circle, however they may be drawn, are always divided so that the rectangle constructed with the segments of the one is equal to that constructed with the segments of the other. The question now is: Does this law lie in the circle or in the understanding? That is, does this figure, independently of the understanding, contain in itself the ground of the law; or does the understanding, having constructed according to its concepts (of the equality of the radii) the figure itself, introduce into it this law of the chords intersecting in geometrical proportion? When we follow the proofs of this law, we soon perceive that it can only be derived from the condition on which the understanding founds the construction of this figure, viz., the equality of the radii. But if we enlarge this concept to pursue further the unity of manifold properties of geometrical figures under common laws and consider the circle as a conic section, which of course is subject to the same fundamental conditions of construction as other conic sections, we shall find that all the chords which intersect within the circle, ellipse, parabola, and hyperbola always intersect so that the rectangles of their segments are not indeed equal but always bear a constant ratio to one another. If we proceed still further to the fundamental doctrines of physical astronomy, we find a physical law of reciprocal attraction extending over the whole material nature, the rule of which is that it decreases inversely as the square of the distance from each attracting point, just as the spherical surfaces through which this force diffuses itself increase; and this law seems to be necessarily inherent in the very nature of things, so that it is usually propounded as cognizable *a priori*. Simple as the sources of this law are, merely resting upon the relation of spherical surfaces of different radii, its consequence is so excellent with regard to the variety and regularity of its agreement that not only are all possible orbits of the celestial bodies conic sections, but such a relation of these orbits to each other results that no other law of attraction than that of the inverse square of the distance can be thought as fit for a cosmical system.

Here, accordingly, is nature resting on laws which the understanding cognizes *a priori*, and chiefly from universal principles of the determination of space. Now I ask: do the laws of nature lie in space, and does the understanding learn them by merely endeavoring to find out the enormous wealth of meaning that lies in space; or do they inhere in the understanding and in the way in which it determines space according to the conditions of the synthetic unity in which its concepts are all centered? Space is something so uniform and as to all particular properties so indeterminate that we should certainly not seek a store of laws of nature in it. Whereas that which determines space to assume the form of a circle, or the figures of a cone and a sphere is the understanding, so far as it contains the ground of the unity of their constructions. The mere universal form of intuition, called space, must therefore be the substratum of all intuitions determinable to particular objects; and in it, of course, the condition of the possibility and of the variety of these intuitions lies. But the unity of the objects is entirely determined by the understanding, and according to conditions which lie in its own nature; and thus the understanding is the origin of the universal order of nature, in that it comprehends all appearances under its own laws and thereby brings about, in an *a priori* way, experience (as to its form), by means of which whatever is to be cognized only by experience is necessarily subjected to its laws. For we are not concerned with the nature of things in themselves, which is independent of the conditions both of our sensibility and our understanding, but with nature as an object of possible experience; and in this case the understanding, because it makes experience possible, thereby insists that the sensuous world is either not an object of experience at all, or else is nature.

Notes

(Notes are Kant's except for those in brackets, which were added by the translator.)

1. I freely grant that these examples do not represent such judgments of perception as ever could become judgments of experience, even though a concept of the understanding were superadded, because they refer merely to feeling, which everybody knows to be merely subjective and which, of course, can never be attributed to the object and, consequently, never can become objective. I only wished to give here an example of a judgment that is merely subjectively valid, containing no ground for necessary universal validity and thereby for a relation to the object. An example of the judgments of perception which become judgments of experience by superadded concepts of the understanding will be given in the next note.

2. As an easier example, we may take the following: when the sun shines on the stone, it grows warm. This judgment, however often I and others may have perceived it, is a mere judgment of perception and contains no necessity; perceptions are only usually conjoined in this manner. But if I say: the sun warms the stone, I add to the perception a concept of the understanding, viz., that of cause, which necessarily connects with the concept of sunshine that of heat, and the synthetic judgment becomes of necessity universally valid, viz., objective, and is converted from a perception into experience.

3. This name seems preferable to the term *particularia*, which is used for these judgments in logic. For the latter already contains the thought that they are not universal. But when I start from unity (in singular judgments) and proceed to totality, I must not [even indirectly and negatively] include any reference to totality. I think plurality merely without totality, and not the exclusion of totality. This is necessary, if the logical moments are to underlie the pure concepts of the understanding. In logical usage one may leave things as they were.

4. [See last sentence of §23.]

5. But how does the proposition that judgments of experience contain necessity in the synthesis of perceptions agree with my statement so often before inculcated that experience, as cognition *a posteriori*, can afford contingent judgments only? When I say that experience teaches me something, I mean only the perception that lies in experience—for example, that heat always follows the shining of the sun on a stone; consequently, the proposition of experience is always so far contingent. That this heat necessarily follows the shining of the sun is contained indeed in the judgment of experience (by means of the concept of cause), yet is a fact not learned by experience; for, conversely, experience is first of all generated by this addition of the concept of the understanding (of cause) to perception. How perception attains this addition may be seen by referring in the *Critique* itself to the section on the transcendental faculty of judgment, B 176 *et seq.*

6. The three following paragraphs will hardly be understood unless reference be made to what the *Critique* itself says on the subject of the principles; they will, however, be of service in giving a general view of the principles, and in fixing the attention on the main moments. [See *Critique*, B 187–294.]

7. [The Axioms of Intuition. See *Critique*, B 202–207.]

8. [The Anticipations of Perception. See *ibid.*, B 207–218.]

9. [The Analogies of Experience. See *ibid.*, B 218–265.]

10. [The Postulates of Empirical Thought. See *ibid.*, B 265–294.]

11. Heat and light are in a small space just as large, as to degree, as in a large one; in like manner the internal representations, pain, consciousness in general, whether they last a short or a long time, need not vary as to the degree. Hence the quantity is here in a point and in a moment just as great as in any space or time, however great. Degrees are thus quantities not in intuition but in mere sensation (or the quantity of the content of an intuition). Hence they can only he estimated quantitatively by the relation of 1 to 0, viz.,

by their capability of decreasing by infinite intermediate degrees to disappearance, or of increasing from naught through infinite gradations to a determinate sensation in a certain time. *Quantitas qualitatis est gradus* [the quantity of quality is degree].

12. [The principles of sensibility (space and time). See *Critique of Pure Reason*, B 33–B 73.]

13. We speak of the "intelligible world," not (as the usual expression is) "intellectual world." For cognitions are intellectual through the understanding and refer to our world of sense also, but objects, insofar as they can be represented merely by the understanding, and to which none of our sensible intuitions can refer, are termed "intelligible." But as some possible intuition must correspond to every object, we would have to think an understanding that intuits things immediately; but of such we have not the least concept, nor of *beings of the understanding* to which it should be applied.

14. Crusius alone thought of a compromise: that a spirit who can neither err nor deceive implanted these laws in us originally. But since false principles often intrude themselves, as indeed the very system of this man shows in not a few examples, we are involved in difficulties as to the use of such a principle in the absence of sure criteria to distinguish the genuine origin from the spurious, for we never can know certainly what the spirit of truth or the father of lies may have instilled into us.

Natural History: Catastrophism and Uniformitarianism

CHAPTER 24

Introduction to "Catastrophism and Uniformitarianism"

The nineteenth-century debate over "catastrophism" and "uniformitarianism" in geology had a significant influence on the development of, and philosophical commentary on, the life sciences. The terms were coined by Whewell, who also provided one of the first thematic analyses of the debate.

The third chapter of Michael Ruse's *The Darwinian Revolution* (1999) contains a detailed and rich dicussion of the debate and its implications; much of the discussion below is drawn from Ruse's analysis and from the original texts.

Catastrophism is the view that "Much of the past might have been like the present, but every now and then the uniform course of nature is shattered" (Ruse 1999, 38). Uniformitarianism, on the other hand, is the view that "past geological phenomena" can be explained "not only in terms of causes of the same *kind* now operating but also in terms of causes of the same *degree*" (Ruse 1999, 40).

Catastrophists included Georges Cuvier (1769–1832) and Joachim Barrande (1799–1883), and probably George-Louis de Buffon (1707–1788) in France; William Buckland (1784–1856), Adam Sedgwick (1785–1873), and Whewell himself in England; and Louis Agassiz (1807–1873) in Switzerland and the United States.

Uniformitarians included Charles Lyell (1797–1875), John Playfair (1748–1819), and James Hutton (1726–1797), and Charles Darwin (1809–1882) in England.

Ruse points out that the debate between the "intellectual ancestors of the uniformitarians," the "Vulcanists," and the ancestors of the catastrophists,

377

the "Neptunists," is germane to the catastrophist-uniformitarian debate (p. 37). The Vulcanists included James Hutton, whose work was popularized by John Playfair; their view was that

> geological formations come from a combination of weathering and heat. Wind and rain and such eventually lead to the deposition of silt on the ocean bed; heat from within the earth, combined with terrific pressure, causes this detritus to fuse into solid rock; then through heat expansion and volcanic action geological formations are thrown up, completing the cycle. [. . .] The Vulcanists' vision of the earth as subject to a constantly repeating cycle implies its great age. With reason, Hutton's best-known geological claim was that "We find no vestige of a beginning,—no prospect of an end."
> (final quotation Hutton 1795, I:200; Ruse 1999, 37)

According to the Neptunists:

> water and precipitation determined virtually everything. At one point the whole earth had been under water; then, bit by bit, the various earth formations were precipitated out. Only recently did some coal deposits catch fire, thus bringing about volcanoes that formed certain localized igneous rocks. Essentially, the earth's rocks are sedimentary. Although by the beginning of the nineteenth century all serious geologists were beginning to see the earth as fairly old, the Neptunists did not believe it was nearly so old as presupposed by Vulcanism. Moreover, unlike the Vulcanists, the Neptunists saw the world historically: for them it had an identifiable beginning and direction.
> (Ruse 1999, 37)

Cuvier defended what became the classic account of catastrophism, influenced by his own version of Neptunism. He argued that the geological record of organisms and meteorological phenomena was evidence for floodings and other catastrophic events, which he refers to as "revolutions," that "both in nature and cause" were unlike any process going on at the present time (Ruse 1999, 38).

In the selection printed here, Cuvier remarks,

> We are therefore forcibly led to believe, not only that the sea has at one period or another covered all our plains, but that it must have remained there for a long time, and in a state of tranquillity; which circumstance was necessary for the formation of deposits so extensive, so thick, in part so solid, and containing exuviæ so perfectly preserved.

[. . .] Hence it is evident that the basin or reservoir containing the sea has undergone some change at least, either in extent, or in situation, or in both.

Comparisons of geological strata and fossils yield conclusions about "revolutions" that have taken place, whether in the position and extent of bodies of water, or in the relative position of geological strata, or in "changes in condition" of animals and geological formations:

> §6 Life, therefore, has been often disturbed on this earth by terrible events—calamities which, at their commencement, have perhaps moved and overturned to a great depth the entire outer crust of the globe, but which, since these first commotions, have uniformly acted at a less depth and less generally. Numberless living beings have been the victims of these catastrophes; some have been destroyed by sudden inundations, others have been laid dry in consequence of the bottom of the seas being instantaneously elevated. Their races even have become extinct, and have left no memorial of them except some small fragment which the naturalist can scarcely recognise.

Cuvier's view was challenged by the uniformitarians, among the first of whom was Charles Lyell. The first volume of Lyell's *Principles of Geology* appeared in July 1830. There, Lyell argued for three related theses:

> **Actualism** "He wanted to explain past geological phenomena in terms of causes of the kind now operating";
> **Uniformitarianism** "He wanted to explain past geological phenomena not only in terms of causes of the same *kind* now operating but also in terms of causes of the same *degree*";
> **Steady-state view of the earth** "There is no sign of any direction or progression in either the inorganic or the organic world."
>
> (Ruse 1999, 40)

As can be seen from the discussion of Neptunism and Vulcanism above, one's position on the steady-state view is significant to the question of whether one sees a beginning or end to the earth. This question was considered to be of philosophical significance as well; it was even taken on by a young Immanuel Kant in *The question of whether the earth is aging* (Kant 2012/various).

Two catastrophists, Adam Sedgwick and Whewell, replied to Lyell's *Principles* (Ruse 1999, 44–48). As Ruse traces the history, the influence of Lyell is key to the development of Darwin's theory of geology; Ruse explains

Darwin's researches in South America in 1831 as cementing his status as a "Lyellian geologist" who accepts all three of the Lyellian principles above (pp. 48–56).

A question in the case of geology that arises in the texts printed here is the question of the evaluation of the geological record. The question of the incompleteness or completeness of the geological record, of its adequacy or inadequacy for supporting uniformitarian or catastrophist conclusions, is paramount in these texts. De Buffon early on refers to the difficulty of acquiring geological and biological specimens, and remarks,

> when specimens of everything that inhabits the earth have been collected; when, after much difficulty, examples of all things that are found scattered so profusely on the earth have been brought together in one location; and when for the first time this storehouse filled with things diverse, new, and strange is viewed, the first sensation that results is bewilderment, mixed with admiration. [. . .]

Nonetheless, de Buffon observes, many develop a methodology that tries to apply uniform reasoning to such distinct and diverse phenomena:

> We are naturally led to imagine that there is a kind of order and uniformity throughout nature. And when the works of nature are only cursorily examined, it appears at first that she has always worked upon the same plan. Since we ourselves know only one way of arriving at a conclusion, we persuade ourselves that nature creates and carries out everything by the same means and by similar operations. This manner of thinking causes us to invent an infinity of false connections between the things nature produces. Plants have been compared with animals, and minerals have been supposedly observed to vegetate. Their quite different organization and their quite distinct means of operation have often been reduced to the same form. The common matrix of these things so unlike each other lies less in nature than in the narrow mind of those who have poorly conceived her, and who know as little about appraising the strength of a truth as they do about the proper limits of comparative analogy.

De Buffon's point does not necessitate a rejection of uniformitarianism (or, better, actualism) more generally, but de Buffon goes on to argue that "the Creator" can vary causes and processes at will, which appears to be at least an indirect support for catastrophism.

Whewell argues more directly, on the grounds of the incompleteness of the evidence, against actualism and uniformitarianism. According to

Whewell, we do not have sufficient evidence to conclude that the causes and processes of the past exactly resemble the causes and processes of the present; as he puts it in the text printed here, to so conclude would result in illicitly "cramping our speculations" a priori.

Playfair responds to Cuvier, de Buffon, Whewell, and others, that the geological record is good enough evidence to support uniformitarianism:

> If it be said, that only a small part of the earth's surface has yet been surveyed, and described with such accuracy as is found in the writers just named [Deluc, Saussure, Bergman, Ferber, Dolomieu], it may be answered, that the earth is constructed with such a degree of uniformity, that a tract of no very large extent may afford instances of all the leading facts that we can ever observe in the mineral kingdom. The variety of geological appearances which a traveller meets with, is not at all in proportion to the extent of country he traverses; and if he take in a portion of land sufficient to include primitive and secondary strata, together with mountains, rivers, and plains, and unstratified bodies in veins and in masses, though it be not a very large part of the earth's surface, he may find examples of all the most important facts in the history of fossils. Though the labours of mineralogists have embraced but a small part of the globe, they may therefore have comprehended a very large proportion of the phenomena which it exhibits; and hence a presumption arises, that the outlines, at least, of geology have now been traced with tolerable truth, and are not susceptible of great variation. (§449)

It is instructive to compare Playfair's statements here with Darwin's statements in the conclusion of his text printed here, about the imperfection of the geological record. Both are uniformitarian, but come to distinct conclusions.

In the eleventh chapter of the *Origin of Species* (the sixth edition), Darwin argues for two conclusions:

> Thus, on the theory of descent with modification, the main facts with respect to the mutual affinities of the extinct forms of life to each other and to living forms, are explained in a satisfactory manner. And they are wholly inexplicable on any other view.
>
> On this same theory, it is evident that the fauna during any one great period in the earth's history will be intermediate in general character between that which preceded and that which succeeded it.

With respect to the latter, he remarks

We must, however, allow for the entire extinction of some preceding forms, and in any one region for the immigration of new forms from other regions, and for a large amount of modification during the long and blank intervals between the successive formations. Subject to these allowances, the fauna of each geological period undoubtedly is intermediate in character, between the preceding and succeeding faunas. I need give only one instance, namely, the manner in which the fossils of the Devonian system, when this system was first discovered, were at once recognised by palæontologists as intermediate in character between those of the overlying carboniferous, and underlying Silurian systems. But each fauna is not necessarily exactly intermediate, as unequal intervals of time have elapsed between consecutive formations.

Rudwick's *Devonian Controversy* (1985) is a chronicle of this research. In the concluding *Summary*, Darwin says he has argued that

the geological record is extremely imperfect; that only a small portion of the globe has been geologically explored with care; that only certain classes of organic beings have been largely preserved in a fossil state; that the number both of specimens and of species, preserved in our museums, is absolutely as nothing compared with the number of generations which must have passed away even during a single formation.

Darwin concludes that, while the "imperfection" of the geological record may give rise to objections, the theory of "descent with modification through variation and natural selection" can explain salient facts, including

how new species come in slowly and successively; how species of different classes do not necessarily change together, or at the same rate, or in the same degree; yet in the long run that all undergo modification to some extent.

Darwin argues against catastrophism in the realm of biology and palaeontology, and in favor of actualism and uniformitarianism. Throughout the text, he reiterates that the principles that govern ancient descent of species are the same as those governing present descent by variation and modification.

The debates between Neptunists and Vulcanists, uniformity and catastrophe, had philosophical implications. Ruse 1999 discusses the implications of Herschel's *verae causae* doctrine and Whewell's notion of the consilience of inductions for uniformitarianism and catastrophism (pp. 59–63; for more on Whewell's notion of induction see Part II-A of this book).

For Herschel, the model of a completed scientific theory – that toward which all science should aim – was what today we call a 'hypothetico-deductive system.' [. . .] Herschel "found reference to causes essential. Furthermore, he thought that one ought to refer if possible only to causes of a particular kind – *verae causae* (true causes). But how can we be sure that we have *verae causae*? The answer is simple. We can and should argue analogically from our own experience: 'If the analogy of two phenomena be very close and striking, while, at the same time, the cause of one is very obvious, it becomes scarcely possible to refuse to admit the action of an analogous cause in the other, though not so obvious in itself'."

(final quotation Herschel 1831, 149; Ruse 1999, 56–8)

While he agreed with Herschel's theory of science in many respects, Whewell

did deny Herschel's empiricist doctrine of the *vera causa*, where one argues analogically from the experientially known to the unknown. Whewell (1840, 2:442) felt that this interpretation unduly limits any methodological rule based on it [. . .] Whewell tied his interpretation to what he called the 'consilience of inductions'. In particular, he argued that the mark of the best kind of science – the definitive evidence that one is starting from true axioms – comes when different areas of science are brought together and shown to spring from the same principles.

(Ruse 1999, 58)

Further Research

Many books in this tradition are available freely online, and would be excellent resources for related research projects. The following are a representative sample.

Note: Many of the works that Darwin referred to in the chapter from the *Origin* printed here are available online; some of them are listed below.

Charles Lyell. 1830–33. *Principles of Geology*, in three volumes. London: John Murray.
John Herschel. 1831. *A Preliminary Discourse on the Study of Natural Philosophy*. London: Longman, Rees, Orme, Brown, and Green.
James Hutton. 1795. *Theory of the Earth*. Edinburgh: William Creech.
William Whewell. 1840. *The Philosophy of the Inductive Sciences, founded upon their history*, 2 vols. London: John W. Parker.
Adam Sedgwick. April and July 1825. "On the Origin of Alluvial and Diluvial Formations," *Annals of Philosophy*.

Louis Agassiz. 1866. *Geological Sketches*. Boston: Ticknor and Fields.
Louis Agassiz. 1864. *Methods of Study in Natural History*. Boston: Ticknor and Fields.
William Buckland. 1836. *Geology and Mineralogy*. London: William Pickering.
Hugh Falconer and Proby Thomas Cautley. 1846. *Fauna Antiqua Sivalensis: Being the Fossil Zoology of the Sewalik Hills, in the North of India*. London: Smith, Elder, and Co. [In English.]
Édouard Verneuil and Adolphe d'Archiac. 1842. *Memoir on the Fossils of the Older Deposits in the Rhenish Provinces*. Paris: Bourgogne and Martinet.

References and Further Reading

Agassiz, Louis. 1866. *Geological Sketches*, vol. 1. Boston: Ticknor and Fields.
——. 1874. *The Structure of Animal Life*, 3rd ed. New York: Scribner, Armstrong and Co.
Baer, Karl Ernst von. 1828–1837. *Über Entwickelungsgeschichte der Thiere: Beobachtung und Reflexion*, 2 vols. Königsberg: Bornträger.
——. 1876. "Ueber Darwins Lehre," in his *Studien auf dem Gebiete der Naturwissenschaften*, 2 vols. St. Petersberg: Schmitzdorff. 2: 235–480.
Buckland, William. 1824. *Reliquiae diluvianae, or Observations on the organic remains contained in caves, fissures and diluvial gravel and on other geological phenomena attesting the action of an universal deluge*. London: John Murray.
Costa, James. 2009. *The Annotated Origin: A Facsimile of the First Edition of On the Origin of Species*. Cambridge, Mass.: Harvard University Press.
Danby, T. W. 1873. "Obituary – M. de Verneuil," *Geological Society* 10: 429–430.
Dean, Dennis. 1992. *James Hutton and the History of Geology*. Ithaca, NY: Cornell University Press.
Falconer, Hugh and Cautley, Proby Thomas. 1846. *Fauna Antiqua Sivalensis: Being the Fossil Zoology of the Sewalik Hills, in the North of India*. London: Smith, Elder, and Co. [In English.]
Fisch, Menachem. 1985a. "Necessary and Contingent Truth in William Whewell's Antithetical Theory of Knowledge," *Studies in History and Philosophy of Science* 16: 275–314.
——. 1985b. "Whewell's Consilience of Inductions: An Evaluation," *Philosophy of Science* 52: 239–55.
——. 1991. *William Whewell, Philosopher of Science*. Oxford: Oxford University Press.
Fisch, Menachem and Simon Schaffer (eds.). 1991. *William Whewell: A Composite Portrait*. Oxford: Oxford University Press.
Herschel, John. 1831. *A Preliminary Discourse on the Study of Natural Philosophy*. London: Longman, Rees, Orme, Brown, and Green.
——. 1841. "Whewell on Inductive Sciences," *Quarterly Review* 68: 177–238.
Hodge, Johnathan and Radick, Gregory. 2003. *The Cambridge Companion to Darwin*. Cambridge: Cambridge University Press.
Hutton, James. 1795. *Theory of the Earth*. Edinburgh: William Creech.
Kant, Immanuel. 2012/various. *Natural Science*, ed. Eric Watkins, trans. Beck, Edwards, Reinhardt, Schönfeld, and Watkins. Cambridge: Cambridge University Press.
Love, Alan. (ed.) Forthcoming. *Conceptual Change in Biology: Scientific and Philosophical Perspectives on Evolution and Development*. Boston Studies in Philosophy of Science. Berlin: Springer.
Lyell, Charles. 1830. *Principles of Geology*, vol. I. London: John Murray.
McGrew, Timothy, Alspector-Kelly, Marc, and Allhoff, Fritz. 2009. *Philosophy of Science: An Historical Anthology*. Chichester, UK: Wiley-Blackwell.
Playfair, John. 1802. *Illustrations of the Huttonian Theory of the Earth*. Printed for Cadell and Davies, London, and W. Creech, Edinburgh.

Richards, Robert. 2008. "Karl Ernst von Baer," in *Evolution: The First Four Billion Years*, ed. Michael Ruse and Joseph Travis. Cambridge, Mass.: Harvard University Press.

Rudwick, Martin. 1985. *The Meaning of Fossils: Episodes in the History of Palaeontology*, second ed. Chicago: University of Chicago Press.

——. 2008. *Worlds Before Adam: The Reconstruction of Geohistory in the Age of Reform*. Chicago: The University of Chicago Press.

Rupke, Nicolaas. 1983. *The Great Chain of History: William Buckland and the English School of Geology*. Oxford: Clarendon Press.

Ruse, Michael and Richards, Robert. 2009. *The Cambridge Companion to* The Origin of Species. Cambridge: Cambridge University Press.

Ruse, Michael. 1975. "Darwin's Debt to Philosophy: An Examination of the Influence of the Philosophical Ideas of John F. W. Herschel and William Whewell on the Development of Charles Darwin's Theory of Evolution," *Studies in History and Philosophy of Science*, 6: 159–181.

——. 1976. "The Scientific Methodology of William Whewell," *Centaurus*, 20: 227–257.

——. 1991. "William Whewell: Omniscientist," in M. Fisch and S. Schaffer (eds.), pp. 87–116.

——. 1999. *The Darwinian Revolution*. Chicago: University of Chicago Press.

Sedgwick, Adam. 1825. "On the Origin of Alluvial and Diluvial Formations," *Annals of Philosophy* April and July 1825.

Snyder, Laura. 2006. *Reforming Philosophy: A Victorian Debate on Science and Society*. Chicago: University of Chicago Press.

Verneuil, Édouard and d'Archiac, Adolphe. 1842. *Memoir on the Fossils of the Older Deposits in the Rhenish Provinces*. Paris: Bourgogne and Martinet.

Whewell, William. 1840. *The Philosophy of the Inductive Sciences, founded upon their history*, 2 vols. London: John W. Parker.

Wilson, D. B. 1974. "Herschel and Whewell's Versions of Newtonianism," *Journal of the History of Ideas* 35: 79–97.

CHAPTER 25

Initial Discourse

GEORGE-LOUIS DE BUFFON

Trans. John Lyon in *From Natural History to the History of Nature*, South Bend: University of Notre Dame Press, 1981/1749.

Note from the translator[1] The text used for translation purposes is the "Premier Discours" from the *Histoire naturelle, générale et particulière avec la description du cabinet du Roi* (Deux-Ponts: Sanson, 1785–1790), vol. I, pp. 7–69. This pirated edition was the last edition of Buffon's work to appear during his lifetime. It has been checked against the first edition (Paris: Imprimerie Royale, 1749) vol. I, as found in Jean Piveteau, ed., *Oeuvres philosophiques de Buffon* (Paris: Presses Universitaires de France, 1954). I would like to thank the University of Oklahoma Library for the use of their copy of the Deux-Ponts edition.

Natural history, taken in its fullest extent, is an immense subject. It embraces all objects which the universe displays to us. This prodigious multitude of quadrupeds, birds, fishes, insects, plants, minerals, etc., offers to the curiosity of the human mind a vast spectacle, the totality of which is so grand that it appears, and indeed is, inexhaustible in its details. A single division of natural history, such as the history of insects, or the history of plants, is vast enough to occupy the attention of many men. The objects which these particular branches of natural history present are so multitudinous that the most capable observers, after many years' work, have given only very imperfect rough outlines of those branches to which they have been singularly devoted. However, they have done all that they were capable of doing. And, far from blaming these observers for the trifling advancement of the science to which their work has been devoted, one could not give them too much praise for their assiduity and patience. It is impossible to deny that they possess the very highest qualities, for it takes a peculiar force of genius and courage of mind to be able to envisage nature in the innumerable multitude of its productions without losing one's orientation, and to believe oneself capable of understanding and comparing such productions. It takes a particular predilection to love these things, a predilection beyond that which has as its goal only particular objects. For it can be said that the love of the study of nature supposes two qualities of mind which are apparently in opposition to each other: the grand view of an intense intellectual power which takes in everything at a glance, and the detailed attention of an instinct which concentrates laboriously on a single minute detail.

The first obstacle encountered in the study of natural history comes from this great multiplicity of objects. But the variety of these same objects, and the difficulty of bringing together the various productions of different regions is another apparently insurmountable obstacle to the advancement of our understanding, an obstacle which in fact work alone is unable to surmount. It is only by dint of time, care, expenditure of money, and often by lucky accidents, that one is able to obtain well-preserved specimens of each species of animal, plant, or mineral, and thus form a well-ordered collection of all the works of nature.

But when specimens of everything that inhabits the earth have been collected; when, after much difficulty, examples of all things that are found scattered so profusely on the earth have been brought together in one location; and when for the first time this storehouse filled with things diverse, new, and strange is viewed, the first sensation that results is bewilderment, mixed with admiration. And the initial thought that follows is a humbling self-reflection. It seems unimaginable that, even with time, one could come to the point of distinguishing all these different objects, or that one could succeed not only in distinguishing them by their form, but further by knowing all that pertains to the birth, the generation, the organization, the habits—in a word, all that pertains to the history of each thing in particular. However, as these objects become familiar, after they have been seen often and, so to speak, without any plan, they slowly create lasting impressions, which are soon bound together in our mind by fixed and invariable relationships. Furthermore, despite ourselves, we construct more general views by which we are able to embrace at one and the same time many different objects. And it is thus that we find ourselves in a position to undertake disciplined study, to reflect fruitfully, and to open up for ourselves routes by which we may arrive at useful discoveries.

Thus, a beginning should be made by observing things often and by frequently reexamining them. However necessary attention to the whole may be, here, at the beginning, one may dispense with this responsibility: I mean that scrupulous attention which is always useful when a great number of things are undertaken, and often detrimental to those who are beginning to learn natural history. The essential thing is to fill the heads of such beginners with ideas and facts, and thus prevent them, if possible, from prematurely establishing schemata. For it always happens that through ignorance of certain facts and through a limited stock of ideas, such neophytes use up their energy in false combinations, and load their memories with vague consequences and results contrary to truth, which form in the sequel preconceptions that are difficult to erase. [. . .]

But let us return to the man who would apply himself seriously to the study of nature, and take up again a consideration of the subject at the point at which we let it drop, namely, at the point at which the adept begins to generalize ideas and to form for himself a method of arrangement and systems of explication. It is at this point that he should consult those who are proficient in the field, read solid authors, examine their various methods, and borrow insights wherever he comes upon them. But since it ordinarily happens that one is easily carried away at this point by his affection and taste for certain authors, or for a certain method, and that often, without a sufficiently mature examination, it is easy to adopt a system which is sometimes ill-conceived, it is proper that we give here several preliminary notions about the methods that have been devised in order to facilitate a knowledge of natural history. The methods are very useful, when applied with appropriate restrictions. They shorten the work, assist the memory, and offer to the mind a series of ideas composed indeed of objects which differ among themselves but which nevertheless have certain common relations. These common relations then form stronger impressions than would be the case with discrete objects which have no connection among themselves. Therein lies the utility of the various methods. But the disadvantage here is the tendency to overextend or to unduly constrict the chain of connections, to wish to subject the laws of nature to arbitrary laws, to wish to divide this chain where it is not divisible, and to wish to measure its strength by means of our weak imagination. Another drawback which is no less serious, and which is the contrary of the one just described, is the temptation to restrict oneself to a regime of overly-detailed methods, and thus to wish to judge of the whole by a single instance, to reduce nature to the status of petty systems which are foreign to her, and, from her immense works, to fashion arbitrarily just as many unconnected assemblages of data as there are petty systems. The final disadvantage of such methods is that, in multiplying names and systems, they make the language of science more difficult than science itself.

We are naturally led to imagine that there is a kind of order and uniformity throughout nature. And when the works of nature are only cursorily examined, it appears at first that she has always worked upon the same plan. Since we ourselves know only one way of arriving at a conclusion, we persuade ourselves that nature creates and carries out everything by the same means and by similar operations. This manner of thinking causes us to invent an infinity of false connections between the things nature produces. Plants have been compared with animals, and minerals have been supposedly observed to vegetate. Their quite different organization and their quite distinct means of operation have often been reduced to the same form. The common matrix of these things so unlike each other lies less in nature than in the narrow mind of those who have poorly conceived her, and who know as little about appraising the strength of a truth as they do about the proper limits of comparative analogy. For example, since blood circulates, must it be asserted that the sap of plants circulates also? Or should it be concluded that there is a growth in minerals like that known in plants? Is it proper to proceed from the movement of the blood to that of the sap, and from that to the movement of the petrifying juice? Isn't what we are doing in these cases only bringing the abstractions or our limited mind to bear upon the reality of the works of the Creator, and granting to him, so to speak, only such ideas as we possess on the matter? Nevertheless, such poorly founded statements have been made and are repeated every day. Systems are constructed upon uncertain facts which have never been examined, and which only go to show the penchant men have for wishing to find resemblances between most disparate objects, regularity where variety reigns, and order among those things which they perceive only in a confused manner.

For, when not stopping at superficial knowledge—which only gives us incomplete ideas of the productions and methods of nature—we wish to penetrate further and examine more meticulously the form and behavior of nature's works, it is surprising what variety of design, and what a multiplicity of means we see. The number of the productions of nature, however prodigious, is only the least part of our astonishment. Nature's mechanism, art, resources, even its confusion, fill us with admiration. Dwarfed before that immensity, overwhelmed by the number of wonders, the human mind staggers. The hand of the Creator does not appear to be opened in order to give existence to a certain limited number of species. Rather, it appears as if it might have cast into existence all at once a world of beings some of whom are related to each other, and some not; a world of infinite combinations, some harmonious and some opposed; a world of perpetual destruction and renewal. What an impression of power this spectacle offers us! What sentiments of respect this view of the universe inspires in us for its Author! And what would be the case in this regard if the weak light which guides us became sufficiently keen to allow us to perceive the general order of causes and of the dependence of effects? But the greatest mind, the most powerful genius, will never lift itself to such a pinnacle of knowledge. The first causes of things will remain ever hidden from us, and the general results of these causes will remain as difficult for us to know as the causes themselves. All that is given to us is to perceive certain particular effects, to compare these with each other, to combine them, and, finally, to recognize therein more of an order appropriate to our own nature than one pertaining to the existence of the things which we are considering.

But seeing that this is the only route open to us, and since we have no other means of arriving at a knowledge of the things of nature, it is necessary to follow that route as far as it can lead us. We must gather together all the objects, compare them, study them, and extract from the totality of their connections all the insights which may be able to assist us to see them clearly and to know them better.

The first truth which issues from this serious examination of nature is a truth which perhaps humbles man. This truth is that he ought to classify himself with the animals, to whom his whole material being connects him. The instinct of animals will perhaps appear to man even more certain than his own reason, and their industry more admirable than his arts. Then, examining successively and by order the various objects which compose the universe, and placing himself at the head of all created beings, man will see with astonishment that it is possible to descend by almost imperceptible degrees from the most perfect of creatures to the most formless matter, from the most perfectly formed animal to the most amorphous mineral. He will recognize that these imperceptible nuances are the great work of nature, and will find them not only in the size and shape of things, but in changes, productions, and successions of every sort.

In thoroughly studying this idea, one sees clearly that it is impossible to establish one general system, one perfect method, not only for the whole of natural history, but even for one of its branches. For in order to make a system, an arrangement—in a word, a general method—it is necessary that everything be taken in by it. It is necessary to divide the whole under consideration into different classes, apportion these classes into genera, subdivide these genera into species, and to do all this following a principle of arrangement in which there is of necessity an element of arbitrariness. But nature proceeds by unknown gradations, and, consequently, it is impossible to describe her with full accuracy by such divisions, since she passes from one species to another, and often from one genus to another, by imperceptible nuances. As a result, one finds a great number of intermediate species and mixed objects which it is impossible to categorize and which necessarily upset the project of a general system. This truth is too important for me not to insist on whatever might make it clear and evident.

Take botany, for example, that admirable part of natural history which, by virtue of its utility, has deserved at all times to be the most cultivated. Let us call to mind the principles of all methods which botanists have given us. We shall see with some surprise that they have always had in view the aim of comprehending in their methods generally all species of plants, and that none of them have been completely successful. It always turns out that, in each of these methods, a certain number of plants must be considered anomalous, their species falling between two genera; and it has been impossible to categorize them, because there is no more reason to ascribe them to the one genus than to the other. Indeed, to propose to devise a perfect system is to propose an impossibility. It would necessitate a work which would represent exactly all the works of nature. But, contrary to such hopes, it always happens that, despite all known methods and despite any assistance which can be had from the most enlightened system of botany, species are constantly being discovered which it is not possible to assimilate to any of the genera posited by such systems. Experience accords with reason on this point, and one ought to be convinced that it is not possible to design a general and perfect system in botany. However, it appears that the search for such a general system may be the search for a kind of "philosopher's stone" for botanists, a search which they have pursued with infinite pains and infinite labor. Some have taken forty years, and some fifty, in the creation of their systems, and what has happened in botany is the same as what has happened in chemistry, namely, that in the pursuit of the philosopher's stone—which has not been found—an infinite number of useful things have been discovered. Thus, from wishing to design a general and perfect system in botany, plants and their usages have been studied in more detail and have come to be better known. In this respect it is true that men need an imaginary goal in order to sustain them in their work. For if they had been persuaded that they would do only what in effect they are capable of doing, they would do nothing at all.

This proclivity which botanists have for establishing general systems with pretensions of perfection and methodological rigor is thus poorly founded. Consequently, their labors deliver to us only defective systems which have been successively destroyed, the one by the other, and have undergone the common fate of all systems founded on arbitrary principles. And what has contributed the most to this process of successive destruction is the freedom which botanists have allowed themselves of choosing arbitrarily a single feature of plants as a distinguishing characteristic. Some have established their method on the basis of the configuration of leaves; others on their position; others on the form of the flowers; some on the number of flower petals; other, finally, on the basis of the number of stamens. I would never finish if I wished to report in detail all the systems which have been imagined. But at the present time I wish to speak only of those systems which have had a good reception and have been followed, each in its turn, without sufficient attention having been given to that erroneous principle which all these systems share, namely, the desire to judge a whole or a combination of many wholes on the basis of a single part, and by comparing the differences of such single parts. For to desire to discern the differences of plants using solely the configurations of their leaves or their flowers as criteria is as if one set out to discern the differences between animals by means of the variations in their skins or generative organs. For who does not see that whatever proceeds in such a manner cannot be considered a science? It is at the very most only a convention, an arbitrary language, a means of mutual understanding. But no real cognizance [*connaissance réele*] of things can result from it.

Might I be permitted to speak my mind upon the origin of these various systems and upon the causes which have multiplied them to the point that botany itself is actually easier to apprehend than the nomenclature which is merely its language? May I be permitted to say that it would be preferable for a man to have engraved in his memory the forms of all plants and have clear ideas of them, which is what botany really is, than to memorize all the names which the various systems give to these plants, as a result of which scientific terminology has become more difficult than science itself? Here, then, is how it appears to me that this state of affairs has arisen. In the first place, the members of the plant kingdom were divided according to their various sizes. There are, after this fashion, large trees, small trees, shrubby trees, bushes, large plants, small plants, and herbs. This is the foundation of a classification which itself has subsequently been divided and subdivided according to other relations of size and form in order to give each species a particular character. After classification according to this plan, some people came along who have examined such a distribution, and who said that this method has been on the relative size of plants cannot be maintained, for there are within the same species, such as that of the oaks, great variations in sizes. There are some kinds of oak which rise a hundred feet in height, and others which never grow more than two feet tall. The same is true, allowance being made, of chestnuts, pines, aloes, and of an infinity of other kinds of plants. Thus it is said that the genus of plants ought not be determined by their size, since this distinction is equivocal and uncertain. And, with reason, this method has been abandoned. Next, others have appeared on the scene who, believing they can do better, have said that in order to know plants it is necessary to stick to the most obvious part of them. And, since the leaves are the most obvious feature, one should arrange plants according to the form, size, and position of the leaves. Thus one becomes familiar with another scheme or method, and follows it for awhile. But then it become evident that the leaves of almost all plants vary prodigiously with age and terrain, that their form is no more constant than their size, and that their position is still more uncertain. Thus this method has proven more satisfactory then the preceding one. Finally, someone imagined—Gesner,—I believe—that the Creator had put in the reproductive structures of plants a certain number of different and unvariable characters, and that it was on this assumption that one ought to try to create a system.[2] And, since this idea turns out to be true up to a point, in that the organs of generation of plants are found to have some unique features more constantly than all the other organs of the plant taken separately, there have suddenly arisen many systems of botany, each founded nearly upon the same principle. Among these methods that of M. de Tournefort[3] is the most remarkable, the most ingenious, and the most complete. This illustrious botanist was aware of the shortcomings of a system which could be purely arbitrary. As a man of intellect, he avoided the absurdities which are found in most of the other systems of his contemporaries, and he made his allocations and his exceptions with boundless knowledge and skill. In a word, he had put botany in a position to do without other methods, and he had made it capable of a certain degree of perfection. But there arose another methodologist who, after having praised de Tournefort's system, tried to destroy it in order to establish his own. This same person, having adopted with M. de Tournefort the distinguishing characteristics drawn from fructification, then employed all the organs of generation of plants, and above all the stamens, for the purpose of dividing his genera. Holding in contempt the wise concern of M. de Tournefort not to push nature to the point of confusing, for the sake of his system, the most various objects—like trees and herbs—he put together in the same class the mulberry and the nettle, the tulip and the barberry, the elm and the carrot, the rose and the strawberry, the oak and the bloodwort. Now, isn't this to make sport of nature and of those who study her? And if all that classification were not presented with a certain appearance of mysterious order and wrapped up in Greek and botanical erudition, would one be long in perceiving the ridiculousness of such a system, or rather in pointing out the confusion which results from such a bizarre assemblage? But that is not all; and I am going to persist in this assertion, because it is proper to preserve for M. de Tournefort the glory which he merited by his sensitive and persistent labor, and because it is not necessary that those who have learned botany according to de Tournefort's system should waste their time studying that new system wherein everything is changed, even to the names and surnames of plants. I say, then, that this recent method which brings together in the same class genera of plants which are entirely dissimilar, has, furthermore, independently of its incongruities, essential shortcomings and drawbacks greater than all the methods which have preceded it. As the characters of the genera come to be set by distinctions almost infinitely small, it becomes necessary to proceed to the identification of a tree or plant with a microscope in one's hand. The size, the form, the external appearance, the leaves, all the obvious features are useless for purposes of identification. Nothing is important except the stamens, and if one is unable to see the stamens, one can do nothing, one has seen nothing of significance. This large tree which you see is perhaps only a bloodwort. It is necessary to count its stamens in order to know what it is, and, since its stamens are often so small that they escape the naked eye or the magnifying glass, one must have a microscope. But unfortunately for this system there are plants which do not have stamens. There are also plants in which the number of stamens varies, and therein lies the shortcoming of this method of classification, just as in the others, in spite of the magnifying glass and microscope.[4]

After this frank exposition of the bases upon which the various systems of botany have been constructed, it is easy to see that the great shortcoming here is a metaphysical error in the very principle of such systems. This error involves disregarding the progression of nature, which is always a matter of nuances, and wishing to judge the whole by a single part. This is a manifest error, and one that it is astonishing to find so widespread. For almost all who have systematically named things have employed only a single feature, such as the teeth, the claws, or the spurs, as a means of classifying animals, and the leaves or the flowers in classifying plants, rather than making use of all parts of the organism, and searching out the differences and similarities of complete individual specimens. To refuse to make use of all the features of objects which we are considering is voluntarily to renounce the greatest number of advantages which nature offers us as a means of knowing her. And even if one were assured of finding constant and invariable characters in the several parts taken by themselves, it would not be necessary to restrict thus the knowledge of the productions of nature to a knowledge of these constant characters, which only give particular and very imperfect ideas of the whole organism. And it appears to me that the sole means of constructing an instructive and natural system is to put together whatever is similar and to separate those things which differ. If the individual entities resemble each other exactly, or if the differences between them are so small that they can be perceived only with difficulty, such individuals will be of the same species. If the differences begin to be perceptible, while at the same time there are always many more similarities than differences, such individuals will be of different species, but of the same genus. And if the differences are even more marked, without however exceeding the resemblances, then such individuals will be not only of another species, but even of another genus than the first and the second instances, but of the same class, for they resemble each other more than they differ. But if, on the contrary, the differences exceed the similarities, such individuals are not even of the same class. This is the systematic order which ought to be followed in arranging the productions of nature. Certainly the similarities and differences will be taken not only from one feature but from the whole organism. And likewise this method of inspection will be brought to bear on form, size, external bearing, upon the various parts, upon their number and position, upon the very substance of the thing. Similarly, these elements will be used in large or small number as the occasion necessitates. And these principles will be applied in such a way that if an individual specimen, whatever its nature may be, is so singular as to be always recognizable at first sight, it will be given but one name. But if this specimen has a form in common with another, and differs constantly from it in size, color, substance, or by any other obviously sensible quality, then it is given the same name as the other, to which is added an adjective to mark the difference. And thus one proceeds, putting in as many adjectives as there are differences. By this means one will be certain to express the various attributes of each species, and there need be no fear of falling into the inconveniences of the two restricted methods of which we have spoken, and about which we have discoursed at length. This is so because of a common shortcoming of all systems of botany and natural history and because the systems which have been devised for animals are even more defective than the systems of botany. For, as we have already hinted, there has been a desire to pronounce on the resemblance and difference of animals by employing in such proceedings only the number of fingers or claws, of teeth and breasts—a project which greatly resembles that of recognition by stamens, and which is in effect of the same author. [. . .]

Thus, one ought not regard as fundamental to science the methods that these authors have given us concerning natural history in general or those designed for one of its parts. Such methods should be used only as systems of artificial signs which are agreed upon for purposes of mutual understanding. Actually, they are only arbitrary connections and differing points of view under which the objects of nature have been considered. Only by making use of such methods in this spirit is it possible to draw from them some utility. For although it may not appear very necessary, it might be good if one knew all species of plants whose leaves resemble each other, all those whose flowers look alike, all those which may nourish certain kinds of insects, all those which have a given number of stamens, and all those which have certain excretory glands. The same is true of animals: there might be a point in knowing all animals which have a given number of digits. To speak precisely, each of these methods is only a dictionary in which one may find names arranged according to an order derived from a certain idea, and, consequently, arranged as arbitrarily as the alphabetical. But the advantage which may be had from such arrangements is that, in comparing all the results, one may finally come across the true method, which involves the complete description and exact history of each particular thing.

Here is the principal goal which must be kept in mind: A prefabricated method can be used as a convenience for studying, a means of mutual understanding. But the sole true means of advancing natural science is to labor at the description and history of the various things which are its objects.

Things, in relation to us, are nothing in themselves; nor does giving them a name call them into existence. But they begin to exist for us when we become acquainted with their relations to each other and their properties. Yet even by means of their relations we are unable to give things a definition. Now, a definition such as we can construct verbally is still no more than a very imperfect representation of the thing, and we are never able adequately to define a thing without describing it exactly. This difficulty of forming an adequate definition is found constantly in all systems and in all the epitomes which have been attempted in order to relieve the burden on the memory. [. . .]

With regard to the general order and the method of distribution of the various subjects of natural history, this could be considered purely arbitrary. Consequently, one is certainly free to choose what seems either the most convenient or the most commonly accepted. But before giving reasons which would lead to the adoption of one system rather than another, it is still necessary to make a few reflections by means of which we shall try to make the reader aware of how far the divisions which we have made of natural productions correspond to reality.

In order to recognize this we must dismantle our prejudices for a moment and even abstract from our ideas. Let us imagine a man who indeed has forgotten everything, or who awakens to completely strange surroundings. Let us set this man in a field where animals, birds, fishes, plants, and stones appear successively to his eyes. This man, upon first perceiving them, would distinguish nothing and confound everything. But allow his ideas to become gradually more settled by means of repeated sensations from the same objects, and soon he will form a general idea of animated matter which he will easily distinguish from inaminate matter. And shortly thereafter he will distinguish quite accurately between animated matter and vegetative matter, and he will naturally arrive at that first great division, *Animal, Vegetable,* and *Mineral.* And since at the same time he will have come to a clear idea of those great and quite diverse objects, *Earth, Air,* and *Water,* he will come shortly to form a particular idea of the animals who inhabit the earth, of those who live in the water, and of those who take to the air. And, consequently, he will easily make that second division between *Four-footed Animals, Birds,* and *Fishes.* Likewise, in the vegetable kingdom he will distinguish trees and plants with facility, whether it be by their size, their substance, or their shape. This is what simple observation must necessarily show him, and what with the very least attention he could not fail to recognize. That is what we also must recognize as real, what we must respect as a division given by nature herself. Next, let us put ourselves in the place of that man, or let us suppose that he may have acquired as much knowledge and experience as we have on this matter. He will come to judge the objects of natural history by the connections which they have with his own life. Those which are the most necessary or useful to him will hold the first rank—for example, he will give preference in the order of animals to the horse, the dog, oxen, etc., and he will always know more about those which are most familiar to him. Next, he will occupy himself with those which, without being familiar, nevertheless inhabit the same places and climates as he does—such as deer, hares, and all the wild animals. And only after acquiring all these details will his curiosity lead him to inquire into what the animals of foreign regions may be like—those such as elephants, dromedaries, etc. The case will be the same with fishes, birds, insects, shellfish, plants, minerals, and all the other productions of nature. He will study them in proportion to their usefulness; he will consider them to the extent that they are familiar to him, and he will rank them in his mind relative to the order of his acquaintance with them, because that is indeed the order according to which he experienced them and according to which it is important to him to preserve them.

This order, the most natural of all, is what we believe ought to be followed. Our method of distribution is no more mysterious than that which we have just observed. We start from general divisions such as we have just indicated, divisions which are incontestable. Next we take those objects which interest us the most owing to the connections which they have with us. From this point we pass little by little to those which are most distant, and which are foreign to us. And we believe that this simple and natural manner of considering things is preferable to more recondite and complex methods because there is not one of them, whether of those which have been constructed, or of all those which might be constructed, which would not have more of an arbitrary element in them than our method. From every point of view, it is easier, more agreeable, and more useful to consider things in relation to us rather than from another point of view. [. . .]

However, [...] it is not necessary to imagine even today that, in the study of natural history, one ought to limit oneself solely to the making of exact descriptions and the ascertaining of particular facts. This is, in truth, and as has been pointed out, the essential end which ought to be proposed at the outset. But we must try to raise ourselves to something greater and still more worthy of our efforts, namely: the combination of observations, the generalization of facts, linking them together by the power of analogies, and the effort to arrive at a high degree of knowledge. From this level we can judge that particular effects depend upon more general ones; we can compare nature with herself in her vast operations; and, finally, we are able to open new routes for the further perfection of the various branches of natural philosophy. A vast memory, assiduity, and attention suffice to arrive at the first end. But more is needed here. General views, a steady eye, and a process of reasoning informed more by reflection than by study are what is called for. Finally, that quality of mind is needed which makes us capable of grasping distant relationships, bringing them together, and making out of them a body of reasoned ideas after having precisely determined their nearness to truth and weighed their probabilities.

Here there is need for a methodical approach to guide the mind, not for that [artificial method] of which we have spoken, for that only serves to arrange words arbitrarily, but for that method which sustains the very order of things, guides our reasoning, enlightens our views, extends them, and prevents us reasoning, enlightens our views, extends them, and prevents us from being led astray.

The greatest philosophers have felt the need for such a method, and they have indeed attempted to give us principles and samples of it. But some of them have left us only the history of their thoughts, while others have left only the story of their imagination. And if some have risen to the elevated stations of metaphysics from which the principles, connections, and totality of the sciences can be viewed, none of them have communicated their ideas to us on these subjects or given us any advice concerning them; and the manner of properly conducting one's mind in the sciences is yet to be found. In the absence of precepts, examples have been substituted; in place of principles, definitions have been used; instead of authenticated facts, risky suppositions have been supplied by guesswork.

Even in our own century, when the sciences seem to be cultivated with care, I believe that it is easy to perceive that philosophy is neglected, perhaps more so than in any other, century. The skills which one would like to call scientific have taken its place. The methods of calculus and geometry, those of botany and natural history—formulas, in a word, and dictionaries—occupy almost everyone. We think that we know more because we have increased the number of symbolic expressions and learned phrases. We pay hardly any attention to the fact that all these skills are only the scaffolding of science, and not science itself. We ought to use them only when we cannot do without them, and we ought always to be careful lest they happen to fail us when we wish to apply them to the edifice of science itself. [. . .]

The phenomena which offer themselves daily to our eyes, which follow one another and repeat themselves without interruption and uniformly, are the foundation of our physical knowledge. It is enough that a thing always happens in the same way for it to become a certainty or a truth for us. All the facts of nature which we have observed, or which we could observe, are just so many truths. And thus we are able to increase the number as much as we please by multiplying our observations. Our science is limited in this case only by the dimensions of the universe.

But when, after having determined the facts through repeated observations; when, after having established new truths through precise experiments, we wish to search out the reasons for these same occurrences, the causes of these effects, we find ourselves suddenly baffled, reduced to trying to deduce effects from more general effects, and obliged to admit that causes are and always will be unknown to us, because our senses, themselves being the effects of causes of which we have no knowledge, can give us ideas only of effects and never of causes. Thus we must be content to call cause a general effect, and must forego hope of knowing anything beyond that.

These general effects are for us the true laws of nature. All the phenomena that we recognize as holding to these laws and depending on them will be so many accountable facts, so many truths understood. Those phenomena which we are unable to associate with these general effects will be simple "occurrences" which we must keep in reserve until such time as a greater number of observations and a more extended experience make us aware of other facts and bring to light their physical cause, that is to say, the general effect from which these particular effects derive. It is here that the union of the two sciences of mathematics and physics might result in great advantages. The one gives the "how many," the other the "how" of things. And since it is a question here of combining and estimating probabilities in order to judge whether an effect depends more on one cause than on another, when you have imagined by physics the how, that is to say, when you have seen that such and such an effect might well depend upon such and such a cause, you then apply mathematics in order to assure yourself as to how often this effect happens in conjunction with its cause. And if you find that the result accords with the observations, the probability that you have guessed correctly is so increased that it becomes a certainty. But in the absence of such corroboration, the relation would have remained a simple probability.

It is true that this union of mathematics and physics can be accomplished only for a very small number of subjects. In order for this to take place it is necessary that the phenomena that we are concerned with explaining be susceptible to being considered in an abstract manner and that their nature be stripped of almost all physical qualities. For mathematics is inapplicable to the extent that such subjects are not simple abstractions. The most beautiful and felicitous use to which this method has ever been applied is to the system of the world. We must admit that if Newton had only given us the physical conformations of his system without having supported them by precise mathematical evaluations they would not have had nearly the same force. But at the same time one ought to be aware that there are very few subjects as simple as this, that is to say, as stripped of physical qualities as the Newtonian universe. For the distance of the planets is so great that it is possible to consider them in reference to each other as being no more than points. And it is possible simultaneously, without being mistaken, to abstract from all the physical qualities of the planets and take into consideration only their force of attraction. Their movements are, moreover, the most regular that we know, and suffer no retardation from resistance. All of this combines to render the explanation of the system of the world a problem in mathematics, for the realization of which fortunately there was needed only one well-conceived physical idea, that idea being to have thought that the force which makes bodies fall to the surface of the earth might well be the same as that which holds the moon in its orbit. [. . .]

Here then is the most delicate and the most important point in the study of the sciences: to know how to distinguish what is real in a subject from what we arbitrarily put there in considering it, to recognize clearly the properties which belong to it and those which we give to it. This appears to me to be the foundation of the true method of leading one's mind in the way of the sciences. And if this principle were always kept in mind, a false step would never be taken. One might thus avoid falling into learned errors which at times are taken as truths. Paradoxes and insoluble problems in the abstract sciences would begin to disappear. The prejudices and the doubts which we ourselves bring to the sciences of the real would become apparent, and agreement would be reached on the metaphysics of sciences. Disputes would cease, and all would unite to advance along the same path following experience. Finally, we would arrive at the knowledge of all the truths which are within the competence of the human mind.

When the subjects are too complicated to allow the advantageous application of calculation and measurement, as is almost always the case with natural history and the physics of the particular, it seems to me that the true method of guiding one's mind in such research is to have recourse to observations, to gather these together, and from them to make new observations in sufficient number to assure the truth of the principal facts, and to use mathematics only for the purpose of estimating the probabilities of the consequences which may be drawn from these facts. Above all, it is necessary to try to generalize these facts and to distinguish well those which are essential from those which are only accessories to the subject under consideration. It is then necessary to tie such facts together by analogies, confirm or destroy certain equivocal points by means of experiment, form one's plan of explication on the basis of the combination of all these connections, and present them in the most natural order. This order can be established in two ways: the first is to ascend from particular effects to more general ones, and the other is to descend from the general to the particular. Both ways are good, and the choice of one or the other depends more on the bent of the author than on the nature of things, which always allows of being treated equally well by either method.

Notes

1. [Notes in square brackets were added by the translator (JL).]
2. [Konrad von Gesner (1516–1565), Swiss polymath and naturalist. (JL)]
3. [Joseph Pitton de Tournefort (1656–1708), French botanist. (JL)]
4. "This system, indeed, namely Linnaeus's, is not only far more vile and inferior to the already-known systems, but is further exceedingly forced, slippery, and fallacious; indeed, I would consider it childish, for it not only brings after itself enormous confusion with regard to the division and denomination of plants, but it is also to be feared that from this would come an almost complete clouding and disruption of the more solid botanical systems" *Vaniloq. Botan. specimen refutatum à Siegesbek*, Petropoli, 1741 (sic).

CHAPTER 26

Preliminary Observations to the Essay on the Theory of the Earth

GEORGES CUVIER

Pages 1–25 (§1–8), Pages 37–52 (§18–21) from *Essay on the Theory of the Earth*, trans. Robert Jameson. Edinburgh: William Blackwood, 1817. Facsimile Reprint available from New York: Arno Press, 1978.

Iᴛ is my object, in the following work, to travel over ground which has as yet been little explored, and to make my reader acquainted with a species of Remains, which, though absolutely necessary for understanding the history of the globe, have been hitherto almost uniformly neglected. [. . .]

The ancient history of the globe, which is the ultimate object of all these researches, is also of itself one of the most curious subjects that can engage the attention of enlightened men; and if they take any interest in examining, in the infancy of our species, the almost obliterated traces of so many nations that have become extinct, they will doubtless take a similar interest in collecting, amidst the darkness which covers the infancy of the globe, the traces of those revolutions which took place anterior to the existence of all nations.

We admire the power by which the human mind has measured the motions of globes which nature seemed to have concealed for ever from our view: Genius and science have burst the limits of space, and a few observations, explained by just reasoning, have unveiled the mechanism of the universe. Would it not also be glorious for man to burst the limits of time, and, by a few observations, to ascertain the history of this world, and the series of events which preceded the birth of the human race? Astronomers, no doubt, have advanced more rapidly than naturalists; and the present period, with respect to the theory of the earth, bears some resemblance to that in which some philosophers thought that the heavens were formed of polished stone, and that the moon was no larger than the Peloponnesus; but, after Anaxagoras, we have had our Copernicuses, and our Keplers, who pointed out the way to Newton; and why should not natural history also have one day its Newton?

2. Plan of this Essay

What I now offer comprehends but a few of the facts which must enter into the composition of this ancient history. But these few are important; many of them are decisive; and I hope that the rigorous methods which I have adopted for the purpose of establishing them, will make them be considered as points so determinately fixed as to admit of no departure from them. Though this hope should only be realised with respect to some of them, I shall think myself sufficiently rewarded for my labour.

In this preliminary discourse I shall describe the whole of the results at which the theory of the earth seems to me to have arrived. I shall mention the relations which connect the history of the fossil bones of land animals with these results, and the considerations which render their history peculiarly important. I shall unfold the principles on which is founded the art of ascertaining these bones, or, in other words, of discovering a genus and of distinguishing a species by a single fragment of bone,—an art on the certainty of which depends that of the whole work. I shall give a rapid sketch of the results to which my researches lead, of the new species and genera which these have been the means of discovering, and of the different strata in which they are found deposited. And as the difference between these species and the species which still exist is bounded by certain limits, I shall show that these limits are a great deal more extensive than those which now distinguish the varieties of the same species; and shall then point out how far these varieties may be owing to the influence of time, of climate, or of domestication.

In this way I shall be prepared to conclude that great events were necessary to produce the more considerable differences which I have discovered: I shall next take notice of the particular modifications which my performance should introduce into the hitherto received opinions respecting the primitive history of the globe; and, last of all, I shall enquire how far the civil and religious history of different nations corresponds with the results of an examination of the physical history of the earth, and with the probabilities afforded by such examination concerning the period at which societies of men had it in their power to take up fixed abodes, to occupy fields susceptible of cultivation, and consequently to assume a settled and durable form.

3. Of the First Appearance of the Earth

When the traveller passes through those fertile plains where gently-flowing streams nourish in their course an abundant vegetation, and where the soil, inhabited by a numerous population, adorned with flourishing villages, opulent cities, and superb monuments, is never disturbed except by the ravages of war and the oppression of tyrants, he is not led to suspect that nature also has had her intestine wars, and that the surface of the globe has been much convulsed by successive revolutions and various catastrophes. But his ideas change as soon as he digs into that soil which presented such a peaceful aspect, or ascends the hills which border the plain; they are expanded, if I may use the expression, in proportion to the expansion of his view; and they begin to embrace the full extent and grandeur of those ancient events to which I have alluded, when he climbs the more elevated chains whose base is skirted by these first hills, or when, by following the beds of the descending torrents, he penetrates into their interior structure, which is thus laid open to his inspection.

4. First Proofs of Revolutions on the Surface of the Globe

The lowest and most level parts of the earth, when penetrated to a very great depth, exhibit nothing but horizontal strata composed of various substances, and containing almost all of them innumerable marine productions. Similar strata, with the same kind of productions, compose the hills even to a great height. Sometimes the shells are so numerous as to constitute the entire body of the stratum. They are almost everywhere in such a perfect state of preservation, that even the smallest of them retain their most delicate parts, their sharpest ridges, and their finest and tenderest processes. They are found in elevations far above the level of every part of the ocean, and in places to which the sea could not be conveyed by any existing cause. They are not only inclosed in loose sand, but are often incrusted and penetrated on all sides by the hardest stones. Every part of the earth, every hemisphere, every continent, every island of any size, exhibits the same phenomenon. We are therefore forcibly led to believe, not only that the sea has at one period or another covered all our plains, but that it must have remained there for a long time, and in a state of tranquillity; which circumstance was necessary for the formation of deposits so extensive, so thick, in part so solid, and containing exuviæ so perfectly preserved.

The time is past for ignorance to assert that these remains of organized bodies are mere *lusus naturæ*,—productions generated in the womb of the earth by its own creative powers. A nice and scrupulous comparison of their forms, of their contexture, and frequently even of their composition, cannot detect the slightest difference between these shells and the shells which still inhabit the sea. They have therefore once lived in the sea, and been deposited by it: the sea consequently must have rested in the places where the deposition has taken place. Hence it is evident that the basin or reservoir containing the sea has undergone some change at least, either in extent, or in situation, or in both. Such is the result of the very first search, and of the most superficial examination.

The traces of revolutions become still more apparent and decisive when we ascend a little higher, and approach nearer to the foot of the great chains of mountains. There are still found many beds of shells; some of these are even larger and more solid; the shells are quite as numerous and as entirely preserved; but they are not of the same species with those which were found in the less elevated regions. The strata which contain them are not so generally horizontal; they have various degrees of inclination, and are sometimes situated vertically. While in the plains and low hills it was necessary to dig deep in order to detect the succession of the strata, here we perceive them by means of the vallies which time or violence has produced, and which disclose their edges to the eye of the observer. At the bottom of these declivities, huge masses of their *debris* are collected, and form round hills, the height of which is augmented by the operation of every thaw and of every storm.

These inclined or vertical strata, which form the ridges of the secondary mountains, do not rest on the horizontal strata of the hills which are situated at their base, and serve as their first steps; but, on the contrary, are situated underneath them. The latter are placed upon the declivities of the former. When we dig through the horizontal strata in the neighbourhood of the inclined strata, the inclined strata are invariably found below. Nay, sometimes, when the inclined strata are not too much elevated, their summit is surmounted by horizontal strata. The inclined strata are therefore more ancient than the horizontal strata. And as they must necessarily have been formed in a horizontal position, they have been subsequently shifted into their inclined or vertical position, and that too before the horizontal strata were placed above them.

Thus the sea, previous to the formation of the horizontal strata, had formed others, which, by some means, have been broken, lifted up, and overturned in a thousand ways. There had therefore been also at least one change in the basin of that sea which preceded ours; it had also experienced at least one revolution; and as several of these inclined strata which it had formed first, are elevated above the level of the horizontal strata which have succeeded and which surround them, this revolution, while it gave them their present inclination, had also caused them to project above the level of the sea, so as to form islands, or at least rocks and inequalities; and this must have happened whether one of their edges was lifted up above the water, or the depression of the opposite edge caused the water to subside. This is the second result, not less obvious, nor less clearly demonstrated, than the first, to every one who will take the trouble of studying carefully the remains by which it is illustrated and proved.

5. Proofs That Such Revolutions Have Been Numerous

If we institute a more detailed comparison between the various strata and those remains of animals which they contain, we shall soon discover still more numerous differences among them, indicating a proportional number of changes in their condition. The sea has not always deposited stony substances of the same kind. It has observed a regular succession as to the nature of its deposits; the more ancient the strata are, so much the more uniform and extensive are they; and the more recent they are, the more limited are they, and the more variation is observed in them at small distances. Thus the great catastrophes which have produced revolutions in the basin of the sea, were preceded, accompanied, and followed by changes in the nature of the fluid and of the substances which it held in solution; and when the surface of the seas came to be divided by islands and projecting ridges, different changes took place in every separate basin.

Amidst these changes of the general fluid, it must have been almost impossible for the same kind of animals to continue to live:—nor did they do so in fact. Their species, and even their genera, change with the strata; and although the same species occasionally recur at small distances, it is generally the case that the shells of the ancient strata have forms peculiar to themselves; that they gradually disappear, till they are not to be seen at all in the recent strata, still less in the existing seas, in which, indeed, we never discover their corresponding species, and where several even of their genera are not to be found; that, on the contrary, the shells of the recent strata resemble, as it respects the genus, those which still exist in the sea; and that in the last-formed and loosest of these strata there are some species which the eye of the most expert naturalist cannot distinguish from those which at present inhabit the ocean.

In animal nature, therefore, there has been a succession of changes corresponding to those which have taken place in the chemical nature of the fluid; and when the sea last receded from our continent, its inhabitants were not very different from those which it still continues to support.

Finally, if we examine with greater care these remains of organized bodies, we shall discover, in the midst even of the most ancient secondary strata, other strata that are crowded with animal or vegetable productions, which belong to the land and to fresh water; and amongst the more recent strata, that is, the strata which are nearest the surface, there are some of them in which land animals are buried under heaps of marine productions. Thus the various catastrophes of our planet have not only caused the different parts of our continent to rise by degrees from the basin of the sea, but it has also frequently happened, that lands which had been laid dry have been again covered by the water, in consequence either of these lands sinking down below the level of the sea, or of the sea being raised above the level of the lands. The particular portions of the earth also which the sea has abandoned by its last retreat, had been laid dry once before, and had at that time produced quadrupeds, birds, plants, and all kinds of terrestrial productions; it had then been inundated by the sea, which has since retired from it, and left it to be occupied by its own proper inhabitants.

The changes which have taken place in the productions of the shelly strata have not, therefore, been entirely owing to a gradual and general retreat of the waters, but to successive irruptions and retreats, the final result of which, however, has been an universal depression of the level of the sea.

6. Proofs That the Revolutions Have Been Sudden

These repeated irruptions and retreats of the sea have neither been slow nor gradual; most of the catastrophes which have occasioned them have been sudden; and this is easily proved, especially with regard to the last of them, the traces of which are most conspicuous. In the northern regions it has left the carcases of some large quadrupeds which the ice had arrested, and which are preserved even to the present day with their skin, their hair, and their flesh. If they had not been frozen as soon as killed they must quickly have been decomposed by putrefaction. But this eternal frost could not have taken possession of the regions which these animals inhabited except by the same cause which destroyed them; this cause, therefore, must have been as sudden as its effect.[1] The breaking to pieces and overturnings of the strata, which happened in former catastrophes, shew plainly enough that they were sudden and violent like the last; and the heaps of *debris* and rounded pebbles which are found in various places among the solid strata, demonstrate the vast force of the motions excited in the mass of waters by these overturnings. Life, therefore, has been often disturbed on this earth by terrible events—calamities which, at their commencement, have perhaps moved and overturned to a great depth the entire outer crust of the globe, but which, since these first commotions, have uniformly acted at a less depth and less generally. Numberless living beings have been the victims of these catastrophes; some have been destroyed by sudden inundations, others have been laid dry in consequence of the bottom of the seas being instantaneously elevated. Their races even have become extinct, and have left no memorial of them except some small fragment which the naturalist can scarcely recognise.

Such are the conclusions which necessarily result from the objects that we meet with at every step of our enquiry, and which we can always verify by examples drawn from almost every country. Every part of the globe bears the impress of these great and terrible events so distinctly, that they must be visible to all who are qualified to read their history in the remains which they have left behind.

But what is still more astonishing and not less certain, there have not been always living creatures on the earth, and it is easy for the observer to discover the period at which animal productions began to be deposited.

7. Proofs of the Occurrence of Revolutions before the Existence of Living Beings

As we ascend to higher points of elevation, and advance towards the lofty summits of the mountains, the remains of marine animals, that multitude of shells we have spoken of, begin very soon to grow rare, and at length disappear altogether. We arrive at strata of a different nature, which contain no vestige at all of living creatures. Nevertheless their crystallization, and even the nature of their strata, shew that they also have been formed in a fluid; their inclined position and their slopes shew that they also have been moved and overturned; the oblique manner in which they sink under the shelly strata shews that they have been formed before these; and the height to which their bare and rugged tops are elevated above all the shelly strata, shews that their summits have never again been covered by the sea since they were raised up out of its bosom.

Such are those primitive or primordial mountains which traverse our continents in various directions, rising above the clouds, separating the basins of the rivers from one another, serving, by means of their eternal snows, as reservoirs for feeding the springs, and forming in some measure the skeleton, or, as it were, the rough framework of the earth.

The sharp peaks and rugged indentations which mark their summits, and strike the eye at a great distance, are so many proofs of the violent manner in which they have been elevated. Their appearance in this respect is very different from that of the rounded mountains and the hills with flat surfaces, whose recently formed masses have always remained in the situation in which they were quietly deposited by the sea which last covered them.

These proofs become more obvious as we approach. The vallies have no longer those gently sloping sides, or those alternately salient and reentrant angles opposite to one another, which seem to indicate the beds of ancient streams. They widen and contract without any general rule; their waters sometimes expand into lakes, and sometimes descend in torrents; and here and there the rocks, suddenly approaching from each side, form transverse dikes, over which the waters fall in cataracts. The shattered strata of these vallies expose their edges on one side, and present on the other side large portions of their surface lying obliquely; they do not correspond in height, but those which on one side form the summit of the declivity, often dip so deep on the other as to be altogether concealed.

Yet, amidst all this confusion, some naturalists have thought that they perceived a certain degree of order prevailing, and that among these immense beds of rocks, broken and overturned though they be, a regular succession is observed, which is nearly the same in all the different chains of mountains. According to them, the granite, which surmounts every other rock, also dips under every other rock; and is the most ancient of any that has yet been discovered in the place assigned it by nature. The central ridges of most of the mountain chains are composed of it; slaty rocks, such as clay slate, granular quartz, (*grès*) and mica slate, rest upon its sides and form lateral chains; granular, foliated limestone, or marble, and other calcareous rocks that do not contain shells, rest upon the slate, forming the exterior ranges, and are the last formations by which this ancient uninhabited sea seems to have prepared itself for the production of its beds of shells.[2]

On all occasion, even in districts that lie at a distance from the great mountain chains, where the more recent strata have been digged through, and the external covering of the earth penetrated to a considerable depth, nearly the same order of stratification has been found as that already described. The crystallized marbles never cover the shelly strata; the granite in mass never rests upon the crystallized marble, except in a few places where it seems to have been formed of granites of newer epochs. In one word, the foregoing arrangement appears to be general, and must therefore depend upon general causes, which have on all occasions exerted the same influence from one extremity of the earth to the other.

Hence, it is impossible to deny, that the waters of the sea have formerly, and for a long time, covered those masses of matter which now constitute our highest mountains; and farther, that these waters, during a long time, did not support any living bodies. Thus, it has not been only since the commencement of animal life that these numerous changes and revolutions have taken place in the constitution of the external covering of our globe: For the masses formed previous to that event have suffered changes, as well as those which have been formed since; they have also suffered violent changes in their positions, and a part of these assuredly took place while they existed alone, and before they were covered over by the shelly masses. The proof of this lies in the overturnings, the disruptions, and the fissures which are observable in their strata, as well as in those of more recent formation, which are there even in greater number and better defined.

But these primitive masses have also suffered other revolutions, posterior to the formation of the secondary strata, and have perhaps given rise to, or at least have partaken of, some portion of the revolutions and changes which these latter strata have experienced. There are actually considerable portions of the primitive strata uncovered, although placed in lower situations than many of the secondary strata; and we cannot conceive how it should have so happened, unless the primitive strata, in these places, had forced themselves into view, after the formation of those which are secondary. In some countries, we find numerous and prodigiously large blocks of primitive substances scattered over the surface of the secondary strata, and separated by deep vallies from the peaks or ridges whence these blocks must have been derived. It is necessary therefore, either that these blocks must have been thrown into those situations by means of eruptions, or that the vallies, which otherwise must have stopped their course, did not exist at the time of their being transported to their present sites.[3]

Thus we have a collection of facts, a series of epochs anterior to the present time, and of which the successive steps may be ascertained with perfect certainty, although the periods which intervened cannot be determined with any degree of precision. These epochs form so many fixed points, answering as rules for directing our enquiries respecting this ancient chronology of the earth.

8. Examination of the Causes Which Act at Present on the Surface of Our Globe

We now propose to examine those changes which still take place on our globe, investigating the causes which continue to operate on its surface, and endeavouring to determine the extent of those effects which they are capable of producing. This portion of the history of the earth is so much the more important, as it has been long considered possible to explain the more ancient revolutions on its surface by means of these still existing causes; in the same manner as it is found easy to explain past events in political history, by an acquaintance with the passions and intrigues of the present day. But we shall presently see that unfortunately this is not the case in physical history; the thread of operation is here broken, the march of nature is changed, and none of the agents that she now employs were sufficient for the production of her ancient works.

There still exist, however, four causes in full activity, which contribute to make alterations on the surface of our earth. These are rains and thaws, which waste down the steep mountains, and occasion their fragments to collect at their bottoms; streams of water, which sweep away these fragments, and afterwards deposit them in places where their current is abated;

the sea which undermines the foundations of elevated coasts, forming steep cliffs in their places, and which throws up hillocks of sand upon flat coasts; and, finally, volcanoes, which pierce through the most solid strata from below, and either elevate or scatter abroad the vast quantity of matter which they eject. [...]

18. Of Astronomical Causes of the Revolutions on the Surface of the Earth

The pole of the earth moves in a circle round the pole of the ecliptic, and its axis is more or less inclined to the plane of the ecliptic; but these two motions, the causes of which are now ascertained, are confined within certain bounds, and are much too limited for the production of those effects which we have stated. Besides, as these motions are exceedingly slow, they are altogether inadequate to account for catastrophes which must necessarily have been sudden.

The same reasoning applies to all other slow motions which have been conceived as causes of the revolutions on the surface of our earth, chosen doubtless in the hope that their existence could not be denied, as it might always be assorted that their extreme slowness rendered them imperceptible. But it is of no importance whether these assumed slow motions be true or false, for they explain nothings, since no cause acting slowly could possibly have produced sudden effects.

Admitting that there was a gradual diminution of the waters; that the sea might take away solid matters from one place and carry them to another; that the temperature of the globe may have diminished or increased; none of these causes could have overthrown our strata; inclosed great quadrupeds with their flesh and skin in ice; laid dry sea-shells in as perfect preservation as if just drawn up alive from the bottom of the ocean; or utterly destroyed many species, and even entire genera, of testaceous animals.

These considerations have presented themselves to most naturalists: And, among those who have endeavoured to explain the present state of the globe, hardly any one has attributed the entire changes it has undergone to slowly operating causes, and still less to causes which continue to act, as it were, under our observation. The necessity to which they were thus reduced, of seeking for causes different from those which we still observe in activity, is the very thing which has forced them to make so many extraordinary suppositions, and to lose themselves in so many erroneous and contradictory speculations, that the very name of their science, as I have elsewhere said, has become ridiculous in the opinion of prejudiced persons, who only see in it the systems which it has exploded, and forget the extensive and important series of facts which it has brought to light and established.[4]

19. Of Former Systems of Geology

During a long time, two events or epochs only, the Creation and the Deluge, were admitted as comprehending the changes which have occurred upon the globe; and all the efforts of geologists were directed to account for the present actual state of the earth, by arbitrarily ascribing to it a certain primitive state, afterwards changed and modified by the deluge, of which also, as to its causes, its operation, and its effects, every one of them entertained his own theory.

Thus, in the opinion of *Burnet*, the whole earth at the first consisted of a uniform light crust, which covered over the abyss of the sea, and which, being broken for the production of the deluge, formed the mountains by its fragments.[5] According to *Woodward*, the deluge was occasioned by a momentary suspension of cohesion among the particles of mineral bodies; the whole mass of the globe was dissolved, and the soft paste became penetrated by, shells.[6] *Scheuchzer* conceived that God raised up the mountains for the purpose of allowing the waters of the deluge to run off, and accordingly selected those portions which contained the greatest abundance of rocks, without which they could not have supported themselves.[7] *Whiston* fancied that the earth was created from the atmosphere of one comet, and that it was deluged by the tail of another.[8] The heat which remained from its first origin, in his opinion, excited the whole antediluvian population, men and animals, to sin, for which they were all drowned in the deluge, excepting the fish, whose passions were apparently less violent.

It is easy to see, that though naturalists might have a range sufficiently wide within the limits prescribed by the book of Genesis, they very soon found themselves in too narrow bounds: and when they had succeeded in converting the six days employed in the work of creation into so many periods of indefinite length, their systems took a flight proportioned to the periods, which they could then dispose of at pleasure.

Even the great *Leibnitz*, as well as *Descartes*, amused his imagination by conceiving the world to be an extinguished sun, or vitrified globe; upon which the vapours condensing in proportion as it cooled, formed the seas, and afterwards deposited calcareous strata.[9]

By *Demaillet*, the globe was conceived to have been covered with water for many thousand years. He supposed that this water had gradually retired; that all the terrestrial animals were originally inhabitants of the sea; that man himself began his career as a fish: And he asserts, that it not uncommon, even now, to meet with fishes in the ocean, which are still only half men, but whose descendants will in time become perfect human beings.[10]

The system of *Buffon* is merely an extension of that before devised by Leibnitz, with the addition only of a comet, which, by a violent blow upon the sun, struck off the mass of our earth in a liquefied state, along with the masses of all the other planets of our system at the same instant. From this supposition, he was enabled to assume positive dates or epochs: As, from the actual temperature of the earth, it could be calculated how long time it had taken to cool so far. And, as all the other planets had come from the sun at the same time, it could also be calculated how many ages were still required for cooling the greater ones, and how far the smaller ones were already frozen.

In the present day, men of bolder imaginations than ever, have employed themselves on this great subject. Some writers have revived and greatly extended the ideas of Demaillet. They suppose that every thing was originally fluid; that this universal fluid gave existence to animals, which were at first of the simplest kind, such as the monads and other infusory microscopic animalcules; that, in process of time, and by acquiring different habits, the races of these animals became complicated, and assumed that diversity of nature and character in which they now exist. It is by all those races of animals that the waters of the ocean have been gradually converted into calcareous earth; while the vegetables, concerning the origin and metamorphoses of which these authors give us no account, have converted a part of the same water into clay; and these two earths, after being stript of the peculiar characters they had received respectively from animal and vegetable life, are resolved by a final analysis into silex: Hence the more ancient mountains are more silicious than the rest. Thus, according to these authors, all the solid particles of our globe owe their existence to animal or vegetable life, and without this our globe would still have continued entirely liquid.[11]

Other writers have preferred the ideas of Kepler, and, like that great astronomer, have considered the globe itself as possessed of living faculties. According to them, it contains a circulating vital fluid. A process of assimilation goes on in it as well as in animated bodies. Every particle of it is alive. It possesses instinct and volition even to the most elementary of its molecules, which attract and repel each other according to sympathies and antipathies. Each kind of mineral substance is capable of converting immense masses of matter into its own peculiar nature, as we convert our aliment into flesh and blood. The mountains are the respiratory organs of the globe, and the schists its organs of secretion. By the latter it decomposes the waters of the sea in order to produce volcanic eruptions. The veins in strata are caries, or abscesses of the mineral kingdom, and the metals are products of rottenness and disease, to which it is owing that almost all of them have so bad a smell.[12]

It must, however, be noticed, that these are what may be termed extreme examples, and that all geologists have not permitted themselves to be carried away by such bold or extravagant conceptions as those we have just cited. Yet, among those who have proceeded with more caution, and have not searched for geological causes beyond the established limits of physical and chemical science, there still remain much diversity and contradiction.

According to one of these writers, every thing has been successively precipitated and deposited, nearly as it exists at present; but the sea, which covered all, has gradually retired.[13]

Another conceives, that the materials of the mountains are incessantly wasted and floated down by the rivers, and carried to the bottom of the ocean, to be there heated under an enormous pressure, and to form strata which shall be violently lifted up at some future period, by the heat that now consolidates and hardens them.[14]

A third supposes the fluid materials of the globe to have been divided among a multitude of successive lakes, placed like the benches of an amphitheatre; which, after having deposited our shelly strata, have successively broken their dikes, to descend and fill the basin of the ocean.[15]

According to a fourth, tides of seven or eight hundred fathoms have carried off from time to time the bottom of the ocean, throwing it up in mountains and hills on the primitive vallies and plains of the continent.[16]

A fifth conceives the various fragments of which the surface of the earth is composed to have fallen successively from heaven, in the manner of meteoric stones, and alleges that they still retain the marks of their origin in the unknown species of animals whose exuviæ they contain.[17]

By a sixth, the globe is supposed to be hollow, and to contain in its cavity a nucleus of loadstone, which is dragged from one pole of the earth to the other by the attraction of comets, changing the centre of gravity, and consequently hurrying the great body of the ocean along with it, so as alternately to drown the two hemispheres.[18] [. . .]

21. Statement of the Nature and Conditions of the Problem to Be Solved

To quit the language of mathematics, it may be asserted, that almost all the authors of these systems, confining their attention to certain difficulties by which they were struck more forcibly than by others, have endeavoured to solve these in a way more or less probable, and have allowed others to remain unnoticed, equally numerous and equally important. For example, the only difficulty with one consisted in explaining the change which had taken place on the level of the seas; with another it consisted in accounting for the solution of all terrestrial substances in the same fluid; and with a third, it consisted in shewing how animals that were natives of the torrid could live under the frigid zone. Exhausting the whole of their ingenuity on these questions, they conceived that they had done every thing that was necessary, when they had contrived some method of answering them; and yet, while they neglected all the other phenomena, they did not always think of determining with precision the measure and extent of those which they attempted to explain. This is peculiarly the case in regard to the secondary stratifications, which constitute, however, the most difficult and most important portion of the problem. It has hardly ever been attempted carefully to ascertain the superpositions of their strata, or the connections of these strata with the species of animals and of plants whose remains they inclose.

Are there certain animals and plants peculiar to certain strata, and not found in others? What are the species that appear first in order, and those which succeed? Do these two kinds of species ever accompany one another? Are there alternations in their appearances; or, in other words, does the first species appear a second time, and does the second species then disappear? Have these animals and plants lived in the places where their exuviæ are found, or have they been brought there from other places? Do all these animals and plants still continue to live in some part of the earth, or have they been totally or partially destroyed? Is there any constant connection between the antiquity of the strata, and the resemblance or non-resemblance of the extraneous fossils, to the animals and plants that still exist? Is there any connection, in regard to climate, between the extraneous fossils and the still living organized bodies which most nearly resemble them? May it be concluded, that the transportation of these living organized bodies, if such a thing ever happened, has taken place from north to south, or from east to west; or was it effected by means that irregularly scattered and mingled them together? And, finally, is it still possible to distinguish the epochs of these transportations, by attentively examining the strata which inclose the remains, or are imprinted by their forms?

If, from the want of sufficient evidence, these questions cannot be satisfactorily answered, how shall we be able to explain the causes of the presently existing state of our globe? It is certain, that so far from any of these points being as yet completely established, naturalists seem to have scarcely any idea of the propriety of investigating facts before they construct their systems. The cause of this strange procedure may be discovered, by considering that all geologists hitherto have either been mere cabinet naturalists, who had themselves hardly paid any attention to the structure of mountains, or mere mineralogists, who had not studied in sufficient detail the innumerable diversity of animals, and the almost infinite complication of their various parts and organs. The former of these have only constructed systems; while the latter have made excellent collections of observations, and have laid the foundations of true geological science, but have been unable to raise and complete the edifice.

Notes

1. The two most remarkable phenomena of this kind, and which must for ever banish all idea of a slow and gradual revolution, are the rhinoceros discovered in 1771 in the banks of the *Vilhoui*, and the elephant recently found by M. Adams near the mouth of the *Lena*. This last retained its flesh and skin, on which was hair of two kinds; one short, fine, and crisped, resembling wool, and the other like long bristles. The flesh was still in such high preservation, that it was eaten by dogs.
2. See Pallas, in his *Memoir on the Formation of Mountains*.
3. The scientific journies of Saussure and Deluc give a prodigious number of instances of this nature.
4. When I formerly mentioned this circumstance, of the science of geology having become ridiculous, I only expressed a well-known truth, without presuming to give my own opinion, as some respectable geologists seem to have believed. If their mistake arose from my expressions having been rather equivocal, I take this opportunity of explaining my meaning.
5. *Telluris Theoria Sacra*. Lond. 1681.
6. *Essay towards the Natural History of the Earth*. Lond. 1702.
7. *Mémoires de l'Academie*, 1708.
8. *A New Theory of the Earth*. Lond. 1708.
9. Leibnitz, *Protogœa. Act Lips.* 1683; *Gott.* 1749.
10. *Telliamed.* [*Telliamed* was the main work of Benoît de Maillet (1656–1738), and is his name spelled backwards.—(LP)]
11. See *La Physique de Rodig.* p. 106. Leipsic, 1801, and *Telliamed*, p. 169. Lamarck has expanded this system at great length, and supported it with much sagacity, in his *Hydrogéologie,* and *Philosophie Zoologique.*
12. M. Patrin has used much ingenuity to establish this view of the subject, in several articles of the *Nouveau Dictionnaire d'Histoire Naturelle.*
13. In his *Geology,* Delamétherie assumes crystallization as the chief cause or agent.
14. Hutton, and Playfair in his *Illustrations of the Huttonian Theory of the Earth.* Edinb. 1802.
15. See Lamanon, in various parts of the *Journal de Physique.*
16. Dolomieu, in the *Journal de Physique.*
17. M. M. de Marschall, in *Researches respecting the Origin and Developement of the present State of the Earth.* Geissen, 1802 [Marschall von Bibenstein, Karl Wilhelm, and Ernst Franz Ludwig Marschall von Bibenstein. 1802. *Untersuchungen über den Ursprung und die Ausbildung der gegenwärtigen Anordnungen des Weltbäudes.* Giessen and Darmstadt.—LP].
18. Bertrand, *Periodical Renewal of the Terrestrial Continents.* Hamburgh, 1799.

CHAPTER 27

Prejudices Relating to the Theory of the Earth

JOHN PLAYFAIR[1]

Pages 510–528, Note XXVI, §133, of *Illustrations of the Huttonian theory of the earth*. Printed for Cadell and Davies, London, and W. Creech, Edinburgh, 1802.

445. Among the prejudices which a new theory of the earth has to overcome, is an opinion, held, or affected to be held, by many, that geological science is not yet ripe for such elevated and difficult speculations. They would, therefore, get rid of these speculations, *by moving the previous question,* and declaring that at present we ought to have no theory at all. We are not yet, they allege, sufficiently acquainted with the phenomena of geology; the subject is so various and extensive, that our knowledge of it must for a long time, perhaps for ever, remain extremely imperfect. And hence it is, that the theories hitherto proposed have succeeded one another with so great rapidity, hardly any of them having been able to last longer than the discovery of a new fact, or a fact unknown when it was invented. It has proved insufficient to connect this fact with the phenomena already known, and has therefore been justly abandoned. In this manner, they say, have passed away the theories of Woodward, Burnet, Whiston, and even of Buffon; and so will pass, in their turn, those of Hutton and Werner.

446. This unfavourable view of geology, ought not, however, to be received without examination; in science, presumption is less hurtful than despair, and inactivity is more dangerous than error.

One reason of the rapid succession of geological theories, is the mistake that has been made as to their object, and the folly of attempting to explain by them the first origin of things. This mistake has led to fanciful speculations that had nothing but their novelty to recommend them, and which, when that charm had ceased, were rejected as mere suppositions, incapable of proof. But if it is once settled, that a theory of the earth ought to have no other aim but to discover the laws that regulate the changes on the surface, or in the interior of the globe, the subject is brought within the sphere either of observation or analogy; and there is no reason to suppose, that man, who has numbered the stars, and measured their forces, shall ultimately prove unequal to this investigation.

447. Again, theories that have a rational object, though they be false or imperfect in their principles, are for the most part approximations to the truth, suited to the information at the time when they were proposed. They are steps, therefore, in the advancement of knowledge, and are terms of a series that must end when the real laws of nature are discovered. It is, on this account, rash to conclude, that in the revolutions of science, what has happened must continue to happen, and because systems have changed rapidly in time past, that they must necessarily do so in time to come.

He who would have reasoned so, and who had seen the ancient physical systems, at first all rivals to one another, and then swallowed up by the Aristotelian; the Aristotelian physics giving way to those of Des Cartes; and the physics of Des Cartes to those of Newton; would have predicted that these last were also, in their turn, to give place to the philosophy of some later period. This is, however, a conclusion that hardly any one will now be bold enough to maintain, after a hundred years of the most scrupulous examination have done nothing but add to the evidence of the NEWTONIAN SYSTEM. It seems certain, therefore, that the rise and fall of theories in times past, does not argue, that the same will happen in the time that is to come.

448. The multifarious and extremely diversified object of geological researches, does, no doubt, render the first steps difficult, and may very well account for the instability hitherto observed in such theories; but the very same thing gives reason for expecting a very high degree of certainty to be ultimately attained in these inquiries.

Where the phenomena are few and simple, there may be several different theories that will explain them in a manner equally satisfactory; and in such cases, the true and the false hypotheses are not easily distinguished from one another. When, on the other hand, the phenomena are greatly varied, the probability is, that among them, some of those *instantiæ crucis* will be found, that exclude every hypothesis but one, and reduce the explanation given to the highest degree of certainty.[2] It was thus, when the phenomena of the heavens were but imperfectly known, and were confined to a few general and simple facts, that the Philolaic could claim no preference to the Ptolemaic system: The former seemed a possible hypothesis; but as it performed nothing that the other did not perform, and was inconsistent with some of our most natural prejudices, it had but few adherents. The invention of the telescope, and the use of more accurate instruments, by multiplying and diversifying the facts, established its credit; and when not only the general laws, but also the inequalities, and disturbances of the planetary motions were understood, all physical hypotheses vanished, like phantoms, before the philosophy of NEWTON. Hence the number, the variety, and even the complication of facts, contribute ultimately to separate truth from falsehood; and the same causes which, in any case, render the first attempts toward a theory difficult, make the final success of such attempts just so much the more probable.

This maxim, however, though a general encouragement to the prosecution of geological inquiries, does not amount to a proof that we are yet arrived at the period when those inquiries may safely assume the form of a theory. But that we are arrived at such a period, appears clear from other circumstances.

449. It cannot be denied, that a great multitude of facts, respecting the mineral kingdom, are now known with considerable precision; and that the many diligent and skilful observers, who have arisen in the course of the last thirty years, have produced a great change in the state of geological knowledge. It is unnecessary to enumerate them all; Ferber, Bergman, De Luc, Saussure, Dolomieu, are those on whom Dr. Hutton chiefly relied; and it is on their observations and his own that his system is founded.[3] If it be said, that only a small part of the earth's surface has yet been surveyed, and described with such accuracy as is found in the writers just named, it may be answered, that the earth is constructed with such a degree of uniformity, that a tract of no very large extent may afford instances of all the leading facts that we can ever observe in the mineral kingdom. The variety of geological appearances which a traveller meets with, is not at all in proportion to the extent of country he traverses; and if he take in a portion of land sufficient to include primitive and secondary strata, together with mountains, rivers, and plains, and unstratified bodies in veins and in masses, though it be not a very large part of the earth's surface, he may find examples of all the most important facts in the history of fossils. Though the labours of mineralogists have embraced but a small part of the globe, they may therefore have comprehended a very large proportion of the phenomena which it exhibits; and hence a presumption arises, that the outlines, at least, of geology have now been traced with tolerable truth, and are not susceptible of great variation.

450. When the phenomena of any class are in general ambiguous, and admit of being explained by different or even opposite theories; if few of

those exclusive facts are known, which admit but of one or a few solutions, then we have no right to expect much from our endeavours to generalize, except the knowledge of the points where our information is most deficient, and to which our observations ought chiefly to be directed. But that many of the exclusive and unambiguous instances are known, in the natural history of the globe, I think is evident from the reasoning in the foregoing pages, where so many examples have occurred of appearances that give the most direct negative to the Neptunian system, and exclude it from the number of possible hypotheses, by which the phenomena of geology can be explained. The abundance of such instances is an infallible sign, that the mass of knowledge is in that state of fermentation, from which the true theory may be expected to emerge.

451. Another indication of the same kind, is the near approach that even the most opposite theories make, in some respects, to one another. There are so many points of contact between them, that they appear to approximate to an ultimate state, in which, however unwillingly, they must at last coincide. That ultimate form, too, which all these theories have a tendency to put on, if I am not deceived, is no other than that of the Huttonian theory.

452. The first example I shall take from the system of Saussure. It is to be regretted, that this excellent geologist has no where given us a complete account of his theory. Some of the leading principles of it are, however, unfolded in the course of his observations, and enable us to form a notion of its general outline. It was evidently far removed from the system of subterraneous heat, and seems, especially in the latter part of the author's life, to have been very much accommodated to the prevailing system of Werner.[4] Nevertheless, with so little affinity between their general views, Saussure and Hutton agree in that most important article which regards the elevation of the strata. Saussure plainly perceived the impossibility of the strata being formed in the vertical situations which so many of them now occupy; and he takes great pains to demonstrate this impossibility, from some facts that have been referred to above. He also believed that this elevation had been given to strata that were originally level, by a force directed upwards, or by the *refoulement* of the beds, not by their falling in, as is the opinion of De Luc and some other of the Neptunists.

Now, whoever admits this principle, and reasons on it consistently, without being afraid to follow it through all its consequences, must unavoidably come very close to the Huttonian theory. He must see, that a power which, acting from below, produced this great effect, can never have belonged to water, unless rarefied into steam by the application of heat. But if it be once admitted that heat resides in the mineral regions, the great objection to Dr. Hutton's system is removed; and the theorist, who was furnished with so active and so powerful an agent, would be very unskilful in the management of his own resources, if he did not employ it in the work of consolidating as well as in that of raising up the strata. A little attention will shew, that it is qualified for both purposes; though insuperable objections must, no doubt, offer themselves, where the effects of compression are not understood. We may safely conclude, then, that the accurate and ingenious Oreologist of Geneva ought to have been a *Plutonist,* in order to give consistency to the principles which he had adopted, and to make them coalesce as parts of one and the same system. If he embraced an opposite opinion, it probably was from feeling the force of those objections that arise from our discovering nothing in the bowels of the earth like the remains left by combustion, or inflammation, at its surface. The secret by which these seeming contradictions are to be reconciled, was unknown to this mineralogist, and he has accordingly decided strongly against the action of fire, even in the case of those unstratified substances that have the greatest affinity to volcanic lava.

453. The theoretical conclusions of another accurate and skilful observer, Dolomieu, furnish a still more remarkable example of a tendency to union between systems professedly hostile to one another.

This ingenious mineralogist, observing the interposition of the basalt between stratified rocks, so that it had not only regular beds of sandstone for its base, but was also covered with beds of the same kind, saw plainly that these appearances were inconsistent with the supposition of common volcanic explosions at the surface. He therefore conceived, that the volcanic eruption had happened at the bottom of the sea, (the level of which, in former ages, had been much higher than at present), and that the materials afterwards deposited on the lava, had been in length of time consolidated into beds of stone. It is evident, that this notion of submarine volcanoes, comes very near, in many respects, to Dr. Hutton's explanation of the same appearances. If the only thing to be accounted for were the phenomenon in question, it cannot be denied that Dolomieu's hypothesis would be perfectly sufficient; but Dr. Hutton, to whom this phenomenon was familiar, and who, like Dolomieu, conceived the basalt to have been in fusion, was convinced that the retreat of the sea was not a fact well attested by geological appearances, and if admitted, was inadequate to account for the facts usually explained by it. He conceived, therefore, that such lava as the preceding had flowed not only at the bottom of the sea, but in the bowels of the earth, and having been forced up through the fissures of rocks already formed, had heaved up some of these rocks, and interposed itself between them. This agrees with the other facts in the natural history both of the basaltes and the strata.

It is plain, that, in this, there is a great approach of the two theories to one another: both maintain the igneous origin of basaltes, and its affinity to lava; both acknowledge that this lava cannot have flowed at the surface, and that the strata which cover it have been formed at the bottom of the sea. They only differ as to the mode in which the submarine or subterraneous volcano produced its effect, and that difference arises merely from the one geologist having generalized more than the other. Dolomieu sought to connect the basalt with the lavas that proceed from volcanic explosions at the surface; Dr. Hutton sought not only to connect these two appearances with one another, but also with the other phenomena of mineralogy, particularly with the veins of basaltes, and the elevation of the strata.

454. In another point, the coincidence of Dolomieu's opinions and Dr. Hutton's is still more striking. The former has remarked, that many of the extinguished volcanoes are in granite countries, and that, nevertheless, the lavas that they have erupted contain no granitic stones. There must be, therefore, says he, something under the granite, and this last is not, at least in all cases, to be considered as the basis of the mineral kingdom, or as the body on which all others rest. In this system, therefore, granite is not always a primordial rock, any more than in Dr. Hutton's.

But Dolomieu makes a still nearer advance to the Huttonian theory; for he supposes, that under the solid and hard crust of the globe, there is a sphere of melted stone, from which this basaltic lava was thrown up. The system of subterraneous heat is here adopted in its utmost extent, and in that form which is considered as the most liable to objection, viz. the existence of it at the present moment, in such a degree as to melt rocks, and keep them in a state of fusion. In this conclusion, the two theories agree perfectly; and if they do so, it is only because the nature of things has forced them into union, notwithstanding the dissimilitude of their fundamental principles.

This ought to be considered as a strong proof, that the phenomena known to mineralogists are sufficient to justify the attempts to form a theory of the earth, and are such as lead to the same conclusions, where there was not only no previous concert, but even a very marked opposition. I have already observed, that there is a greater tendency to agree among geological theories, than among the authors of those theories.

455. Another circumstance worthy of consideration is, that in the search which the Neptunists have made, for facts most favourable to the aqueous formation of minerals, we find hardly any of a kind that was unknown to the author of the system here explained. The appearances on which Werner grounds his opinion with respect to basaltes, and by which he would exclude the action of fire from any share in the formation of it, are all comprehended in the alternation of that rock with beds, or strata obviously of aqueous origin. Now these appearances were well known to Dr. Hutton, and are easily explained by his theory, provided the effects of compression are admitted. From this, and the other circumstances just observed, I am

disposed to think, that the great facts on which every geological system must depend, are now known, and that it is not too bold an anticipation to say, that a theory of the earth, which, explains all the phenomena with which we are at present acquainted, will be found to explain all those that remain to be discovered.

456. The time indeed was, and we are not yet far removed from it, when one of the most important principles involved in Dr. Hutton's theory was not only unknown, but could not be discovered. This was before the causticity produced in limestone by exposure to fire was understood, and when it was not known that it arose from the expulsion of a certain aerial fluid, which before was a component part of the stone. It could not then be perceived, that this aerial part might be retained by pressure, even in spite of the action of fire, and that in a region where great compression existed, the absence of causticity was no proof that great heat had not been applied. The discoveries of Dr. Black, therefore, mark an era, before which men were not qualified to judge of the nature of the powers that had acted in the consolidation of mineral substances.[5] Those discoveries were, indeed, destined to produce a memorable change in chemistry, and in all the branches of knowledge allied to it; and have been the foundation of that brilliant progress, by which a collection of practical rules, and of insulated facts, has in a few years risen to the rank of a very perfect science. But even before they had explained the nature of carbonic gas, and its affinity to calcareous earth, I am not sure but that Dr. Hutton's theory was, at least, partly formed, though it must certainly have remained, even in his own opinion, exposed to great difficulties. His active and penetrating genius soon perceived, in the experiments of his friend, the solution of those difficulties, and formed that happy combination of principles, which has enabled him to explain the most enigmatical appearances in the natural history of the earth.

As we are not yet far removed from the time when our chemical knowledge was too imperfect to admit of a satisfactory explanation of the phenomena of mineralogy, so it is not unlikely that we are approaching to other discoveries that are to throw new light on this science. It would, however, be to argue strangely to say, that we must wait till those discoveries are made before we begin any theoretical reasonings. If this rule were followed, we should not know where the imperfections of our science lay, nor when the remedies were found out, should we be in a condition to avail ourselves of them. Such conduct would not be caution, but timidity, and an excess of prudence fatal to all philosophical inquiry.

457. The truth, indeed, is, that in physical inquiries, the work of theory and observation must go hand in hand, and ought to be carried on at the same time, more especially if the matter is very complicated, for there the clue of theory is necessary to direct the observer. Though a man may begin

to observe without any hypothesis, he cannot continue long without seeing some general conclusion arise; and to this nascent theory it is his business to attend, because, by seeking either to verify or to disprove it, he is led to new experiments, or new observations. He is led also to the very experiments and observations that are of the greatest importance, namely, to those *instantiæ crucis,* which are the *criteria* that naturally present themselves for the trial of every hypothesis. He is conducted to the places where the transitions of nature are most perceptible, and where the absence of former, or the presence of new circumstances, excludes the action of imaginary causes. By this correction of his first opinion, a new approximation is made to the truth; and by the repetition of the same process, certainty is finally obtained. Thus theory and observation mutually assist one another; and the spirit of system, against which there are so many and such just complaints, appears, nevertheless, as the animating principle of inductive investigation. The business of found philosophy is not to extinguish this spirit, but to restrain and direct its efforts.

458. It is therefore hurtful to the progress of physical science to represent observation and theory as standing opposed to one another. Bergman has said, "Observationes veras quàm ingeniosissimas fictiones sequi præstat; naturæ mysteria potius indagare quàm divinare."

If it is meant by this merely to say, that it is better to have facts without theory, than theory without facts, and that it is wiser to inquire into the secrets of nature, than to guess at them, the truth of the maxim will hardly be controverted. But if we are to understand by it, as some may perhaps have done, that all theory is mere fiction, and that the only alternative a philosopher has, is to devote himself to the study of facts unconnected by theory, or of theory unsupported by facts, the maxim is as far from the truth, as I am convinced it is from the real sense of Bergman. Such an opposition between the business of the theorist and the observer, can only occur when the speculations of the former are vague and indistinct, and cannot be so *embodied* as to become visible to the latter. But the philosopher who has ascended to his theory by a regular generalization of facts, and who descends from it again by drawing such palpable conclusions as may be compared with experience, furnishes the infallible means of distinguishing between *perfect science* and *ingenious fiction.* Of a geological theory that has stood this double test of the analytic and synthetic methods, Dr. Hutton has furnished us with an excellent instance, in his explanation of granite. The appearances which he observed in that stone led him to conclude, that it had been melted, and injected while fluid, among the stratified rocks already formed. He then considered, that if this is true, veins of granite must often run from the larger masses of that stone, and penetrate the strata in various directions; and this must be visible at those places where these different kinds of rock come into contact with one another. This led him to search in Arran and Glen-tilt for the phenomena in question; the result, as we have seen, afforded to his theory the fullest confirmation, and to himself the high satisfaction which must ever accompany the success of candid and judicious inquiry.

459. It cannot, however, be denied, that the impartiality of an observer may often be affected by system; but this is a misfortune against which the want of theory is not always a complete security. The partialities in favour of opinions are not more dangerous than the prejudices against them; for such is the spirit of system, and so naturally do all men's notions tend to reduce themselves into some regular form, that the very belief that there can be no theory, becomes a theory itself, and may have no inconsiderable sway over the mind of an observer. Besides, one man may have as much delight in pulling down, as another has in building up, and may choose to display his dexterity in the one occupation as well as in the other. The want of theory, then, does not secure the candour of an observer, and it may very much diminish his skill. The discipline that seems best calculated to promote both, is a thorough knowledge of the methods of inductive investigation; an acquaintance with the history of physical discovery; and the careful study of those sciences in which the rules of philosophizing have been most successfully applied.

Editor's Notes

1. These notes were added by the editor; the original text does not use footnotes here. See, among others, Dean 1992 for the historical background to the figures and themes discussed.
2. The term *instantia crucis* is from Francis Bacon's *Novum Organum*, Book II, Aphorism 36.
3. Johann Jakob Ferber (1743–1790), Torbern Bergman (1735–1784), Jean André Deluc (1727–1817), Horace-Bénédict de Saussure (1740–1799), Déodat de Dolomieu (1750–1801).
4. Abraham Werner (1749–1817), a German geologist.
5. Joseph Black (1728–1799), Scottish chemist.

CHAPTER 28

On the Geological Succession
of Organic Beings

CHARLES DARWIN[1]

Pages 290–315 of *Origin of Species*, 6th ed. London: John Murray. Ch.
XI, "On the Geological Succession of Organic Beings," 1873.

On the slow and successive appearance of new species—On their different rates of change—Species once lost do not reappear—Groups
of species follow the same general rules in their appearance and disappearance as do single species—On extinction—On simultaneous
changes in the forms of life throughout the world—On the affinities
of extinct species to each other and to living species—On the state of
development of ancient forms—On the succession of the same types
within the same areas—Summary of preceding and present chapter.

LET US now see whether the several facts and laws relating to the geological succession of organic beings accord best with the common view of the
immutability of species, or with that of their slow and gradual modification,
through variation and natural selection.

New species have appeared very slowly, one after another, both on the
land and in the waters. Lyell has shown that it is hardly possible to resist
the evidence on this head in the case of the several tertiary stages; and every
year tends to fill up the blanks between the stages, and to make the proportion between the lost and existing forms more gradual. In some of the most
recent beds, though undoubtedly of high antiquity if measured by years,

only one or two species are extinct, and only one or two are new, having appeared there for the first time, either locally, or, as far as we know, on the face of the earth. The secondary formations are more broken; but, as Bronn has remarked, neither the appearance nor disappearance of the many species embedded in each formation has been simultaneous.[2]

Species belonging to different genera and classes have not changed at the same rate, or in the same degree. In the older tertiary beds a few living shells may still be found in the midst of a multitude of extinct forms. Falconer has given a striking instance of a similar fact, for an existing crocodile is associated with many lost mammals and reptiles in the sub-Himalayan deposits.[3] The Silurian Lingula differs but little from the living species of this genus; whereas most of the other Silurian Molluscs and all the Crustaceans have changed greatly. The productions of the land seem to have changed at a quicker rate than those of the sea, of which a striking instance has been observed in Switzerland. There is some reason to believe that organisms high in the scale, change more quickly than those that are low: though there are exceptions to this rule. The amount of organic change, as Pictet has remarked, is not the same in each successive so-called formation.[4] Yet if we compare any but the most closely related formations, all the species will be found to have undergone some change. When a species has once disappeared from the face of the earth, we have no reason to believe that the same identical form ever reappears. The strongest apparent exception to this latter rule is that of the so-called "colonies" of M. Barrande, which intrude for a period in the midst of an older formation, and then allow the pre-existing fauna to reappear; but Lyell's explanation, namely, that it is a case of temporary migration from a distinct geographical province, seems satisfactory.[5]

These several facts accord well with our theory, which includes no fixed law of development, causing all the inhabitants of an area to change abruptly, or simultaneously, or to an equal degree. The process of modification must be slow, and will generally affect only a few species at the same time; for the variability of each species is independent of that of all others. Whether such variations or individual differences as may arise will be accumulated through natural selection in a greater or less degree, thus causing a greater or less amount of permanent modification, will depend on many complex contingencies—on the variations being of a beneficial nature, on the freedom of intercrossing, on the slowly changing physical conditions of the country, on the immigration of new colonists, and on the nature of the other inhabitants with which the varying species come into competition. Hence it is by no means surprising that one species should retain the same identical form much longer than others; or, if changing, should change in a less degree. We find similar relations between the existing inhabitants of distinct countries; for instance, the land-shells and coleopterous insects of

Madeira have come to differ considerably from their nearest allies on the continent of Europe, whereas the marine shells and birds have remained unaltered. We can perhaps understand the apparently quicker rate of change in terrestrial and in more highly organised productions compared with marine and lower productions, by the more complex relations of the higher beings to their organic and inorganic conditions of life, as explained in a former chapter. When many of the inhabitants of any area have become modified and improved, we can understand, on the principle of competition, and from the all-important relations of organism to organism in the struggle for life, that any form which did not become in some degree modified and improved, would be liable to extermination. Hence we see why all the species in the same region do at last, if we look to long enough intervals of time, become modified, for otherwise they would become extinct.

In members of the same class the average amount of change, during long and equal periods of time, may, perhaps, be nearly the same; but as the accumulation of enduring formations, rich in fossils, depends on great masses of sediment being deposited on subsiding areas, our formations have been almost necessarily accumulated at wide and irregularly intermittent intervals of time; consequently the amount of organic change exhibited by the fossils embedded in consecutive formations is not equal. Each formation, on this view, does not mark a new and complete act of creation, but only an occasional scene, taken almost at hazard, in an ever slowly changing drama.

We can clearly understand why a species when once lost should never reappear, even if the very same conditions of life, organic and inorganic, should recur. For though the offspring of one species might be adapted (and no doubt this has occurred in innumerable instances) to fill the place of another species in the economy of nature, and thus supplant it; yet the two forms—the old and the new—would not be identically the same; for both would almost certainly inherit different characters from their distinct progenitors; and organisms already differing would vary in a different manner. For instance, it is possible, if all our fantail pigeons were destroyed, that fanciers might make a new breed hardly distinguishable from the present breed; but if the parent rock-pigeon were likewise destroyed, and under nature we have every reason to believe that parent-forms are generally supplanted and exterminated by their improved offspring, it is incredible that a fantail, identical with the existing breed, could be raised from any other species of pigeon, or even from any other well-established race of the domestic pigeon, for the successive variations would almost certainly be in some degree different, and the newly-formed variety would probably inherit from its progenitor some characteristic differences.

Groups of species, that is, genera and families, follow the same general rules in their appearance and disappearance as do single species, changing more or less quickly, and in a greater or lesser degree. A group, when it has once disappeared, never reappears; that is, its existence, as long as it lasts, is continuous. I am aware that there are some apparent exceptions to this rule, but the exceptions are surprisingly few, so few that E. Forbes, Pictet, and Woodward (though all strongly opposed to such views as I maintain) admit its truth; and the rule strictly accords with the theory. For all the species of the same group, however long it may have lasted, are the modified descendants, one from the other, and all from a common progenitor. In the genus Lingula, for instance, the species which have successively appeared at all ages must have been connected by an unbroken series of generations, from the lowest Silurian stratum to the present day.

We have seen in the last chapter that whole groups of species sometimes falsely appear to have been abruptly developed; and I have attempted to give an explanation of this fact, which if true would be fatal to my views. But such cases are certainly exceptional; the general rule being a gradual increase in number, until the group reaches its maximum, and then, sooner or later, a gradual decrease. If the number of the species included within a genus, or the number of the genera within a family, be represented by a vertical line of varying thickness, ascending through the successive geological formations in which the species are found, the line will sometimes falsely appear to begin at its lower end, not in a sharp point, but abruptly; it then gradually thickens upwards, often keeping of equal thickness for a space, and ultimately thins out in the upper beds, marking the decrease and final extinction of the species. This gradual increase in number of the species of a group is strictly conformable with the theory, for the species of the same genus, and the genera of the same family, can increase only slowly and progressively; the process of modification and the production of a number of allied forms necessarily being a slow and gradual process,—one species first giving rise to two or three varieties, these being slowly converted into species, which in their turn produce by equally slow steps other varieties and species, and so on, like the branching of a great tree from a single stem, till the group becomes large.

On Extinction

We have as yet spoken only incidentally of the disappearance of species and of groups of species. On the theory of natural selection, the extinction of old forms and the production of new and improved forms are intimately connected together. The old notion of all the inhabitants of the earth having been swept away by catastrophes at successive periods is very generally given up, even by those geologists, as Elie de Beaumont, Murchison, Barrande, &c., whose general views would naturally lead them to this conclusion.[6] On the contrary, we have every reason to believe, from the study of the tertiary formations, that species and groups of species gradually disappear, one after another, first from one spot, then from another, and finally from the world. In some few cases, however, as by the breaking of an isthmus and the consequent irruption of a multitude of new inhabitants into an adjoining sea, or by the final subsidence of an island, the process of extinction may have been rapid. Both single species and whole groups of species last for very unequal periods; some groups, as we have seen, have endured from the earliest known dawn of life to the present day; some have disappeared before the close of the palæozoic period. No fixed law seems to determine the length of time during which any single species or any single genus endures. There is reason to believe that the extinction of a whole group of species is generally a slower process than their production: if their appearance and disappearance be represented, as before, by a vertical line of varying thickness the line is found to taper more gradually at its upper end, which marks the progress of extermination, than at its lower end, which marks the first appearance and the early increase in number of the species. In some cases, however, the extermination of whole groups, as of ammonites, towards the close of the secondary period, has been wonderfully sudden.

The extinction of species has been involved in the most gratuitous mystery. Some authors have even supposed that, as the individual has a definite length of life, so have species a definite duration. No one can have marvelled more than I have done at the extinction of species. When I found in La Plata the tooth of a horse embedded with the remains of Mastodon, Megatherium, Toxodon, and other extinct monsters, which all co-existed with still living shells at a very late geological period, I was filled with astonishment; for, seeing that the horse, since its introduction by the Spaniards into South America, has run wild over the whole country and has increased in numbers at an unparalleled rate, I asked myself what could so recently have exterminated the former horse under conditions of life apparently so favourable. But my astonishment was groundless. Professor Owen soon perceived that the tooth, though so like that of the existing horse, belonged to an extinct species.[7] Had this horse been still living, but in some degree rare, no naturalist would have felt the least surprise at its rarity; for rarity is the attribute of a vast number of species of all classes, in all countries. If we ask ourselves why this or that species is rare, we answer that something is unfavourable in its conditions of life; but what that something is, we can hardly ever tell. On the supposition of the fossil horse still existing as a rare species, we might have felt certain, from the analogy of all other mammals, even of the slow-breeding elephant, and from the history of the naturalisation of the domestic horse in South America, that under more favourable conditions it would in a very few years have stocked the whole continent.

But we could not have told what the unfavourable conditions were which checked its increase, whether some one or several contingencies, and at what period of the horse's life, and in what degree, they severally acted. If the conditions had gone on, however slowly, becoming less and less favourable, we assuredly should not have perceived the fact, yet the fossil horse would certainly have become rarer and rarer, and finally extinct;—its place being seized on by some more successful competitor.

It is most difficult always to remember that the increase of every creature is constantly being checked by unperceived hostile agencies; and that these same unperceived agencies are amply sufficient to cause rarity, and finally extinction. So little is this subject understood, that I have heard surprise repeatedly expressed at such great monsters as the Mastodon and the more ancient Dinosaurians having become extinct; as if mere bodily strength gave victory in the battle of life. Mere size, on the contrary, would in some cases determine, as has been remarked by Owen, quicker extermination from the greater amount of requisite food. Before man inhabited India or Africa, some cause must have checked the continued increase of the existing elephant. A highly capable judge, Dr. Falconer, believes that it is chiefly insects which, from incessantly harassing and weakening the elephant in India, check its increase; and this was Bruce's conclusion with respect to the African elephant in Abyssinia.[8] It is certain that insects and blood-sucking bats determine the existence of the larger naturalised quadrupeds in several parts of S. America.

We see in many cases in the more recent tertiary formations, that rarity precedes extinction; and we know that this has been the progress of events with those animals which have been exterminated, either locally or wholly, through man's agency. I may repeat what I published in 1845, namely, that to admit that species generally become rare before they become extinct—to feel no surprise at the rarity of a species, and yet to marvel greatly when the species ceases to exist, is much the same as to admit that sickness in the individual is the forerunner of death—to feel no surprise at sickness, but, when the sick man dies, to wonder and to suspect that he died by some deed of violence.

The theory of natural selection is grounded on the belief that each new variety, and ultimately each new species, is produced and maintained by having some advantage over those with which it comes into competition; and the consequent extinction of the less-favoured forms almost inevitably follows. It is the same with our domestic productions; when a new and slightly improved variety has been raised, it at first supplants the less improved varieties in the same neighbourhood; when much improved it is transported far and near, like our short-horn cattle, and takes the place of other breeds in other countries. Thus the appearance of new forms and the disappearance of old forms, both those naturally and those artificially produced, are bound together. In flourishing groups, the number of new specific forms which have been produced within a given time has at some periods probably been greater than the number of the old specific forms which have been exterminated; but we know that species have not gone on indefinitely increasing, at least during the later geological epochs, so that, looking to later times, we may believe that the production of new forms has caused the extinction of about the same number of old forms.

The competition will generally be most severe, as formerly explained and illustrated by examples, between the forms which are most like each other in all respects. Hence the improved and modified descendants of a species will generally cause the extermination of the parent-species; and if many new forms have been developed from any one species, the nearest allies of that species, *i.e.* the species of the same genus, will be the most liable to extermination. Thus, as I believe, a number of new species descended from one species, that is a new genus, comes to supplant an old genus, belonging to the same family. But it must often have happened that a new species belonging to some one group has seized on the place occupied by a species belonging to a distinct group, and thus have caused its extermination. If many allied forms be developed from the successful intruder, many will have to yield their places; and it will generally be the allied forms, which will suffer from some inherited inferiority in common. But whether it be species belonging to the same or to a distinct class, which have yielded their places to other modified and improved species, a few of the sufferers may often be preserved for a long time, from being fitted to some peculiar line of life, or from inhabiting some distant and isolated station, where they will have escaped severe competition. For instance, some species of Trigonia, a great genus of shells in the secondary formations, survive in the Australian seas; and a few members of the great and almost extinct group of Ganoid fishes still inhabit our fresh waters. Therefore the utter extinction of a group is generally, as we have seen, a slower process than its production.

With respect to the apparently sudden extermination of whole families or orders, as of Trilobites at the close of the palæozoic period and of Ammonites at the close of the secondary period, we must remember what has been already said on the probable wide intervals of time between our consecutive formations; and in these intervals there may have been much slow extermination. Moreover, when, by sudden immigration or by unusually rapid development, many species of a new group have taken possession of an area, many of the older species will have been exterminated in a correspondingly rapid manner; and the forms which thus yield their places will commonly be allied, for they will partake of the same inferiority in common.

Thus, as it seems to me, the manner in which single species and whole groups of species become extinct accords well with the theory of natural selection. We need not marvel at extinction; if we must marvel, let it be at our own presumption in imagining for a moment that we understand the many complex contingencies on which the existence of each species depends. If we forget for an instant, that each species tends to increase inordinately, and that some check is always in action, yet seldom perceived by us, the whole economy of nature will be utterly obscured. Whenever we can precisely say why this species is more abundant in individuals than that; why this species and not another can be naturalised in a given country; then, and not until then, we may justly feel surprise why we cannot account for the extinction of any particular species or group of species.

On the Forms of Life changing almost simultaneously throughout the World

Scarcely any palæontological discovery is more striking than the fact, that the forms of life change almost simultaneously throughout the world. Thus our European Chalk formation can be recognised in many distant regions, under the most different climates, where not a fragment of the mineral chalk itself can be found; namely, in North America, in equatorial South America, in Tierra del Fuego, at the Cape of Good Hope, and in the peninsula of India. For at these distant points, the organic remains in certain beds present an unmistakeable resemblance to those of the Chalk. It is not that the same species are met with; for in some cases not one species is identically the same, but they belong to the same families, genera, and sections of genera, and sometimes are similarly characterised in such trifling points as mere superficial sculpture. Moreover, other forms, which are not found in the Chalk of Europe but which occur in the formations either above or below, occur in the same order at these distant points of the world. In the several successive palæozoic formations of Russia, Western Europe, and North America, a similar parallelism in the forms of life has been observed by several authors: so it is, according to Lyell, with the European and North American tertiary deposits. Even if the few fossil species which are common to the Old and New Worlds were kept wholly out of view, the general parallelism in the successive forms of life, in the palæozoic and tertiary stages, would still be manifest, and the several formations could be easily correlated.

These observations, however, relate to the marine inhabitants of the world: we have not sufficient data to judge whether the productions of the land and of fresh water at distant points change in the same parallel manner. We may doubt whether they have thus changed: if the Megatherium, Mylodon, Macrauchenia, and Toxodon had been brought to Europe from La Plata, without any information in regard to their geological position, no one would have suspected that they had co-existed with sea-shells all still living; but as these anomalous monsters co-existed with the Mastodon and Horse, it might at least have been inferred that they had lived during one of the later tertiary stages.

When the marine forms of life are spoken of as having changed simultaneously throughout the world, it must not be supposed that this expression relates to the same year, or to the same century, or even that it has a very strict geological sense; for if all the marine animals now living in Europe, and all those that lived in Europe during the pleistocene period (a very remote period as measured by years, including the whole glacial epoch) were compared with those now existing in South America or in Australia, the most skilful naturalist would hardly be able to say whether the present or the pleistocene inhabitants of Europe resembled most closely those of the southern hemisphere. So, again, several highly competent observers maintain that the existing productions of the United States are more closely related to those which lived in Europe during certain late tertiary stages, than to the present inhabitants of Europe; and if this be so, it is evident that fossiliferous beds now deposited on the shores of North America would hereafter be liable to be classed with somewhat older European beds. Nevertheless, looking to a remotely future epoch, there can be little doubt that all the more modern *marine* formations, namely, the upper pliocene, the pleistocene and strictly modern beds, of Europe, North and South America, and Australia, from containing fossil remains in some degree allied, and from not including those forms which are found only in the older underlying deposits, would be correctly ranked as simultaneous in a geological sense.

The fact of the forms of life changing simultaneously, in the above large sense, at distant parts of the world, has greatly struck those admirable observers, MM. de Verneuil and d'Archiac.[9] After referring to the parallelism of the palæozoic forms of life in various parts of Europe, they add,

If, struck by this strange sequence, we turn our attention to North America, and there discover a series of analogous phenomena, it will appear certain that all these modifications of species, their extinction, and the introduction of new ones, cannot be owing to mere changes in marine currents or other causes more or less local and temporary, but depend on general laws which govern the whole animal kingdom.

M. Barrande has made forcible remarks to precisely the same effect. It is, indeed, quite futile to look to changes of currents, climate, or other physical conditions, as the cause of these great mutations in the forms of life throughout the world, under the most different climates. We must, as Barrande has remarked, look to some special law. We shall see this more clearly when we treat of the present distribution of organic beings, and find how slight is the relation between the physical conditions of various countries and the nature of their inhabitants.

This great fact of the parallel succession of the forms of life throughout the world, is explicable on the theory of natural selection. New species are formed by having some advantage over older forms; and the forms, which are already dominant, or have some advantage over the other forms in their own country, give birth to the greatest number of new varieties or incipient species. We have distinct evidence on this head, in the plants which are dominant, that is, which are commonest and most widely diffused, producing the greatest number of new varieties. It is also natural that the dominant, varying, and far-spreading species, which have already invaded to a certain extent the territories of other species, should be those which would have the best chance of spreading still further, and of giving rise in new countries to other new varieties and species. The process of diffusion would often be very slow, depending on climatal and geographical changes, on strange accidents, and on the gradual acclimatisation of new species to the various climates through which they might have to pass, but in the course of time the dominant forms would generally succeed in spreading and would ultimately prevail. The diffusion would, it is probable, be slower with the terrestrial inhabitants of distinct continents than with the marine inhabitants of the continuous sea. We might therefore expect to find, as we do find, a less strict degree of parallelism in the succession of the productions of the land than with those of the sea.

Thus, as it seems to me, the parallel, and, taken in a large sense, simultaneous, succession of the same forms of life throughout the world, accords well with the principle of new species having been formed by dominant species spreading widely and varying; the new species thus produced being themselves dominant, owing to their having had some advantage over their already dominant parents, as well as over other species, and again spreading, varying, and producing new forms. The old forms which are beaten and which yield their places to the new and victorious forms, will generally be allied in groups, from inheriting some inferiority in common; and therefore, as new and improved groups spread throughout the world, old groups disappear from the world; and the succession of forms everywhere tends to correspond both in their first appearance and final disappearance.

There is one other remark connected with this subject worth making. I have given my reasons for believing that most of our great formations, rich in fossils, were deposited during periods of subsidence; and that blank intervals of vast duration, as far as fossils are concerned, occurred during the periods when the bed of the sea was either stationary or rising, and likewise when sediment was not thrown down quickly enough to embed and preserve organic remains. During these long and blank intervals I suppose that the inhabitants of each region underwent a considerable amount of modification and extinction, and that there was much migration from other parts of the world. As we have reason to believe that large areas are affected by the same movement, it is probable that strictly contemporaneous formations have often been accumulated over very wide spaces in the same quarter of the world; but we are very far from having any right to conclude that this has invariably been the case, and that large areas have invariably been affected by the same movements. When two formations have been deposited in two regions during nearly, but not exactly, the same period, we should find in both, from the causes explained in the foregoing paragraphs, the same general succession in the forms of life; but the species would not exactly correspond; for there will have been a little more time in the one region than in the other for modification, extinction, and immigration.

I suspect that cases of this nature occur in Europe. Mr. Prestwich, in his admirable Memoirs on the eocene deposits of England and France, is able to draw a close general parallelism between the successive stages in the two countries; but when he compares certain stages in England with those in France, although he finds in both a curious accordance in the numbers of the species belonging to the same genera, yet the species themselves differ in a manner very difficult to account for considering the proximity of the two areas,—unless, indeed, it be assumed that an isthmus separated two seas inhabited by distinct, but contemporaneous, faunas.[10] Lyell has made similar observations on some of the later tertiary formations. Barrande, also, shows that there is a striking general parellelism in the successive Silurian deposits of Bohemia and Scandinavia; nevertheless he finds a surprising amount of difference in the species. If the several formations in these regions have not been deposited during the same exact periods,—a formation in one region often corresponding with a blank interval in the other,—and if in both regions the species have gone on slowly changing during the accumulation of the several formations and during the long intervals of time between them; in this case the several formations in the two regions could be arranged in the same order, in accordance with the general succession of the forms of life, and the order would falsely appear to be strictly parallel; nevertheless the species would not be all the same in the apparently corresponding stages in the two regions.

On the Affinities of Extinct Species to Each Other, and to Living Forms

Let us now look to the mutual affinities of extinct and living species. All fall into a few grand classes; and this fact is at once explained on the principle of descent. The more ancient any form is, the more, as a general rule, it differs from living forms. But, as Buckland long ago remarked, extinct species can all be classed either in still existing groups, or between them. That the extinct forms of life help to fill up the intervals between existing genera, families, and orders, is certainly true; but as this statement has often been ignored or even denied, it may be well to make some remarks on this subject, and to give some instances. If we confine our attention either to the living or to the extinct species of the same class, the series is far less perfect than if we combine both into one general system. In the writings of Professor Owen we continually meet with the expression of generalised forms, as applied to extinct animals; and in the writings of Agassiz, of prophetic or synthetic types; and these terms imply that such forms are in fact intermediate or connecting links. Another distinguished palæontologist, M. Gaudry, has shown in the most striking manner that many of the fossil mammals discovered by him in Attica serve to break down the intervals between existing genera.[11] Cuvier ranked the Ruminants and Pachyderms, as two of the most distinct orders of mammals; but so many fossil links have been disentombed that Owen has had to alter the whole classification, and has placed certain pachyderms in the same sub-order with ruminants; for example, he dissolves by gradations the apparently wide interval between the pig and the camel. The Ungulata or hoofed quadrupeds are now divided into the even-toed or odd-toed divisions; but the Macrauchenia of S. America connects to a certain extent these two grand divisions. No one will deny that the Hipparion is intermediate between the existing horse and certain older ungulate forms. What a wonderful connecting link in the chain of mammals is the Typotherium from S. America, as the name given to it by Professor Gervais expresses, and which cannot be placed in any existing order. The Sirenia form a very distinct group of mammals, and one of the most remarkable peculiarities in the existing dugong and lamentin is the entire absence of hind limbs, without even a rudiment being left; but the extinct Halitherium had, according to Professor Flower, an ossified thigh-bone "articulated to a well-defined acetabulum in the pelvis," and it thus makes some approach to ordinary hoofed quadrupeds, to which the Sirenia are in other respects allied. The cetaceans or whales are widely different from all other mammals, but the tertiary Zeuglodon and Squalodon, which have been placed by some naturalists in an order by themselves, are considered by Professor Huxley to be undoubtedly cetaceans, "and to constitute connecting links with the aquatic carnivora."[12]

Even the wide interval between birds and reptiles has been shown by the naturalist just quoted to be partially bridged over in the most unexpected manner, on the one hand, by the ostrich and extinct Archeopteryx, and on the other hand, by the Compsognathus, one of the Dinosaurians—that group which includes the most gigantic of all terrestrial reptiles. Turning to the Invertebrata, Barrande asserts, and a higher authority could not be named, that he is every day taught that, although palæozoic animals can certainly be classed under existing groups, yet that at this ancient period the groups were not so distinctly separated from each other as they now are.

Some writers have objected to any extinct species, or group of species, being considered as intermediate between any two living species, or groups of species. If by this term it is meant that an extinct form is directly interme- diate in all its characters between two living forms or groups, the objection is probably valid. But in a natural classification many fossil species certainly stand between living species, and some extinct genera between living gen- era, even between genera belonging to distinct families. The most common case, especially with respect to very distinct groups, such as fish and reptiles, seems to be, that, supposing them to be distinguished at the present day by a score of characters, the ancient members are separated by a somewhat lesser number of characters; so that the two groups formerly made a somewhat nearer approach to each other than they now do.

It is a common belief that the more ancient a form is, by so much the more it tends to connect by some of its characters groups now widely sepa- rated from each other. This remark no doubt must be restricted to those groups which have undergone much change in the course of geological ages; and it would be difficult to prove the truth of the proposition, for every now and then even a living animal, as the Lepidosiren, is discovered having affinities directed towards very distinct groups. Yet if we compare the older Reptiles and Batrachians, the older Fish, the older Cephalopods, and the eocene Mammals, with the more recent members of the same classes, we must admit that there is truth in the remark.

Let us see how far these several facts and inferences accord with the theory of descent with modification. As the subject is somewhat complex, I must request the reader to turn to the diagram in the fourth chapter.

We may suppose that the numbered letters in italics represent genera, and the dotted lines diverging from them the species in each genus. The diagram is much too simple, too few genera and too few species being given, but this is unimportant for us. The horizontal lines may represent succes- sive geological formations, and all the forms beneath the uppermost line may be considered as extinct. The three existing genera a^{14}, q^{14}, p^{14}, will form a small family; b^{14} and f^{14} a closely allied family or sub-family; and o^{14}, e^{14}, m^{14}, a third family. These three families, together with the many extinct

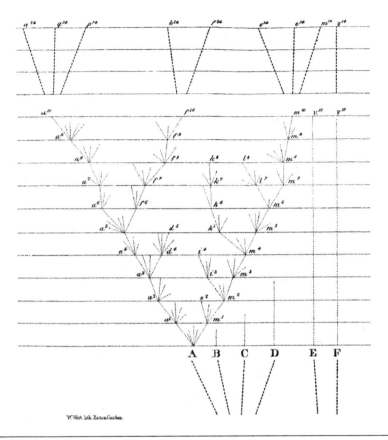

Figure 28.1

genera on the several lines of descent diverging from the parent-form (A) will form an order; for all will have inherited something in common from their ancient progenitor. On the principle of the continued tendency to divergence of character, which was formerly illustrated by this diagram, the more recent any form is, the more it will generally differ from its ancient progenitor. Hence we can understand the rule that the most ancient fossils differ most from existing forms. We must not, however, assume that divergence of character is a necessary contingency; it depends solely on the descendants from a species being thus enabled to seize on many and different places in the economy of nature. Therefore it is quite possible, as we have seen in the case of some Silurian forms, that a species might go on being slightly modified in relation to its slightly altered conditions of life, and yet retain throughout a vast period the same general characteristics. This is represented in the diagram by the letter F^{14}.

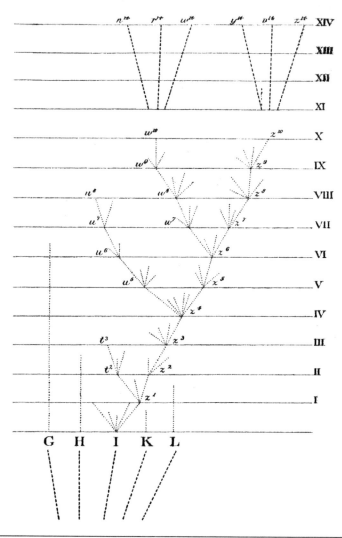

Figure 28.2

All the many forms, extinct and recent, descended from (A), make, as before remarked, one order; and this order, from the continued effects of extinction and divergence of character, has become divided into several sub-families and families, some of which are supposed to have perished at different periods, and some to have endured to the present day.

By looking at the diagram we can see that if many of the extinct forms supposed to be imbedded in the successive formations, were discovered at several points low down in the series, the three existing families on the uppermost line would be rendered less distinct from each other. If, for instance, the genera a^1, a^5, a^{10}, f^8, m^3, m^6, m^9, were disinterred, these three families would be so closely linked together that they probably would have to be united into one great family, in nearly the same manner as has occurred with ruminants and certain pachyderms. Yet he who objected to consider as intermediate the extinct genera, which thus link together the living genera of three families, would be partly justified, for they are intermediate, not directly, but only by a long and circuitous course through many widely different forms. If many extinct forms were to be discovered above one of the middle horizontal lines or geological formations—for instance, above No. VI.—but none from beneath this line, then only two of the families (those on the left hand, a^{14}, &c., and b^{14}, &c.) would have to be united into one; and there would remain two families, which would be less distinct from each other than they were before the discovery of the fossils. So again if the three families formed of eight genera (a^{14} to m^{14}), on the uppermost line, be supposed to differ from each other by half-a-dozen important characters, then the families which existed at the period marked VI. would certainly have differed from each other by a less number of characters; for they would at this early stage of descent have diverged in a less degree from their common progenitor. Thus it comes that ancient and extinct genera are often in a greater or less degree intermediate in character between their modified descendants, or between their collateral relations.

Under nature the process will be far more complicated than is represented in the diagram; for the groups will have been more numerous; they will have endured for extremely unequal lengths of time, and will have been modified in various degrees. As we possess only the last volume of the geological record, and that in a very broken condition, we have no right to expect, except in rare cases, to fill up the wide intervals in the natural system, and thus to unite distinct families or orders. All that we have a right to expect is, that those groups which have, within known geological periods, undergone much modification, should in the older formations make some slight approach to each other; so that the older members should differ less from each other in some of their characters than do the existing members of the same groups; and this by the concurrent evidence of our best palæontologists is frequently the case.

Thus, on the theory of descent with modification, the main facts with respect to the mutual affinities of the extinct forms of life to each other and to living forms, are explained in a satisfactory manner. And they are wholly inexplicable on any other view.

On this same theory, it is evident that the fauna during any one great period in the earth's history will be intermediate in general character between that which preceded and that which succeeded it. Thus the species which lived at the sixth great stage of descent in the diagram are the modified offspring of those which lived at the fifth stage, and are the parents of those which became still more modified at the seventh stage; hence they could hardly fail to be nearly intermediate in character between the forms of life above and below. We must, however, allow for the entire extinction of some preceding forms, and in any one region for the immigration of new forms from other regions, and for a large amount of modification during the long and blank intervals between the successive formations. Subject to these allowances, the fauna of each geological period undoubtedly is intermediate in character, between the preceding and succeeding faunas. I need give only one instance, namely, the manner in which the fossils of the Devonian system, when this system was first discovered, were at once recognised by palæontologists as intermediate in character between those of the overlying carboniferous, and underlying Silurian systems. But each fauna is not necessarily exactly intermediate, as unequal intervals of time have elapsed between consecutive formations.

It is no real objection to the truth of the statement that the fauna of each period as a whole is nearly intermediate in character between the preceding and succeeding faunas, that certain general offer exceptions to the rule. For instance, the species of mastodons and elephants, when arranged by Dr. Falconer in two series,—in the first place according to their mutual affinities, and in the second place according to their periods of existence,—do not accord in arrangement. The species extreme in character are not the oldest or the most recent; nor are those which are intermediate in character, intermediate in age. But supposing for an instant, in this and other such cases, that the record of the first appearance and disappearance of the species was complete, which is far from the case, we have no reason to believe that forms successively produced necessarily endure for corresponding lengths of time. A very ancient form may occasionally have lasted much longer than a form elsewhere subsequently produced, especially in the case of terrestrial productions inhabiting separated districts. To compare small things with great; if the principal living and extinct races of the domestic pigeon were arranged in serial affinity, this arrangement would not closely accord with the order in time of their production, and even less with the order of their disappearance; for the parent rock-pigeon still lives; and many varieties between the rock-pigeon and the carrier have become extinct; and carriers which are extreme in the important character of length of beak originated earlier than short-beaked tumblers, which are at the opposite end of the series in this respect.

Closely connected with the statement, that the organic remains from an intermediate formation are in some degree intermediate in character, is the fact, insisted on by all palæontologists, that fossils from two consecutive formations are far more closely related to each other, than are the fossils from two remote formations. Pictet gives as a well-known instance, the general resemblance of the organic remains from the several stages of the Chalk formation, though the species are distinct in each stage. This fact alone, from its generality, seems to have shaken Professor Pictet in his belief in the immutability of species. He who is acquainted with the distribution of existing species over the globe, will not attempt to account for the close resemblance of distinct species in closely consecutive formations, by the physical conditions of the ancient areas having remained nearly the same. Let it be remembered that the forms of life, at least those inhabiting the sea, have changed almost simultaneously throughout the world, and therefore under the most different climates and conditions. Consider the prodigious vicissitudes of climate during the pleistocene period, which includes the whole glacial epoch, and note how little the specific forms of the inhabitants of the sea have been affected.

On the theory of descent, the full meaning of the fossil remains from closely consecutive formations being closely related, though ranked as distinct species, is obvious. As the accumulation of each formation has often been interrupted, and as long blank intervals have intervened between successive formations, we ought not to expect to find, as I attempted to show in the last chapter, in any one or in any two formations, all the intermediate varieties between the species which appeared at the commencement and close of these periods: but we ought to find after intervals, very long as measured by years, but only moderately long as measured geologically, closely allied forms, or, as they have been called by some authors, representative species; and these assuredly we do find. We find, in short, such evidence of the slow and scarcely sensible mutations of specific forms, as we have the right to expect.

On the State of Development of Ancient Compared with Living Forms

We have seen in the fourth chapter that the degree of differentiation and specialisation of the parts in organic beings, when arrived at maturity, is the best standard, as yet suggested, of their degree of perfection or highness. We have also seen that, as the specialisation of parts is an advantage to each being, so natural selection will tend to render the organisation of each being more specialised and perfect, and in this sense higher; not but that it may leave many creatures with simple and unimproved structures fitted for simple conditions of life, and in some cases will even degrade or simplify the organisation, yet leaving such degraded beings better fitted for their new walks of life. In another and more general manner, new species become superior to their predecessors; for they have to beat in the struggle for life all the older forms, with which they come into close competition. We may therefore conclude that if under a nearly similar climate the eocene inhabitants of the world could be put into competition with the existing inhabitants, the former would be beaten and exterminated by the latter, as would the secondary by the eocene, and the palæozoic by the secondary forms. So that by this fundamental test of victory in the battle for life, as well as by the standard of the specialisation of organs, modern forms ought, on the theory of natural selection, to stand higher than ancient forms. Is this the case? A large majority of palæontologists would answer in the affirmative; and it seems that this answer must be admitted as true, though difficult of proof.

It is no valid objection to this conclusion, that certain Brachiopods have been but slightly modified from an extremely remote geological epoch; and that certain land and fresh-water shells have remained nearly the same, from the time when, as far as is known, they first appeared. It is not an insuperable difficulty that Foraminifera have not, as insisted on by Dr. Carpenter, progressed in organisation since even the Laurentian epoch; for some organisms would have to remain fitted for simple conditions of life, and what could be better fitted for this end than these lowly organised Protozoa?[13] Such objections as the above would be fatal to my view, if it included advance in organisation as a necessary contingent. They would likewise be fatal, if the above Foraminifera, for instance, could be proved to have first come into existence during the Laurentian epoch, or the above Brachiopods during the Cambrian formation; for in this case, there would not have been time sufficient for the development of these organisms up to the standard which they had then reached. When advanced up to any given point, there is no necessity, on the theory of natural selection, for their further continued progress; though they will, during each successive age, have to be slightly modified, so as to hold their places in relation to slight changes in their conditions. The foregoing objections hinge on the question whether we really know how old the world is, and at what period the various forms of life first appeared; and this may well be disputed.

The problem whether organisation on the whole has advanced is in many ways excessively intricate. The geological record, at all times imperfect, does not extend far enough back, to show with unmistakeable clearness that within the known history of the world organisation has largely advanced. Even at the present day, looking to members of the same class, naturalists are not unanimous which forms ought to be ranked as highest: thus, some look at the selaceans or sharks, from their approach in some important points of structure to reptiles, as the highest fish; others look at the teleosteans as the highest. The ganoids stand intermediate between the selaceans and teleosteans; the latter at the present day are largely preponderant in number; but formerly selaceans and ganoids alone existed; and in this case, according to the standard of highness chosen, so will it be said that fishes have advanced or retrograded in organisation. To attempt to compare members of distinct types in the scale of highness seems hopeless; who will decide whether a cuttle-fish be higher than a bee—that insect which the great Von Baer believed to be "in fact more highly organised than a fish, although upon another type"?[14] In the complex struggle for life it is quite credible that crustaceans, not very high in their own class, might beat cephalopods, the highest molluscs; and such crustaceans, though not highly developed, would stand very high in the scale of invertebrate animals, if judged by the most decisive of all trials—the law of battle. Besides these inherent difficulties in deciding which forms are the most advanced in organisation, we ought not solely to compare the highest members of a class at any two periods—though undoubtedly this is one and perhaps the most important element in striking a balance—but we ought to compare all the members, high and low, at the two periods. At an ancient epoch the highest and lowest molluscoidal animals, namely, cephalopods and brachiopods, swarmed in numbers; at the present time both groups are greatly reduced, whilst others, intermediate in organisation, have largely increased; consequently some naturalists maintain that molluscs were formerly more highly developed than at present; but a stronger case can be made out on the opposite side, by considering the vast reduction of brachiopods, and the fact that our existing cephalopods, though few in number, are more highly organised than their ancient representatives. We ought also to compare the relative proportional numbers at any two periods of the high and low classes throughout the world: if, for instance, at the present day fifty thousand kinds of vertebrate animals exist, and if we knew that at some former period only ten thousand kinds existed, we ought to look at this increase in number in the highest class, which implies a great displacement of lower forms, as a decided advance in the organisation of the world. We thus see how hopelessly difficult it is to compare with perfect fairness, under such extremely complex relations, the standard of organisation of the imperfectly-known faunas of successive periods.

We shall appreciate this difficulty more clearly, by looking to certain existing faunas and floras. From the extraordinary manner in which European productions have recently spread over New Zealand, and have seized on places which must have been previously occupied by the indigenes, we must believe, that if all the animals and plants of Great Britain were set free in New Zealand, a multitude of British forms would in the course of time become thoroughly naturalised there, and would exterminate many of the natives. On the other hand, from the fact that hardly a single inhabitant of the southern hemisphere has become wild in any part of Europe, we may well doubt whether, if all the productions of New Zealand were set free in Great Britain, any considerable number would be enabled to seize on places now occupied by our native plants and animals. Under this point of view, the productions of Great Britain stand much higher in the scale than those of New Zealand. Yet the most skilful naturalist, from an examination of the species of the two countries, could not have foreseen this result.

Agassiz and several other highly competent judges insist that ancient animals resemble to a certain extent the embryos of recent animals belonging to the same classes; and that the geological succession of extinct forms is nearly parallel with the embryological development of existing forms. This view accords admirably well with our theory. In a future chapter I shall attempt to show that the adult differs from its embryo, owing to variations having supervened at a not early age, and having been inherited at a corresponding age. This process, whilst it leaves the embryo almost unaltered, continually adds, in the course of successive generations, more and more difference to the adult. Thus the embryo comes to be left as a sort of picture, preserved by nature, of the former and less modified condition of the species. This view may be true, and yet may never be capable of proof. Seeing, for instance, that the oldest known mammals, reptiles, and fishes strictly belong to their proper classes, though some of these old forms are in a slight degree less distinct from each other than are the typical members of the same groups at the present day, it would be vain to look for animals having the common embryological character of the Vertebrata, until beds rich in fossils are discovered far beneath the lowest Cambrian strata—a discovery of which the chance is small.

On the Succession of the Same Types within the Same Areas, during the Later Tertiary Periods

Mr. Clift many years ago showed that the fossil mammals from the Australian caves were closely allied to the living marsupials of that continent.[15] In South America, a similar relationship is manifest, even to an uneducated eye, in the gigantic pieces of armour, like those of the armadillo, found in several parts of La Plata; and Professor Owen has shown in the most striking manner that most of the fossil mammals, buried there in such numbers, are related to South American types. This relationship is even more clearly seen in the wonderful collection of fossil bones made by MM. Lund and Clausen in the caves of Brazil.[16] I was so much impressed with these facts that I strongly insisted, in 1839 and 1845, on this "law of the succession of types,"—on "this wonderful relationship in the same continent between the dead and the living." Professor Owen has subsequently extended the same generalisation to the mammals of the Old World. We see the same law in this author's restorations of the extinct and gigantic birds of New Zealand. We see it also in the birds of the caves of Brazil. Mr. Woodward has shown that the same law holds good with sea-shells, but, from the wide distribution of most molluscs, it is not well displayed by them. Other cases could be added, as the relation between the extinct and living land-shells of Madeira; and between the extinct and living brackish water-shells of the Aralo-Caspian Sea.

Now what does this remarkable law of the succession of the same types within the same areas mean? He would be a bold man who, after comparing the present climate of Australia and of parts of South America, under the same latitude, would attempt to account, on the one hand through dissimilar physical conditions, for the dissimilarity of the inhabitants of these two continents; and, on the other hand through similarity of conditions, for the uniformity of the same types in each continent during the later tertiary periods. Nor can it be pretended that it is an immutable law that marsupials should have been chiefly or solely produced in Australia; or that Edentata and other American types should have been solely produced in South America. For we know that Europe in ancient times was peopled by numerous marsupials; and I have shown in the publications above alluded to, that in America the law of distribution of terrestrial mammals was formerly different from what it now is. North America formerly partook strongly of the present character of the southern half of the continent; and the southern half was formerly more closely allied, than it is at present, to the northern half. In a similar manner we know, from Falconer and Cautley's discoveries, that Northern India was formerly more closely related in its mammals to Africa than it is at the present time.[17] Analogous facts could be given in relation to the distribution of marine animals.

On the theory of descent with modification, the great law of the long enduring, but not immutable, succession of the same types within the same areas, is at once explained; for the inhabitants of each quarter of the world will obviously tend to leave in that quarter, during the next succeeding period of time, closely allied though in some degree modified descendants. If the inhabitants of one continent formerly differed greatly from those of another continent, so will their modified descendants still differ in nearly the same manner and degree. But after very long intervals of time, and after great geographical changes, permitting much intermigration, the feebler will yield to the more dominant forms, and there will be nothing immutable in the distribution of organic beings.

It may be asked in ridicule, whether I suppose that the megatherium and other allied huge monsters, which formerly lived in South America, have left behind them the sloth, armadillo, and anteater, as their degenerate descendants. This cannot for an instant be admitted. These huge animals have become wholly extinct, and have left no progeny. But in the caves of

Brazil, there are many extinct species which are closely allied in size and in all other characters to the species still living in South America; and some of these fossils may have been the actual progenitors of the living species. It must not be forgotten that, on our theory, all the species of the same genus are the descendants of some one species; so that, if six genera, each having eight species, be found in one geological formation, and in a succeeding formation there be six other allied or representative genera each with the same number of species, then we may conclude that generally only one species of each of the older genera has left modified descendants, which constitute the new genera containing the several species; the other seven species of each old genus having died out and left no progeny. Or, and this will be a far commoner case, two or three species in two or three alone of the six older genera will be the parents of the new genera: the other species and the other whole genera having become utterly extinct. In failing orders, with the genera and species decreasing in numbers as is the case with the Edentata of South America, still fewer genera and species will leave modified blood-descendants.

Summary of the Preceding and Present Chapters

I have attempted to show that the geological record is extremely imperfect; that only a small portion of the globe has been geologically explored with care; that only certain classes of organic beings have been largely preserved in a fossil state; that the number both of specimens and of species, preserved in our museums, is absolutely as nothing compared with the number of generations which must have passed away even during a single formation; that, owing to subsidence being almost necessary for the accumulation of deposits rich in fossil species of many kinds, and thick enough to outlast future degradation, great intervals of time must have elapsed between most of our successive formations; that there has probably been more extinction during the periods of subsidence, and more variation during the periods of elevation, and during the latter the record will have been least perfectly kept; that each single formation has not been continuously deposited; that the duration of each formation is, probably, short compared with the average duration of specific forms; that migration has played an important part in the first appearance of new forms in any one area and formation; that widely ranging species are those which have varied most frequently, and have oftenest given rise to new species; that varieties have at first been local; and lastly, although each species must have passed through numerous transitional stages, it is probable that the periods, during which each underwent modification, though many and long as measured by years, have been short in comparison with the periods during which each remained in an unchanged condition. These causes, taken conjointly, will to a large extent explain why—though we do find many links—we do not find interminable varieties, connecting together all extinct and existing forms by the finest graduated steps. It should also be constantly borne in mind that any linking variety between two forms, which might be found, would be ranked, unless the whole chain could be perfectly restored, as a new and distinct species; for it is not pretended that we have any sure criterion by which species and varieties can be discriminated.

He who rejects this view of the imperfection of the geological record, will rightly reject the whole theory. For he may ask in vain where are the numberless transitional links which must formerly have connected the closely allied or representative species, found in the successive stages of the same great formation? He may disbelieve in the immense intervals of time which must have elapsed between our consecutive formations; he may overlook how important a part migration has played, when the formations of any one great region, as those of Europe, are considered; he may urge the apparent, but often falsely apparent, sudden coming in of whole groups of species. He may ask where are the remains of those infinitely numerous organisms which must have existed long before the Cambrian system was deposited? We now know that at least one animal did then exist; but I can answer this last question only by supposing that where our oceans now extend they have extended for an enormous period, and where our oscillating continents now stand they have stood since the commencement of the Cambrian system; but that, long before that epoch, the world presented a widely different aspect; and that the older continents, formed of formations older than any known to us, exist now only as remnants in a metamorphosed condition, or lie still buried under the ocean.

Passing from these difficulties, the other great leading facts in palæontology agree admirably with the theory of descent with modification through variation and natural selection. We can thus understand how it is that new species come in slowly and successively; how species of different classes do not necessarily change together, or at the same rate, or in the same degree; yet in the long run that all undergo modification to some extent. The extinction of old forms is the almost inevitable consequence of the production of new forms. We can understand why when a species has once disappeared it never reappears. Groups of species increase in numbers slowly, and endure for unequal periods of time; for the process of modification is necessarily slow, and depends on many complex contingencies. The dominant species belonging to large and dominant groups tend to leave many modified descendants, which form new sub-groups and groups. As these are formed, the species of the less vigorous groups, from their inferiority inherited from a common progenitor, tend to become extinct together, and to leave no modified offspring on the face of the earth. But the utter extinction of a whole group of species has sometimes been a slow process, from the survival of a few descendants, lingering in protected and isolated situations. When a group has once wholly disappeared, it does not reappear; for the link of generation has been broken.

We can understand how it is that dominant forms which spread widely and yield the greatest number of varieties tend to people the world with allied, but modified, descendants; and these will generally succeed in displacing the groups which are their inferiors in the struggle for existence. Hence, after long intervals of time, the productions of the world appear to have changed simultaneously.

We can understand how it is that all the forms of life, ancient and recent, make together a few grand classes. We can understand, from the continued tendency to divergence of character, why the more ancient a form is, the more it generally differs from those now living; why ancient and extinct forms often tend to fill up gaps between existing forms, sometimes blending two groups, previously classed as distinct, into one; but more commonly bringing them only a little closer together. The more ancient a form is, the more often it stands in some degree intermediate between groups now distinct; for the more ancient a form is, the more nearly it will be related to, and consequently resemble, the common progenitor of groups, since become widely divergent. Extinct forms are seldom directly intermediate between existing forms; but are intermediate only by a long and circuitous course through other extinct and different forms. We can clearly see why the organic remains of closely consecutive formations are closely allied; for they are closely linked together by generation. We can clearly see why the remains of an intermediate formation are intermediate in character.

The inhabitants of the world at each successive period in its history have beaten their predecessors in the race for life, and are, in so far, higher in the scale, and their structure has generally become more specialised; and this may account for the common belief held by so many palæontologists, that organisation on the whole has progressed. Extinct and ancient animals resemble to a certain extent the embryos of the more recent animals belonging to the same classes, and this wonderful fact receives a simple explanation according to our views. The succession of the same types of structure within the same areas during the later geological periods ceases to be mysterious, and is intelligible on the principle of inheritance.

If then the geological record be as imperfect as many believe, and it may at least be asserted that the record cannot be proved to be much more perfect, the main objections to the theory of natural selection are greatly diminished or disappear. On the other hand, all the chief laws of palæontology plainly proclaim, as it seems to me, that species have been produced by ordinary generation: old forms having been supplanted by new and improved forms of life, the products of Variation and the Survival of the Fittest.

Editor's Notes

1. These notes were added by the editor, and are not features of the original text. See Costa 2009 for additional and detailed annotations to the text.
2. Heinrich Georg Bronn (1800–1862), a German paleontologist. Bronn translated Darwin's *Origin* into German.
3. Hugh Falconer (1808–1865), a Scottish geologist, botanist, palaeontologist, and paleoanthropologist.
4. Francois-Jules Pictet de la Rive (1809–1872), a French zoologist and palaeontologist.
5. Joachim Barrande (1799–1883), a French geologist and palaeontologist. Barrande wrote five volumes of the *Défense des colonies*, which appeared in 1861, 1862, 1865, 1870, and 1881.
6. Léonce Élie de Beaumont (1798–1874) and Sir Roderick Impey Murchison (1792–1871).
7. Sir Richard Owen (1804–1892), British biologist, anatomist, and palaeontologist.
8. Possibly James Bruce (1730–1794), a Scottish traveller.
9. Édouard de Verneuil (1805–1873) and Adolphe d'Archiac (1802–1868), French geologists and palaeontologists. Verneuil and d'Archiac "devoted much attention to the study of the Devonian rocks and fossils of the Bas-Boulonnais and Rhenish provinces" (Danby 1873, 429). These were communicated to the Geological Society of London as Verneuil and d'Archiac 1842. See Rudwick 1985, 387 and *passim*.
10. Sir Joseph Prestwich (1812–1896), British palaeontologist.
11. Jean Albert Gaudry (1827–1908), French geologist and palaeontologist.
12. Thomas Henry Huxley (1825–1895), British comparative anatomist.
13. William Carpenter (1813–1885), British zoologist and physiologist.
14. Karl Ernst von Baer (1792–1876), born in Estonia to a German family; naturalist, embryologist, and geologist, among other accomplishments. See Richards 2008 for an analysis of von Baer's contributions. The citation here probably is from Baer 1828–1837.
15. William Clift (1775–1849), a British naturalist.
16. Peter Lund (1801–1880) and Peter Clausen (1804–1855), Danish naturalists and palaeontologists.
17. Proby Thomas Cautley (1802–1871), British engineer and palaeontologist. See Falconer and Cautley 1846, also Costa 2009, 340.

CHAPTER 29

Of the Doctrine of Catastrophes and the Doctrine of Uniformity

WILLIAM WHEWELL

Pages 665–680, Vol. 1, book X, Chap. 3, of *Philosophy of the Inductive Sciences*. Second edition, London: John W. Parker, 1847 .

1. *Doctrine of Catastrophes.*—I HAVE already shown, in the *History of Geology*, that the attempts to frame a theory of the earth have brought into view two completely opposite opinions:—one, which represents the course of nature as *uniform* through all ages, the causes which produce change having had the same intensity in former times which they have at the present day;—the other opinion, which sees, in the present condition of things, evidences of *catastrophes*;—changes of a more sweeping kind, and produced by more powerful agencies than those which occur in recent times. Geologists who held the latter opinion, maintained that the forces which have elevated the Alps or the Andes to their present height could not have been any forces which are now in action: they pointed to vast masses of strata hundreds of miles long, thousands of feet thick, thrown into highly-inclined positions, fractured, dislocated, crushed: they remarked that upon the shattered edges of such strata they found enormous accumulations of fragments and rubbish, rounded by the action of water, so as to denote ages of violent aqueous action: they conceived that they saw instances in which whole mountains of rock in a state of igneous fusion, must have burst the earth's crust from below: they found that in the course of the revolutions by which one stratum of rock was placed upon another, the whole collection

of animal species which tenanted the earth and the seas had been removed, and a new set of living things introduced in its place: finally, they found, above all the strata, vast masses of sand and gravel containing bones of animals, and apparently the work of a mighty deluge. With all these proofs before their eyes, they thought it impossible not to judge that the agents of change by which the world was urged from one condition to another till it reached its present state must have been more violent, more powerful, than any which we see at work around us. They conceived that the evidence of "catastrophes" was irresistible.

2. *Doctrine of Uniformity.*—I need not here repeat the narrative (given in the *History*[1]) of the process by which this formidable array of proofs was, in the minds of some eminent geologists, weakened, and at last overcome. This was done by showing that the sudden breaks in the succession of strata were apparent only, the discontinuity of the series which occurred in one country being removed by terms interposed in another locality:—by urging that the total effect produced by existing causes, taking into account the accumulated result of long periods, is far greater than a casual speculator would think possible:—by making it appear that there are in many parts of the world evidences of a slow and imperceptible rising of the land since it was the habitation of now existing species:—by proving that it is not universally true that the strata separated in time by supposed catastrophes contain distinct species of animals:—by pointing out the limited fields of the supposed diluvial action:—and finally, by remarking that though the *creation* of species is a mystery, the *extinction* of species is going on in our own day. Hypotheses were suggested, too, by which it was conceived that the change of climate might be explained, which, as the consideration of the fossil remains seemed to show, must have taken place between the ancient and the modern times. In this manner the whole evidence of catastrophes was explained away: the notion of a series of paroxysms of violence in the causes of change was represented as a delusion arising from our contemplating short periods only, in the action of present causes: length of time was called in to take the place of intensity of force: and it was declared that Geology need not despair of accounting for the revolutions of the earth, as Astronomy accounts for the revolutions of the heavens, by the universal action of causes which are close at hand to us, operating through time and space without variation or decay.

An antagonism of opinions, somewhat of the same kind as this, will be found to manifest itself in the other Palætiological Sciences as well as in Geology; and it will be instructive to endeavour to balance these opposite doctrines. I will mention some of the considerations which bear upon the subject in its general form.

3. *Is Uniformity Probable A Priori?*—The doctrine of Uniformity in the course of nature has sometimes been represented by its adherents as possessing a great degree of *à priori* probability. It is highly unphilosophical, it has been urged, to assume that the causes of the geological events of former times were of a different kind from causes now in action, if causes of this latter kind can in any way be made to explain the facts. The analogy of all other sciences compels us, it was said, to explain phenomena by known, not by unknown, causes. And on these grounds the geological teacher recommended "an earnest and patient endeavour to reconcile the indications of former change with the evidence of gradual mutations now in progress."[2]

But on this we may remark, that if by *known* causes we mean causes acting with the same intensity which they have had during historical times, the restriction is altogether arbitrary and groundless. Let it be granted, for instance, that many parts of the earth's surface are now undergoing an imperceptible rise. It is not pretended that the rate of this elevation is rigorously uniform; what, then, are the limits of its velocity? Why may it not increase so as to assume that character of violence which we may term a *catastrophe* with reference to all changes hitherto recorded? Why may not the rate of elevation be such that we may conceive the strata to assume *suddenly* a position nearly vertical? and is it, in fact, easy to conceive a position of strata nearly vertical, a position which occurs so frequently, to be *gradually* assumed? In cases where the strata are nearly vertical, as in the Isle of Wight, and hundreds of other places, or where they are actually inverted, as sometimes occurs, are not the causes which have produced the effect as truly *known* causes, as those which have raised the coasts where we trace the former beach in an elevated terrace? If the latter case proves *slow* elevation, does not the former case prove *rapid* elevation? In neither case have we any measure of the time employed in the change; but does not the very nature of the results enable us to discern, that if one was gradual, the other was comparatively sudden?

The causes which are now elevating a portion of Scandinavia can be called known *causes*, only because we know the *effect*. Are not the causes which have elevated the Alps and the Andes known causes in the same sense? We know nothing in either case which confines the intensity of the force within any limit, or prescribes to it any law of uniformity. Why, then, should we make a merit of cramping our speculations by such assumptions? Whether the causes of change do act uniformly;—whether they oscillate only within narrow limits;—whether their intensity in former times was nearly the same as it now is;—these are precisely the questions which we wish Science to answer to us impartially and truly: where is then the wisdom of "an earnest and patient endeavour" to secure an *affirmative* reply?

Thus I conceive that the assertion of an *à priori* claim to probability and philosophical spirit in favour of the doctrine of uniformity, is quite untenable. We must learn from an examination of all the facts, and not from any assumption of our own, whether the course of nature be uniform. The limit of intensity being really unknown, catastrophes are just as probable as uniformity. If a volcano may repose for a thousand years, and then break out and destroy a city; why may not another volcano repose for ten thousand years, and then destroy a continent; or if a continent, why not the whole habitable surface of the earth?

4. *Cycle of Uniformity Indefinite.*—But this argument may be put in another form. When it is said that the course of nature is uniform, the assertion is not intended to exclude certain smaller variations of violence and rest, such as we have just spoken of;—alternations of activity and repose in volcanoes; or earthquakes, deluges, and storms, interposed in a more tranquil state of things. With regard to such occurrences, terrible as they appear at the time, they may not much affect the average rate of change; there may be a *cycle*, though an irregular one, of rapid and slow change; and if such cycles go on succeeding each other, we may still call the order of nature uniform, notwithstanding the periods of violence which it involves. The maximum and minimum intensities of the forces of mutation alternate with one another; and we may estimate the average course of nature as that which corresponds to something between the two extremes.

But if we thus attempt to maintain the uniformity of nature by representing it as a series of *cycles,* we find that we cannot discover, in this conception, any solid ground for excluding catastrophes. What is the length of that cycle, the repetition of which constitutes uniformity? What interval from the maximum to the minimum does it admit of? We may take for our cycle a hundred or a thousand years, but evidently such a proceeding is altogether arbitrary. We may mark our cycles by the greatest known paroxysms of volcanic and terremotive agency, but this procedure is no less indefinite and inconclusive than the other.

But further; since the cycle in which violence and repose alternate is thus indefinite in its length and in its range of activity, what ground have we for assuming more than *one* such cycle, extending from the origin of things to the present time? Why may we not suppose the maximum force of the causes of change to have taken place at the earliest period, and the tendency towards the minimum to have gone on ever since? Or instead of only one cycle, there may have been several, but of such length that our historical period forms a portion only of the last;—the feeblest portion of the latest cycle. And thus violence and repose may alternate upon a scale of time and intensity so large, that man's experience supplies no evidence enabling him to estimate the amount. The course of things is *uniform*, to an Intelligence which can embrace the succession of several cycles, but it is *catastrophic* to the contemplation of man, whose survey can grasp a part only of one cycle. And thus the hypothesis of uniformity, since it cannot exclude degrees of change, nor limit the range of these degrees, nor define the interval of their recurrence, cannot possess any essential simplicity which, previous to inquiry, gives it a claim upon our assent superior to that of the opposite catastrophic hypothesis.

5. *Uniformitarian Arguments are Negative Only.*—There is an opposite tendency in the mode of maintaining the catastrophist and the uniformitarian opinions, which depends upon their fundamental principles, and shows itself in all the controversies between them. The Catastrophist is affirmative, the Uniformitarian is negative in his assertions: the former is constantly attempting to construct a theory; the latter delights in demolishing all theories. The one is constantly bringing fresh evidence of some great past event, or series of events, of a striking and definite kind; his antagonist is at every step explaining away the evidence, and showing that it proves nothing. One geologist adduces his proofs of a vast universal deluge; but another endeavours to show that the proofs do not establish either the universality or the vastness of such an event. The inclined broken edges of a certain formation, covered with their own fragments, beneath superjacent horizontal deposits, are at one time supposed to prove a catastrophic breaking up of the earlier strata; but this opinion is controverted by showing that the same formations, when pursued into other countries, exhibit a uniform gradation from the lower to the upper, with no trace of violence. Extensive and lofty elevations of the coast, continents of igneous rock, at first appear to indicate operations far more gigantic than those which now occur; but attempts are soon made to show that time only is wanting to enable the present age to rival the past in the production of such changes. Each new fact adduced by the catastrophist is at first striking and apparently convincing; but as it becomes familiar, it strikes the imagination less powerfully; and the uniformitarian, constantly labouring to produce some imitation of it by the machinery which he has so well studied, at last in every case seems to himself to succeed, so far as to destroy the effect of his opponent's evidence.

This is so with regard to more remote, as well as with regard to immediate evidences of change. When it is ascertained that in every part of the earth's crust the temperature increases as we descend below the surface, at first this fact seems to indicate a central heat: and a central heat naturally suggests an earlier state of the mass, in which it was incandescent, and from which it is now cooling. But this original incandescence of the globe of the earth is manifestly an entire violation of the present course of things; it belongs to the catastrophist view, and the advocates of uniformity have to explain it away. Accordingly, one of them holds that this increase of heat in descending below the surface may very possibly not go on all the way to the center. The heat which increases at first as we descend, may, he conceives, afterwards decrease; and he suggests causes which may have produced such a succession of hotter and colder shells within the mass of the earth. I have mentioned this suggestion in the *History of Geology*; and have given my reasons for believing it altogether untenable.[3] Other persons also, desirous of reconciling this subterraneous heat with the tenet of uniformity, have offered another suggestion:—that the warmth or incandescence of the interior parts of the earth does not arise out of an originally hot condition from which it is gradually cooling, but results from chemical action constantly going on among the materials of the earth's substance. And thus new attempts are perpetually making, to escape from the cogency of the reasonings which send us towards an original state of things different from the present. Those who theorize concerning an origin go on building up the fabric of their speculations, while those who think such theories unphilosophical, ever and anon dig away the foundation of this structure. As we have already said, the uniformitarian's doctrines are a collection of negatives.

This is so entirely the case, that the uniformitarian would for the most part shrink from maintaining as positive tenets the explanations which he so willingly uses as instruments of controversy. He puts forward his suggestions as difficulties, but he will not stand by them as doctrines. And this is in accordance with his general tendency; for any of his hypotheses, if insisted upon as positive theories, would be found inconsistent with the assertion of uniformity. For example, the nebular hypothesis appears to give to the history of the heavens an aspect which obliterates all special acts of creation, for, according to that hypothesis, new planetary systems are constantly forming; but when asserted as the origin of our own solar system, it brings with it an original incandescence, and an origin of the organic world. And if, instead of using the chemical theory of subterraneous heat to neutralize the evidence of original incandescence, we assert it as a positive tenet, we can no longer maintain the infinite past duration of the earth; for chemical forces, as well as mechanical, tend to equilibrium; and that condition once attained, their efficacy ceases. Chemical affinities tend to form new compounds; and though, when many and various elements are mingled together, the play of synthesis and analysis may go on for a long time, it must at last end. If, for instance, a large portion of the earth's mass were originally pure potassium, we can imagine violent igneous action to go on so long as any part remained unoxidized; but when the oxidation of the whole has once taken place, this action must be at an end; for there is in the hypothesis no agency which can reproduce the deoxidized metal. Thus a perpetual motion is impossible in chemistry, as it is in mechanics; and a theory of constant change continued through infinite time, is untenable when asserted upon chemical, no less than upon mechanical principles. And thus the Skepticism of the uniformitarian is of force only so long as it is employed against the Dogmatism of the catastrophist. When the Doubts are erected into Dogmas, they are no longer consistent with the tenet of Uniformity. When the Negations become Affirmations, the Negation of an Origin vanishes also.

6. *Uniformity in the Organic World.*—In speaking of the violent and sudden changes which constitute catastrophes, our thoughts naturally turn at first to great *mechanical* and *physical* effects;—ruptures and displacements of strata; extensive submersions and emersions of land; rapid changes of temperature. But the catastrophes which we have to consider in geology affect the *organic* as well as the inorganic world. The sudden extinction of one collection of species, and the introduction of another in their place, is a catastrophe, even if unaccompanied by mechanical violence. Accordingly, the antagonism of the catastrophist and uniformitarian school has shown itself in this department of the subject, as well as in the other. When geologists had first discovered that the successive strata are each distinguished by appropriate organic fossils, they assumed at once that each of these collections of living things belonged to a separate creation. But this conclusion, as I have already said, Mr. Lyell has attempted to invalidate, by proving that in the existing order of things, some species become extinct; and by suggesting it as possible, that in the same order it may be true that new species are from time to time produced, even in the present course of nature. And in this, as in the other part of the subject, he calls in the aid of vast periods of time, in order that the violence of the changes may be softened down: and he appears disposed to believe that the actual extinction and creation of species may be so slow as to excite no more notice than it has hitherto obtained; and yet may be rapid enough, considering the immensity of geological periods, to produce such a succession of different collections of species as we find in the strata of the earth's surface.

7. *Origin of the Present Organic World.*—The last great event in the history of the vegetable and animal kingdoms was that by which their various tribes were placed in their present seats. And we may form various hypotheses with regard to the sudden or gradual manner in which we may suppose this distribution to have taken place. We may assume that at the beginning of the present order of things, a stock of each species was placed in the vegetable or animal *province* to which it belongs, by some cause out of the common order of nature; or we may take a uniformitarian view of the subject, and suppose that the provinces of the organic world derived their population from some anterior state of things by the operation of natural causes.

Nothing has been pointed out in the existing order of things which has any analogy or resemblance, of any valid kind, to that creative energy which must be exerted in the production of a new species. And to assume the introduction of new species as "a part of the order of nature," without pointing out any natural fact with which such an event can be classed, would be to reject creation, by an arbitrary act. Hence, even on natural grounds, the most intelligible view of the history of the animal and vegetable kingdoms seems to be, that each period which is marked by a distinct collection of species forms a cycle; and that at the beginning of each such cycle a creative power was exerted, of a kind to which there was nothing at all analogous in the succeeding part of the same cycle. If it be urged that in some cases the same species, or the same genus, runs through two geological formations, which must, on other grounds, be referred to different cycles of creative energy, we may reply that the creation of many new species does not imply the extinction of all the old ones.

Thus we are led by our reasonings to this view, that the present order of things was commenced by an act of creative power entirely different to any agency which has been exerted since. None of the influences which have modified the present races of animals and plants since they were placed in their habitations on the earth's surface can have had any efficacy in producing them at first. We are necessarily driven to assume, as the beginning of the present cycle of organic nature, an event not included in the course of nature. And we may remark that this necessity is the more cogent, precisely because other cycles have preceded the present.

8. *Nebular Origin of the Solar System.*—If we attempt to apply the same antithesis of opinion (the doctrines of Catastrophe and Uniformity,) to the other subjects of palætiological sciences, we shall be led to similar conclusions. Thus, if we turn our attention to astronomical palætiology, we perceive that the nebular hypothesis has a uniformitarian tendency. According to this hypothesis the formation of this our system of sun, planets, and satellites, was a process of the same kind as those which are still going on in the heavens. One after another, nebulæ condense into separate masses, which begin to revolve about each other by mechanical necessity, and form systems of which our solar system is a finished example. But we may remark, that the uniformitarian doctrine on this subject rests on most unstable foundations. We have as yet only very vague and imperfect reasonings to show that by such condensation a *material* system such as ours could result; and the introduction of *organized* beings into such a material system is utterly out of the reach of our philosophy. Here again, therefore, we are led to regard the present order of the world as pointing towards an origin altogether of a different kind from anything which our material science can grasp.

9. *Origin of Languages.*—We may venture to say that we should be led to the same conclusion once more, if we were to take into our consideration those palætiological sciences which are beyond the domain of matter; for instance, the history of languages. We may explain many of the differences and changes which we become acquainted with, by referring to the action of causes of change which still operate. But what glossologist will venture to declare that the efficacy of such causes has been uniform;—that the influences which mould a language, or make one language differ from others of the same stock, operated formerly with no more efficacy than they exercise now. "Where," as has elsewhere been asked, "do we now find a language in the process of formation, unfolding itself in inflexions, terminations, changes of vowels by grammatical relations, such as characterize the oldest known languages?" Again, as another proof how little the history of languages suggests to the philosophical glossologist the persuasion of a uniform action of the causes of change, I may refer to the conjecture of Dr. Prichard, that the varieties of language produced by the separation of one stock into several, have been greater and greater as we go backwards in history:—that[4] the formation of sister dialects from a common language, (as the Scandinavian, German, and Saxon dialects from the Teutonic, or the Gaelic, Erse and Welsh from the Celtic,) belongs to the first millennium before the Christian era; while the formation of cognate languages of the same family, as the Sanskrit, Latin, Greek and Gothic, must be placed at least two thousand years before that era; and at a still earlier period took place the separation of the great families themselves, the Indo-European, Semitic, and others, in which it is now difficult to trace the features of a common origin. No hypothesis except one of this kind will explain the existence of the families, groups, and dialects of languages, which we find in existence. Yet this is an entirely different view from that which the hypothesis of the uniform progress of change would give. And thus, in the earliest stages of man's career, the revolutions of language must have been, even by the evidence of the theoretical history of language itself, of an order altogether different from any which have taken place within the recent history of man. And we may add, that as the early stages of the progress of language must have widely different from those later ones of which we can in some measure trace the natural causes, we cannot place the origin of language in any point of view in which it comes under the jurisdiction of natural causation at all.

10. *No Natural Origin Discoverable.*—We are thus led by a survey of several of the palætiological sciences to a confirmation of the principle formerly asserted,[5] That in no palætiological science has man been able to arrive at a beginning which is homogeneous with the known course of events. We can in such sciences often go very far back;—determine many of the remote

circumstances of the past series of events;—ascend to a point which seems to be near the origin;—and limit the hypotheses respecting the origin itself: but philosophers never have demonstrated, and, so far as we can judge, probably never will be able to demonstrate, what was that primitive state of things from which the progressive course of the world took its first departure. In all these paths of research, when we travel far backwards, the aspect of the earlier portions becomes very different from that of the advanced part on which we now stand; but in all cases the path is lost in obscurity as it is traced backwards towards its starting point: it. becomes not only invisible, but unimaginable; it is not only an interruption, but an abyss, which interposes itself between us and any intelligible beginning of things.

Notes

1. *Hist. Ind Sci.*, B. xviii. c. viii. sect 2 [Whewell's own *History of the Inductive Sciences.*—LP].
2. Lyell, 4th Ed. B. iv. c. l, p. 328 [*Principles of Geology*—LP].
3. *Hist. Ind. Sci.*, B. xviii. c. v. sect 5, and note.
4. *Researches*, 11. 224 [James Prichard. *Researches into the Physical History of Mankind.*—LP].
5. *Hist. Ind. Sci.*, B. xviii. c. vi. sect. 5.

Index